Estimating and Bidding for Heavy Construction

Stuart H. Bartholomew
California State University, Chico

Prentice Hall
Upper Saddle River, New Jersey   *Columbus, Ohio*

Library of Congress Cataloging-in-Publication Data

Bartholomew, Stuart H.

    Estimating and bidding for heavy construction/Stuart H.
Bartholomew

      p.   cm.

    Includes bibliographical references and index.

    ISBN 0-13-598327-4

    1. Civil engineering—Estimates—United States.   2. Letting of
contracts—United States.   I. Title.

TA183.B37   2000   99–29570

624'.029'9—dc21   CIP

Cover photo: *Ewing Galloway*
Editor: *Ed Francis*
Production Editor: *Christine M. Buckendahl*
Production Coordinator: *Kathleen M. Lafferty/Roaring Mountain Editorial Services*
Design Coordinator: *Karrie Converse-Jones*
Cover Designer: *Dan Eckel*
Production Manager: *Patricia A. Tonneman*
Marketing Manager: *Chris Bracken*

This book was set in Times Roman by Carlisle Communications Ltd., and was printed and bound
by R. R. Donnelley & Sons Company. The cover was printed by Phoenix Color Corp.

Printed in the United States of America

10 9 8 7 6 5 4 3 2 1

ISBN: 0-13-598327-4

Prentice-Hall International (UK) Limited, *London*
Prentice-Hall of Australia Pty. Limited, *Sydney*
Prentice-Hall of Canada, Inc., *Toronto*
Prentice-Hall Hispanoamericana, S.A., *Mexico*
Prentice-Hall of India Private Limited, *New Delhi*
Prentice-Hall of Japan, Inc., *Tokyo*
Prentice-Hall (Singapore) Pte. Ltd., *Singapore*
Editora Prentice-Hall do Brasil, Ltda., *Rio de Janiero*

Foreword

Estimating and Bidding for Heavy Construction opens with the statement that the "life blood of a successful construction contracting company is the ability to obtain contracts that can be executed for a profit under intensely competitive conditions in a complex, difficult industry." This is the key to success in a nutshell. In heavy civil work, there is only one chance to "get it right." If the work is priced correctly—using locally realistic labor and materials costs—the project can usually be built for a profit. The need for precise, current, and local information is one of the reasons the more successful heavy civil companies today are largely regional, midsized firms. They have a management core who, because they are very close to the business, are experts in obtaining and properly using local job-specific data when pricing and performing the work. They have not grown too large to stay in touch.

At the same time, to survive under the stress of competition, larger firms in the heavy construction industry have undergone substantial consolidation into bigger, more comprehensive companies while growing more global in seeking and prosecuting work, usually as joint ventures with firms having a strong local presence. This strategy works if it keeps the venture in touch with the work itself—with local labor productivity and practices, and with current materials and techniques. Today, technology has become a very complex moving target. The workforce is of necessity more educated and more sophisticated than in the past. Projects are larger and more demanding. The result is severe strains on the ability of such evolving, mobile, and complex companies to successfully tender for large heavy civil works.

My experience on heavy civil projects over the years validates the author's detailed, practical, straightforward approach. From early work on the 600-ft-high earth and rock fill Mica Dam project in British Columbia to the equipment and earth-moving intensity of the Tennessee-Tombigbee waterway in the southeastern United States to port, rail, and coal mining project experience on the Cerejon project in Colombia, I can vouch for the importance of this book's emphasis on such matters as accurate quantity takeoffs and load-and-haul calculations. Mistakes on projects such as these can mean the difference between a substantial profit or a huge, company-destroying loss.

On the Channel Tunnel construction between England and France, I managed a joint venture of five British and five French companies. This project, with its three

52-km-long undersea tunnels outfitted with a state-of-the-art rail transportation system including unique, specially designed rolling stock, brought home to me the need for accuracy and experience in estimating. Nothing can be taken for granted on a project like this, including the ability to integrate specialties such as electrical and mechanical practice into the overall project concept.

As the author emphasizes, projects such as the Channel Tunnel could not have been successfully planned or built without the extensive use of computers. For example, materials had to be described in both English and French, and prices were quoted in pounds, francs, dollars, and other currencies where required. As the author also observes, however, without experience and informed attention to the functions the computer is intended to perform, even the best computer system is useless.

Although designed as a textbook for upper-division college students in construction management, construction technology, or civil engineering, this book will prove to be a useful reference for companies with a heavy civil construction component. Its orderly approach to estimating strategies and methods provides a solid model and checklist for those on "the front lines." The direct, practical, no-nonsense approach adds an overlay of seriousness that is quite welcome to anyone who has ever had to struggle with a poorly bid project and its associated aftermath of recalculation, rework, and, often, costly litigation.

Although years of on-the-job experience are needed to develop a reliably consistent and accurate estimator who can cover the range of conditions encountered on a large project, this book provides the basic knowledge needed to begin. For this purpose, the division into four blocks of chapters seems quite appropriate. The first three chapters provide a general introduction to any type of construction, and the next three chapters complete the description of fundamental methodologies for developing a project's direct costs.

Because of the extent and variety of subject matter, the following four chapters are devoted to representative types of construction that an estimator might encounter. Here, the discussion of drill and blast, mass diagrams, load-and-haul, and concrete operations opens a window into the many examples of heavy construction work operations that are the components of more complex topics such as types of tunnels, dams, and megaprojects in general, all of which combine many specific disciplines. The last three chapters describe subjects particular to heavy construction such as indirect costs and equipment mobilization and demobilization costs and end with a discussion of the ethical, legal, and business issues that are specific to heavy construction.

Estimating and Bidding for Heavy Construction presents a timely, straightforward approach to the realities of contemporary construction practice. It is extremely important that individuals with solid theoretical backgrounds who have also worked at "at the sharp end of the business" continue, as this author has done, to pass their knowledge and experience on to those now entering the industry.

Jack K. Lemley
Boise, Idaho

Preface

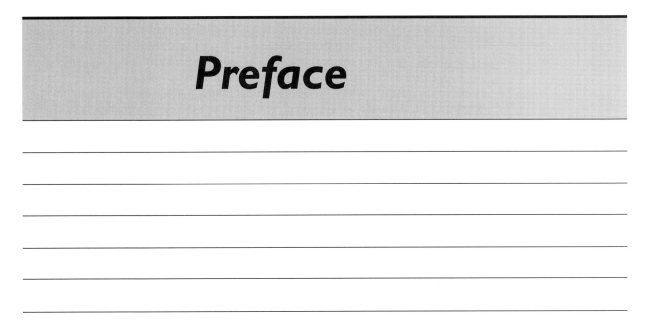

When I started teaching heavy construction estimating and bidding in 1985, I soon realized that there were no comprehensive textbooks on the subject suitable for classroom instruction. Excellent books contained reams of material on various areas of heavy construction, but none made clear in an integrated way the basic principles of cost estimating and bidding that I felt would have been of value to me when I began my own career in construction.

Estimating and Bidding for Heavy Construction is the result of my efforts to identify these principles and to clarify the organizational and analytical processes involved in their implementation. This book is intended as a textbook for students in a four-year baccalaureate program in construction management, construction technology, or civil engineering programs with a construction option. It is assumed that such students are computer literate and have previously completed courses in the ownership and operation of construction equipment, plan reading, soils and concrete technology, the design of formwork and other temporary structures, and basic cost accounting. The focus here is on the development of a construction contractor's competitive bid for the acquisition of a major heavy construction contract to be performed under fixed price and time commercial terms in today's competitive business environment.

Chapters 1 through 3 stress general matters pertaining to any construction estimate, the nature of the various kinds of costs involved, estimate formats, and the major elements comprising a contractor's total bid (both an explanation of what these elements are and how they are developed). Chapter 4 is devoted entirely to the takeoff process as it applies to major heavy construction work, and Chapter 5 covers the details of the basic unit cost database that must be assembled at the inception of any heavy construction estimate. Chapter 6 then explains the pricing process whereby the takeoff information is transformed into actual costs by means of the systematic and organized use of the unit cost database. Thus, the first six chapters comprise the fundamental methodologies involved in developing the direct costs for any kind of heavy construction work.

As Chapter 4 emphasizes, the takeoff process for a heavy construction project cannot be completed without a well-developed plan for the performance of the work that in turn involves what can best be described as heavy construction operations analysis. How is the job going to be built? How are the various work operations to be sequenced? What construction methods are going to be employed? The need to answer these kinds of questions at the time of the estimate distinguishes heavy construction work from residential, commercial, or industrial building construction and from various forms of specialty contracting. The required processes cannot be explained without the use of examples. I found it difficult to choose what particular kinds of construction activity to use as examples. I finally settled on drill and blast operations (Chapter 7); mass diagrams (Chapter 8); earth and rock load, haul, and compact operations (Chapter 9); and structural concrete operations (Chapter 10). These subjects are important operations performed by heavy construction prime contractors with their own forces, and all involve the interaction of work crews with heavy construction equipment, the purchase of expendable and permanent materials, and little subcontracted work. The processes illustrated by these examples also apply to piledriving and ground support operations; marine cofferdam work; underground construction; asphalt and concrete paving; and sand, gravel, and rock quarry operations. Any or all of these areas could have been included in a much larger book.

The last three chapters are devoted to subjects unique to heavy construction. Chapter 11 emphasizes the kinds of indirect costs that heavy construction contractors typically incur and how these costs are determined. Chapter 12 deals with the construction equipment reconciliation and adjustment process, and Chapter 13 addresses the pricing of schedule-of-bid-items bid forms, a process involving ethical, legal, and business issues found nowhere else in the industry and a subject that I have found to be widely misunderstood.

The Key Words and Concepts listed at the beginning of each chapter will prove helpful as a study guide and as a means of testing recollection as the reader completes each chapter. Also, each chapter concludes with questions for testing comprehension of the material provided and problems to be used as laboratory exercises for university estimating courses. Although the solution to these problems would normally be carried out in a supervised classroom setting, they are also suitable for self-study. The short bibliography at the end of the book contains a list of excellent books on construction equipment, methods, and technology, all relevant subjects that this book touches on only to the extent necessary to provide context for illustrating the cost estimating and bidding process.

Computer applications in cost estimating have become indispensable in today's world, and I have frequently referred to their use. In addition, several specific applications illustrate the practical elimination of otherwise onerous number crunching, and a floppy disk containing the programs used and instructions for their use are provided. My intended focus, however, is on the basic analytical and computational processes involved, whether performed by a computer or manually. In my view, an estimator should never use a computer program for any process that he or she does not fully understand and cannot comfortably perform manually.

I would like to thank the reviewers of my manuscript: Edward L. Bernstein, Alabama A&M University; John Fredley, University of Washington; James L. Otter, Pittsburgh State University; and Sandra Weber, Arizona State University. I would also like to thank Ed Francis, senior editor of Career and Technology at Prentice Hall, for his constant encouragement during the writing of this book and to the entire Prentice Hall editorial staff for their help during publication production. On a different level, my sincere thanks are also due to the bosses, mentors, coworkers, and friends in the industry who over more years than I sometimes care to remember have taught me my trade and, in more recent years, to my students at California State University, Chico, who have helped me develop ways to teach this interesting and important subject.

Stuart H. Bartholomew
Chico, California

Contents

CHAPTER 6   PRICING OUT THE DIRECT COST ESTIMATE   114

CHAPTER 7   DRILL-AND-BLAST OPERATIONS   132

CHAPTER 8   MASS DIAGRAMS   171

1

Preliminary Considerations

Key Words and Concepts

Conceptual estimates
Estimates to evaluate alternative designs
Engineer's estimate to evaluate bids
Contractor's bid estimate
Estimates to evaluate alternative construction
   means and/or methods
Contract change and/or breach of contract damages
   estimates
Estimate cost-generating mechanism

Accuracy of cost estimates
Methods of cost expression (horizontal format)
Single-value/multiple-value estimates
Scope of the estimate (vertical format)
Estimating methodology
Unit price estimating
Unit labor-hour estimating
Crew and equipment work item analysis

The life blood of a successful construction contracting company is the ability to obtain contracts that can be executed at a profit under intensely competitive conditions in a complex, difficult industry. Central to the ability to obtain such contracts is the further required ability to forecast, or estimate in advance, the probable cost of construction performance. Nowhere is this ability more important than in the heavy construction segment of the industry, the focus area of this book.

Preparing competent construction cost estimates usually involves considerable expenditure of time and effort, depending on the scope and purpose for which the estimate is made. Before studying the techniques and procedures involved, it therefore seems sensible to consider the reasons and needs for construction cost estimates and the types of estimate made in response to these reasons and needs. In the chronology of typical construction projects, a number of different types of cost estimates are usually generated in the following sequence:

1. Conceptual estimates for project feasibility studies
2. Estimates to evaluate alternative designs
3. Engineer's (owner's) estimate to evaluate contractor's competitive bids or cost proposals for negotiated projects
4. Contractor's estimate as the basis for a competitive bid or cost proposal
5. Contractor's estimates to evaluate alternative methods or schemes for construction after award of construction contract
6. Estimates to evaluate the cost of contract changes and/or monetary damages in the event of breach of contract

GENERAL TYPES OF CONSTRUCTION COST ESTIMATES

Following are typical construction cost estimates listed by their more or less generic names, along with descriptions of the situations creating a need for each estimate type. This discussion includes most estimate types likely to be encountered.

Conceptual Estimates for Project Feasibility Studies

When either a private or public owner determines the need for an infrastructure or other facility, the owner obviously needs to know the possible cost range for the contemplated project. Once the fundamental scope parameters for the project have been established, the owner normally engages an engineering planning organization to prepare a conceptual estimate for the project based on those scope parameters. The engineering planning organization then prepares the conceptual estimate in broad terms, relying on past similar project cost experience applied to a rough quantification of project scope measures, such as square footages of buildings and other structures, miles of roadway, or other similar measures of project scope. The estimator necessarily limits estimating time and effort, recognizing that more accurate cost estimates will be prepared later, when definitive details of the project have been defined.

Estimates to Evaluate Alternative Designs

After the owner has engaged a design professional (architect/engineer, or A/E) to design the project and create the plans and specifications, that organization usually studies alternative design solutions for various project features. To intelligently choose between such alternatives, the design professional must consider probable construction cost. Thus, A/Es need to quickly evaluate costs for the various alternatives. Cost estimates for this purpose are necessarily limited in scope and are usually based on the designer's perceived notions of cost, which are derived from reviewing construction contractor's unit bid costs obtained from abstracts of competitively bid projects of a similar nature. Such estimates usually involve tabulating quantities of physical construction work required for each alternative design and then applying unit costs at the construction contractor bid level (obtained from previous bid abstracts). At this point, the estimator will not consider the method or technique of actual construction.

Engineer's (Owner's) Estimate to Evaluate Contractor Competitive Bids or Cost Proposals for Negotiated Work

Once the project design approaches completion (say, reaches the 80 to 90 percent completion level), the A/E typically will prepare an estimate intended to reflect a probable figure for the lowest responsible, responsive bid expected for competitively bid projects or a realistic contract price expected to be agreed by negotiation with interested contractors for projects that are not competitively bid. Such estimates sometimes involve no more than compiling a list (or takeoff) of the quantities of work or services the contractor is expected to provide and then assigning a unit contract price to each such quantity. Values of the assigned unit prices are generally obtained from past bid abstracts on competitively bid work or from past experience on negotiated construction work of the type and complexity at hand. The estimator may or may not take into consideration the particular methods or techniques to be used to accomplish the construction work. If the A/E or the owner engages a cost-estimating professional to prepare the estimate, the estimator will probably use a somewhat more sophisticated estimating methodology.

Contractor's Estimate for a Competitive Bid or Price Proposal for Negotiated Work

Once the design has been completed, the owner will call for either competitive bids or proposals leading to a negotiated contract. This is typically done by means of an advertised *Invitation for Bids* (IFB) or *Request for Proposals* (RFP). Contractors interested in obtaining the construction work either bid competitively in compliance with terms and conditions established by the owner, or they submit proposals expected to lead to a negotiated contract for the work, as the case may be. Both avenues to obtaining a construction contract require the contractor to make an accurate appraisal of the probable costs of performing the required work and meeting all other requirements of the potential contract. Ordinarily, competing contractors will be furnished a more or less complete set of contract documents upon which to base their bids or cost proposals. These documents include drawings and specifications defining the precise requirements of the construction work to be put in place, plus explicit statements of the general and special conditions under which the proposed work must be performed. The highly competitive nature of this scenario creates a complex and difficult business environment in which the construction contractor's ability to survive (let alone prosper) depends on the ability to accurately forecast probable construction costs. **Although this preliminary chapter approaches the subject of contractor's estimates somewhat more generally, this book focuses on the principles by which such cost estimates are made for the heavy construction segment of the industry.**

Contractor's Estimates to Evaluate Alternative Means and/or Methods of Construction

During preparation of the contractor's bid estimate (or estimate for a price proposal for negotiated work), the estimator probably would have evaluated alternative means and/or methods of performing the work and made appropriate choices. However, once contractors obtain the construction contract, they usually continue to study the project to evaluate further alternatives or schemes of construction not previously considered. Most construction contracts provide by their terms that, as long as construction drawing and specification requirements are met, the contractor has the right to choose the means and/or methods of construction. This also implies the prerogative to make changes in construction means and methods once the contract has been initiated. My experience with contracts I have personally managed and/or for which I have been responsible, and with contracts managed by competitors and clients, is that once the contract has been obtained, the search for cost-saving alternative means and methods intensifies. Estimators typically use the same general procedures and techniques they employed for the contractor's bid estimate for this purpose as well.

Contract Change Estimates and/or Estimates to Evaluate Breach of Contract Damages

Changes in the contract work to be performed are inevitable during the course of construction for any contract. Changes in the drawings and specifications may be made to accommodate some project alteration desired by the owner or, particularly in heavy construction, may be due to the contractor encountering differing site conditions.[1] Contract changes are also frequently required because of delays imposed on the project due to various causes not contemplated by the parties when the contract was formed. For whatever cause, changes in the details of the work to be put in place, or in the conditions under which the work must be performed, create the legal requirement that the contract price and stipulated time period for contract work performance must be equitably adjusted to account for the extra cost and required time of performance due to the change. In this situation, it is usually the contractor who is required to propose the appropriate change in contract price and time to the owner, who in turn will evaluate the proposal and negotiate with the contractor to arrive at a mutually agreed change in contract price and time. The eventual outcome of such "change order" negotiations will involve a mutual examination of the contractor's estimate of the cost and time impact of the change and comparison with a similar estimate made by the owner's engineer. Similarly, the adjudication of claims resulting from breaches of contract by either party also involve determining an appropriate change in contract time or price or both. In all the above cases, cost and time estimates are required. Whether the contractor, the owner's engineer, or a consultant on behalf of the owner makes these estimates, he or she usually will prepare them using estimating principles similar to those utilized for the contractor's original bid estimate.

Do all these types of estimates have common characteristics, and if so, what points apply to all of them? And what are the major points of difference?

---

[1]Physical site conditions different from those indicated in the original contract documents or different from those commonly encountered in similar projects.

POINTS COMMON TO ALL ESTIMATES

All the estimates discussed above are similar in two major respects. All depend on the same basic mechanism to generate estimated costs, and, being estimates, all share the characteristic of impreciseness.

Cost-Generating Mechanism

One way or another, all estimated costs depend on associating or connecting quantities of units to some perceived monetary and time impact for accomplishing a single unit. Usually the quantities involve various kinds of physical construction work to be put in place on the project or physical work in removing something that presently exists on the project site. For instance, such items as cubic yards of concrete, pounds of reinforcing steel, or cubic yards of engineered embankment are all quantities of construction work to be put in place at the site. On the other hand, foundation excavation or clearing and grubbing of trees or other vegetation forms at the site represent quantities of things existing on the site that must be removed to construct the project. Sometimes the quantities are for nontangible units such as an amount of elapsed time. For instance, when determining the cost for a construction superintendent to oversee the project, the quantity involved will be the length of time the superintendent is required at the project site. No estimate for the cost of anything can escape this basic connection between quantities of units and the perceived monetary and time impact of accomplishing a single unit.

Accuracy of Cost Estimates

All estimates are imprecise. The only certainty about any estimate is that it is probably either high or low. One of the most astute estimating professionals I have encountered once said that a competent estimate for a heavy engineered construction project could best be described as a long series of probable errors where those on the high side compensated for those on the low side, resulting in an overall figure that was realistic. When correct principles are applied, the result in heavy construction estimates will usually be considerably better than the above statement implies, depending on the particular risks and unknowns the estimator must deal with when making the estimate; but, even under the best of circumstances, no estimate will be precise.

RESPECTS IN WHICH ESTIMATES DIFFER

The foregoing discussion considered points common to the various generic estimate types. The following discussion focuses on ways in which these estimates vary from one another due to the different needs each estimate is expected to meet.

Method of Cost Expression (Horizontal Format)

Line item and total costs for many of the earlier enumerated estimates are usually expressed as single values. Consider for instance a situation in which it was desired to determine the cost for a small pumphouse. The first step would be to determine (or take

Description	Unit	Quantity	Unit price, $	Total cost, $
Foundation excavation	cy	550	12.50	6,875
Concrete	cy	175	225.00	39,375
Reinforcing steel	lb	15,000	0.40	6,000
Built-up roofing	sy	60	50.00	3,000
Galvanized grating	sf	50	15.00	750
Total cost				56,000

FIGURE 1-1   Single-value estimate for pumphouse.

off) the physical quantities of various kinds of construction work required for the pumphouse. Assume the results of this determination were as follows:

Foundation excavation	550 cy
Structural concrete	175 cy
Reinforcing steel	15,000 lb
Built-up roofing	60 sy
Galvanized grating	50 sf

Assume further that expected unit prices for the various items of work (determined on some basis or another) were believed to be

Foundation excavation	$ 12.50
Concrete	$225.00
Reinforcing steel	$ 0.40
Built-up roofing	$ 50.00
Galvanized grating	$ 15.00

Based on the above assumptions, a **single-value estimate** for the cost of the pumphouse could be expressed as shown in Figure 1-1.

On the other hand, assume that you desire to express the cost of the pumphouse as a **multiple-value estimate.** In this case, it would be necessary to have more detail for the expected unit costs. Assume, for purposes of illustration, that the expected unit costs were believed to consist of the following:

Description	Labor	Materials	Equipment	Total
Foundation excavation	6.00	—	6.50	12.50
Concrete	120.00	60.00	45.00	225.00
Reinforcing steel	0.10	0.20	0.10	0.40
Built-up roofing	20.00	30.00	—	50.00
Galvanized grating	5.00	10.00	—	15.00

Based on this breakdown for the unit costs, the multiple-value estimate for the pumphouse would be expressed as shown in Figure 1-2.

The total estimated cost of the pumphouse in both the single and multiple-value estimate is the same. However, the multiple-value estimate provides more insight into the

| Description | Unit | Quantity | Labor, $ | | Materials, $ | | Equipment, $ | | Total cost, $ |
			Unit price	Total	Unit price	Total	Unit price	Total	
Found. excavation	cy	550	6.00	3,300			6.50	3,575	6,875
Concrete	cy	175	120.00	21,000	60.00	10,500	45.00	7,875	39,375
Reinforcing steel	lb	15,000	0.10	1,500	0.20	3,000	0.10	1,500	6,000
Built-up roofing	sy	60	20.00	1,200	30.00	1,800			3,000
Galvanized grating	sf	50	5.00	250	10.00	500			750
Total cost				27,250		15,800		12,950	56,000

FIGURE 1-2 Multiple-value estimate for pumphouse.

nature of the expected costs; the breakdown into labor, materials, and costs for the use of construction equipment is now provided, as is the total estimated cost for each line item. In the single-value estimate, only the total estimated cost for each line item is developed.

Single-value estimates usually suffice for conceptual estimates, estimates to evaluate alternative designs, and sometimes even the engineer's (owner's) estimate to evaluate contractor's competitive bids. The balance of the estimates discussed earlier in this chapter would normally be made as multiple-value estimates. Construction contractors seldom, if ever, estimate costs in the single-value format.

Scope of the Estimate (Vertical Format)

The vertical format of the estimate, or its line item scope, shows a breakdown of whatever costs are being estimated. For instance, an estimate may be prepared for some small isolated part of a construction project or for the entire project. Further, regardless of the part of the project for which the cost is being estimated (isolated part versus entire project), the scope of the cost estimate may be confined to **direct costs**[2] only, or the estimate may include both direct costs and **indirect costs.**[3] Obviously, an estimate is meaningless if the reader of the estimate is unaware of the intended scope.

Estimating Methodology

In addition to the method of cost expression and scope of the estimate, as discussed above, cost estimates may be prepared using one of several estimating methodologies. The methodologies in common use are more fully discussed below.

ESTIMATING METHODOLOGIES

Construction cost estimates may be prepared by any one of three common methodologies. The particular methodology used will depend on the purpose for which the estimate is made and on the type of entity making the estimate. For instance, planners and

---

[2]Costs associated with discrete separate physical parts of the project.

[3]More general types of project costs that cannot be identified with specific discrete parts of the project.

A/Es employ different methodologies than construction contractors do, and individual construction contractors employ various methodologies, depending on the type of construction they typically perform. A detailed examination of these estimating methodologies follows.

Unit Price Estimating

In unit price estimating, the estimator first determines quantities for various kinds of construction work or other project elements that are expected to generate a cost. Then the estimator will price each quantity by applying a unit cost perceived to be accurate for one unit of that quantity. Both Figures 1-1 and 1-2 illustrate unit price estimating. Conceptual estimates, estimates to evaluate alternative designs, engineer or owner estimates to evaluate contractor competitive bids, and some contractor estimates for competitive bids or cost proposals, or for other purposes after contracts have been entered into are commonly made using this methodology. The obvious advantage of unit pricing is its simplicity, which is the reason it appeals to planners and design engineers. The advantage of simplicity also appeals to many construction contractors, particularly specialty contractors who operate within a restricted geographical area. For instance, contractors exclusively performing pile-driving work, or paving and grading contractors, fit into this category, as do specialty trade subcontractors such as electrical, piping installation, heating/ventilating/air conditioning, and mechanical installation contractors.

Unit price estimates are acceptable only when the estimator uses a database of known unit prices for various kinds of work that is reliable and obtainable for the work being priced. Contractors who use this method typically maintain accurate records of previously performed work of a similar nature from which they can derive unit prices for each particular kind of work.

The disadvantage of unit price estimating is that grossly erroneous estimates will result if the unit prices applied do not represent those that can be obtained when the work is actually performed. Unit prices obtained for a given kind of construction work under one set of applicable labor rates and one set of conditions prevailing in a particular geographical area are inapplicable in other geographical areas where the labor rates and work practices are different. Clearly, the reliability of unit prices depends on where the previous experience upon which the unit prices are based was obtained and on the working conditions that prevailed at that time and place. Unit price estimating is seldom employed by contractors operating in the heavy engineered construction segment of the industry, except possibly for minor elements of a cost estimate.

Unit Labor-Hour Estimating

The unit labor-hour estimating method is similar to unit price estimating except that, instead of a monetary unit price, the number of labor-hours perceived to be required per unit of quantity for a line item of work is applied to the total quantity for the line item to yield the total labor-hours of labor. The total labor-hours of labor for the line items of whatever is being estimated is then converted into dollars and cents by applying a weighted average hourly wage rate believed to represent the mix of construction trades typically involved in performing that particular type of work. In this way, the estimator will obtain a total value for labor cost for the project or portion of the project being estimated. This estimating methodology is primarily utilized in situations where labor is the predominate component of cost expected to be incurred. Costs for the materials and for the use of construction equipment are obtained by applying unit price estimating.

The total cost of the project (or portion of the project) being estimated is then obtained by adding the labor component cost to the materials and equipment usage costs.

Specialty trade contractors, such as electrical, piping, and heating/ventilating/air conditioning contractors, typically utilize unit labor-hour estimating. Figure 1-3 illustrates a unit labor-hour estimate for a single run of stainless steel threaded and coupled pipe from a piping contractor's estimate.

The principal advantage of unit labor-hour estimating is that it identifies the influence of the weighted average hourly labor rate on the total estimated labor cost. The direct connection of labor rate to labor cost is not obvious when unit price estimating is used. Unit labor-hour estimating also has the advantage of simplicity in providing a convenient means of connecting past labor productivity experience with expected labor productivity for similar new work to be performed. The disadvantages are the same as those applying to unit price estimating. If the unit labor-hour figures used were obtained under conditions that differ considerably from conditions that will prevail in the project being estimated, serious error will result. Additionally, as in unit price estimating, there is no apparent connection between the means and methods of construction and the expected costs of the future work. This is a serious disadvantage for estimators who apply these methods to cost estimates for heavy engineered construction.

Crew and Equipment Work-Item Analysis

Crew and equipment work-item analysis is the methodology most commonly employed by construction contractors in the heavy engineered construction segment of the industry. Because planners and A/Es usually lack the construction background to estimate costs by this method, it will normally not be utilized for conceptual estimates or estimates to evaluate alternative designs. However, if the owner or the owner's engineer engages a consultant with construction contracting experience to prepare an estimate, the consultant probably would utilize this methodology.

The crew and equipment work-item analysis method can be illustrated by a simple example. Assume that a pile-driving contractor wanted to make a direct cost estimate

Description	Unit	Quantity	Labor @ $22.00/Labor-Hour			Materials, $		Total cost, $
			Labor-hour factor	Labor-hour	Total, $	Unit price	Total	
Stainless steel pipe, T&C								
2 in. pipe	lf	100	0.30	30		12.00	1,200	
1 in. pipe	lf	125	0.17	22		6.50	813	
3/8 in. pipe	lf	50	0.13	7		3.50	175	
2 in. ell	Each	3	0.60	2		22.00	66	
1 in. ell	Each	2	0.34	1		9.00	18	
3/8 in. ell	Each	2	0.13	1		6.00	12	
Reducer 2 in.–1 in.	Each	1	0.60	1		27.00	27	
Reducer 1 in.–3/8 in.	Each	1	0.34	1		15.00	15	
2 in. valve	Each	2	0.70	2		125.00	250	
3/8 in. valve	Each	2	0.30	1		55.00	110	
Total	–	–	–	68	1,496	–	2,686	4,182

FIGURE 1-3 Unit labor-hour estimate for run of stainless steel threaded and coupled pipe.

for a simple pile-driving project for the foundation of a high-rise office building. Further assume that the engineer's drawings and specifications define the work to be performed as driving 575 steel bearing piles, each 60 ft long. The piles were specified to be 14HP73 bearing piles.[4] Assume further that, based on past experience and study of the soil borings included with the bid documents, the contractor concluded that the following crew could reasonably be expected to achieve the production rates indicated:[5]

Required crew	Production
One 60-ton crawler crane with operator and oiler One set pile-driving leads One Del Mag D-22 pile hammer	Two days required for unloading bearing piles and assembling pile driver; one day required for breakdown of pile driver and removing from site
One pile driver foreman Four pile drivers	Average number of bearing piles driven per day = 18

Assume further that the hourly labor rates, the hourly rates reflecting the value of construction equipment to be used in the work, and the unit material cost for the bearing pile material were as follows:

Classification	Labor rates, $/hr[1]
Pile driver foreman	15.00
Pile driver	12.00
Crane operator	18.00
Crane oiler	12.00
Equipment description	Equipment use rates, $/hr[2]
60 ton crawler crane	30.00
Pile-driving leads	10.00
D-22 hammer	20.00
Material cost, 14HP73 bearing pile, FOB jobsite[3] = $0.25/lb	

[1]Rates listed are deemed to include all applicable labor fringes, insurance, and taxes.

[2]Rates listed include repair and service labor, equipment operating expense, and equipment ownership costs.

[3]"FOB jobsite" means that delivery or freight costs to the site of the work are included in the unit material price.

The labor rates and material cost for the bearing piling is always readily obtainable information. The hourly use rates for construction equipment would be available either from the contractor's records for similar work previously performed or from published equipment use rate manuals, several of which list reasonably reliable use rate figures.

Based on all the above, a crew and equipment work-item analysis type of cost estimate can be readily set out as shown in Figure 1-4.

The obvious disadvantage to the crew and equipment work-item analysis method is that it is more time-consuming than unit price or unit labor-hour estimating. How-

---

[4]Bearing piles of 14-in. deep sections weighing 73 pounds per foot.

[5]Conclusory information of this type would come from supervision and management personnel of an experienced pile-driving contractor who had previously operated in the area of the project.

Description	Unit	Quantity	Labor, $		Material, $		Equipment, $		Total, $	
			Unit	Total	Unit	Total	Unit	Total	Unit	Total
Pile-Driving Crew and Equipment for One Working Day:										
1 60 T crawler crane	hr	8					30.00	240		240
1 set of leads	hr	8					10.00	80		80
1 D-22 hammer	hr	8					20.00	160		160
1 pile driver foreman	hr	8	15.00	120						120
4 pile drivers	hr	32	12.00	384						384
1 crane operator	hr	8	18.00	144						144
1 oiler	hr	8	12.00	96						96
Crew & equipment per day	–	–	–	744	–	–	–	480	–	1,224
A. Move in and set up	Days	2	744	1,488			480	960	1,244	2,448
B. Handle and drive: 575/18	Days	32	744	23,808			480	15,360	1,244	39,168
C. Breakdown & move out	Days	1	744	744			480	480	1,244	1,224
D. Buy pile material: (575)(60)(730)	lb	2,518,500	–	–	0.250	629,625	–	–	–	629,625
Total cost (575)(60 ft/each)	lf	34,500	0.75	26,040	18.25	629,625	0.49	16,800	19.49	672,465

FIGURE 1-4 Example of crew and equipment work-item analysis estimate applied to a simple bearing pile estimate.

ever, this disadvantage is more than overcome by the method's advantages. By using this approach, the estimator can deal directly with the fundamental events contributing to the cost of construction.

In unit cost or unit labor-hour estimating, the underlying cost drivers are not apparent or utilized in developing the costs. The details of the labor force involved in the work, the equipment to be utilized, the expected productivity, the length of time the work will take, and more particularly, **the level of risk** involved, are all unknowns.

When the crew and equipment analysis method of estimating is applied, answers to the above questions are readily apparent. Details of the crew, the equipment used, and the productivity upon which the estimate is based appear directly in the body of the estimate. And more importantly, this method provides a direct means to evaluate the level of risk implicit in the estimate. For instance, in the simple example cited, if the contractor was concerned about the productivity level utilized in the estimate (an average of 18 piles per day), the monetary effect of a lesser productivity could be quickly calculated. If the contractor believed that in the most adverse scenario, productivity would fall to a 12-piles-per-day average, the monetary consequences would be

$$\text{Additional days required} = \frac{575}{12} - \frac{575}{18} = 48 - 32 = 16 \text{ days}$$

$$\text{Cost overrun} = (16 \text{ days})\left(\frac{\$1,224}{\text{day}}\right) = \$19,584$$

Knowing the monetary value of risk implicit in the estimate, the contractor can (and should) reflect this knowledge in the contingency and markup consideration that should be part of every competitive bid or cost proposal. In this respect, the great advantage of the crew and equipment work-item analysis estimating methodology is clear. This benefit is particularly important in the field of cost estimating for heavy engineered construction projects where great risk in correctly estimating productivity often exists.

CONCLUSION

This chapter on preliminary considerations focused on the types of cost estimates commonly made in the construction industry and on the purpose for which these estimates are made. The points all estimates have in common and the points in which they differ were also discussed. In addition to a major focus on estimating methodology (a common point of difference), other points of difference discussed included the method of cost expression (horizontal format) and the scope of the estimate (vertical format). The next two chapters focus on these later topics.

Throughout the discussion in this chapter the point was made that, although the discussion was general, applying to estimates for any type of project, this book will focus on cost estimates made by and for the heavy engineered construction segment of the industry.

QUESTIONS AND PROBLEMS

1. What are the six separate purposes for which cost estimates are made that were discussed in this chapter?

2. What are two points of similarity and three ways in which cost estimates may differ?

3. Name and explain the advantages and disadvantages of the three separate estimating methodologies discussed in this chapter.

4. You have already seen the effect of assumed work operation productivity on total estimated cost for the pile-driving project represented by Figure 1-4. Assuming that local work rules permitted the pile-driving crew to work without the oiler on the crane and with one less pile driver, and you believed that new productivity would be unaffected by operating in this manner, determine the savings in total direct cost.

5. Evaluate the savings in direct cost by reducing the hourly cost for the crawler crane by $5/hr.

6. Based on the discussion in the text on the effect of assumed crew productivity and on your evaluations in questions 4 and 5, and assuming these kinds of variations represent the level of differences in assumptions one might encounter in estimating work, rate the effect in importance of these separate considerations. Which appears to be most important, which next most important, etc.?

7. If the total cost represented by Figure 1-4 had been estimated by applying unit costs from past experience to the linear feet of piling to be driven, could you easily evaluate the effect of differences in assumptions represented by questions 4 and 5?

2

Horizontal Format: The Method of Cost Expression

Key Words and Concepts

Horizontal display of costs
"Bare" labor
Shift differential
Overtime premium
Portal-to-portal pay
Working-through-lunch pay
Travel and subsistence pay
Union fringes
Employer-provided fringes
Payroll taxes
Payroll insurance
Permanent materials
Expendable materials
Subcontracts
Equipment operating expense

Fuel, oil, and grease (FOG)
Repair parts
Tires
Third-party service
Repair and service labor
Equipment rental
Third-party rental
Equipment depreciation
Operated and maintained equipment
Cost volatility
Productivity-related costs
Relation of cost profile to risk
Relation of risk to contingency allowance and
  profit

This chapter discusses the manner in which the elements of cost for any line entry in any part of an estimate are expressed horizontally across the estimate page. Obviously, such a discussion is relevant only to multivalue estimates. The particular way costs are expressed relates to the general type of contractor preparing the estimate as well as to individual preferences of contractors operating within the same segment of the industry. Also, the breakdown of the total project costs displayed at the conclusion of the estimate plays a major role in identifying the degree of risk associated with the project. This in turn dictates the amount of money that should be included in the bid or cost proposal for contingencies and profit.

The **horizontal display of costs** across the estimate page (that is, the number of columns in which costs are segregated and accumulated) varies considerably, even between contractors who historically pursue the same types of projects. Although the following discussion will be framed around one particular format of cost breakdown that has proven effective in practice, it is not meant to imply that this format is necessarily superior to other formats in common use. However, because it is detailed, it provides a convenient basis for a more complete discussion and will be consistently referred to in this chapter and throughout the remainder of this book.

In this format, the elements of cost for any line entry, listed for the time being in an arbitrary order, consist of the following:

- Labor (L)
- Permanent materials (PM)
- Expendable materials (EM)
- Subcontracts (S)
- Construction equipment operating expense (EOE)
- Construction equipment repair and service labor (RL)
- Construction equipment rental, depreciation, or write-off (R)

A detailed discussion of each element follows.

LABOR COSTS

Most contractors find it convenient to carry all costs associated with construction labor in one vertical column. When this is done, one must recognize that the dollar figures recorded, although all are labor-type costs, may originate in a number of different ways.

"Bare" Labor Pay

"Bare" labor pay represents money paid directly to the contractor's employees, either hourly paid or salaried, including any withholdings employers are required to accrue from wages paid for remittal to the state or federal government (the employee's portion of social security taxes, for instance). Conceptually, this element of labor cost represents wages paid for work performed during the day shift of the standard work week consisting of 8 hours per day, Monday through Friday.

Shift Differential

Hourly paid construction workers receive additional payment when working on shifts other than the normal day shift. Typically, the day shift extends from 8:00 A.M. to 4:30 P.M. with a half hour off for lunch. The swing shift extends from 4:30 P.M. to 00:30 A.M.

the following day with one-half hour for lunch, and the graveyard shift extends from 00:30 A.M. to 8:00 A.M. with a half hour off for lunch. This results in a total of 8 hr of actual work being performed on the day shift, $7\frac{1}{2}$ hr on the swing shift, and 7 hr on the graveyard shift. However, workers on each of the three shifts receive pay for 8 hr of work at the straight-time rate, thus, in effect, creating a pay differential for the two night shifts. This shift differential pay, expressed as a percentage of straight-time pay for each of the three shifts, is illustrated by the following calculation:

Day Shift:

8.0 hr actual work performed

8.0 hr pay received at straight-time rate

Shift differential $= 0$

Swing Shift:

7.5 hr actual work performed

8.0 hr pay received at straight-time rate

$$\text{Shift differential} = \left(\frac{\text{hours paid} - \text{hours worked}}{\text{hours worked}}\right) \times 100$$

$$= \left(\frac{8.0 - 7.5}{7.5}\right) \times 100$$

$$= \left(\frac{0.5}{7.5}\right) \times 100$$

$$= 6.67\%$$

Graveyard Shift:

7.0 hr actual work performed

8.0 hr pay received at straight-time rate

$$\text{Shift differential} = \left(\frac{8.0 - 7.0}{7.0}\right) \times 100$$

$$= 14.29\%$$

Note the following:

Hours worked day shift	$+ \frac{1}{2}$ hr lunch $= 8.0 + 0.5 = 8.5$
Hours worked swing shift	$+ \frac{1}{2}$ hr lunch $= 7.5 + 0.5 = 8.0$
Hours worked graveyard shift	$+ \frac{1}{2}$ hr lunch $= 7.0 + 0.5 = \underline{7.5}$
Total elapsed hours for a three-shift day	$= 24.0$ hr

The above scenario is typical for shift work, although starting and stopping times for the various shifts vary considerably in different parts of the country, depending on local area practice.

Do not be concerned at this point with how money for shift differential pay finds its way into the labor column during estimate preparation. The mechanics are explained in Chapter 6.

Overtime Premium

The cost of the premium paid workers for work performed on an overtime basis is similar to the premium paid for shift differential. Both can be thought of as percentage additions to bare labor pay at straight-time wages. Overtime rates are commonly stated as time and one half (a 50% premium), double time (a 100% premium), or triple time (a 200% premium). These overtime rates are paid for all time worked outside of and above the hours of the normal work shift (whether the normal shift is day, swing, or graveyard shift).

In this text, the following notation will be used to designate the overtime premium rate:

1.5/1.5/2.0:   Time and one-half on weekdays (regardless of shift)
                Time and one-half on Saturdays (regardless of shift)
                Double time on Sundays and holidays (regardless of shift)
2.0/2.0/3.0:   Double time on weekdays
                Double time on Saturdays
                Triple time on Sundays and holidays

Thus, with overtime at 1.5/1.5/2.0, 2 hr of overtime work on each of the three weekday shifts would result in the following effective percentage increase to base wages paid for the hours actually worked on the shift:

$$\text{Premium for 2 hr overtime on day shift} = (1.5 \times 2) - 2.0 = 1.0 \text{ hr}$$

$$= \left(\frac{1.0 \text{ hr}}{10.0 \text{ hr}}\right) \times 100 = 10.00\%$$

$$\text{Premium for 2 hr overtime on swing shift} = \left(\frac{1.0 \text{ hr}}{9.5 \text{ hr}}\right) \times 100 = 10.53\%$$

$$\text{Premium for 2 hr overtime on graveyard shift} = \left(\frac{1.0 \text{ hr}}{9.0 \text{ hr}}\right) \times 100 = 11.11\%$$

Again, do not be concerned at this point with the way in which money for overtime premiums is incorporated into the labor column. The mechanism to accomplish that will also be explained in Chapter 6.

Portal-to-Portal Pay

For the majority of construction projects, workers report to work at a designated location reasonably close to the point at which the work will be performed. This will be true, for instance, in commercial building as well as other projects, including many heavy construction projects. However, some heavy construction projects require the transport of workers some distance from the point to which they are directed to report for work to the location at which the actual construction work will be performed. Similarly, at the end of the work shift, the workers must be transported from the work location back to the point at which they reported for work.

A typical heavy construction project example would be bridge substructure construction across a major body of water, where the bridge piers at which the work will be performed lie in the water some distance from the location on shore where the workers are directed to report for work. A crew boat must pick up the work crews and distribute them at various pier locations on the water at the beginning of the shift, and at the end of the shift the crew boat must pick up and transport the workers back to the point on the shore at which they reported for work. The contractor must pay the workers at the straight-time hourly wage rate for the period of transport out to the work location and for returning from work at the end of the shift. During these periods, no productive work is being accomplished. The cost that has been incurred with no commensurate work being performed must be recovered by loading it into the hours of the shift during which work is being performed. The convenient way to do that is to treat the pay for nonwork hours as a percentage addition to the pay for hours during which work is being performed.

Another situation common to heavy construction occurs in underground work, where a tunnel crew must be transported at the beginning of the shift from the tunnel portal to the heading where the crew will work and must be transported from the heading to the tunnel portal at the end of the shift. Typical underground practice is for the contractor employer to pay the crew for travel from the portal to the heading at the start of the shift and for the crew to travel from the heading to the portal at the end of the shift on their own time. The following hypothetical example illustrates the magnitude of portal-to-portal pay costs typically incurred in underground work:

Assume that an average of 30 min (0.50 hr) is required to travel to and from the heading and that the crews will change shifts at the heading rather than at the portal. If the crews do not work through the $\frac{1}{2}$ hr lunch period, this means that the day shift will work 8 hr at the heading, the swing shift, 7.5 hr at the heading, and the graveyard shift, 7 hr at the heading. Under these circumstances, the premium percentages that must be added to the base labor costs can be calculated as follows:

$$\text{Day-shift premium pay} = 0.50 \text{ hr @ overtime rate}$$

$$= \frac{(0.50 \text{ hr.})(1.5)}{8.0 \text{ hr}} \times 100 = 9.38\%$$

$$\text{Swing-shift premium pay} = \frac{(0.5 \text{ hr})(1.5)}{7.50 \text{ hr}} \times 100 = 10.00\%$$

$$\text{Graveyard-shift premium pay} = \frac{(0.5 \text{ hr})(1.5)}{7.00 \text{ hr}} \times 100 = 10.71\%$$

$$\text{Third-shift premium pay} = \frac{(3 \times 0.5 \text{ hr})(1.5)}{(8.0 \text{ hr} + 7.5 \text{ hr} + 7.0 \text{ hr})} \times 100$$

$$= \left(\frac{2.25 \text{ hr}}{22.5 \text{ hr}}\right) \times 100$$

$$= 10.00\%$$

The manner in which these calculated monies are actually incorporated into the estimate labor column will also be explained in Chapter 6.

Working-through-Lunch Pay

In underground projects, work time at the tunnel heading is so important that the contractor usually will maximize it to the extent possible. In typical tunnel practice, the crew will work continuously without shutting down the heading for the lunch period. Crew members eat lunch "on the fly"; that is, one or two workers drop out of the crew at a time to eat their lunch, requiring the crew to work short-handed until all have eaten.

In return, the crew receives pay for the $\frac{1}{2}$ hr lunch period at the overtime rate. This cost is carried in the labor column.

Travel and Subsistence Pay

Heavy construction projects are frequently located in remote areas some distance from population centers where the workers live. In former days, contractor employers usually established a labor camp at remote locations, providing housing and food for the work force. This is no longer a prevalent practice in the domestic United States due to great improvement in the highway and road system, which makes daily travel over considerable distances feasible. Today, workers travel to and from work or stay at a motel or other temporary facility near the jobsite during the work week, some distance from their normal residence. Under these circumstances, the contractor employer typically pays a daily amount to compensate workers for travel or additional living expenses. These costs are also carried in the estimate labor column.

Union Fringes

Construction work performed by union employees (the segment of the industry operating under collective bargaining agreements with union organizations) requires that the contractor employer pay into union trust funds for various categories of benefits. Typically, trust funds are established by the collective bargaining agreement for worker vacation, health and welfare, pension, and so on. The particular trust funds and the dollar amount to be contributed to each fund per hour worked or hour paid (depending on the wording of the collective bargaining agreement) will be set forth in the particular agreement applying to the work. Union fringes amounting to 18 to 20% of base pay are not unusual. These payments are carried in the estimate labor column.

Employer-Provided Fringes

The "open-shop" or "merit-shop" segments of the industry do not operate under collective bargaining agreements with labor unions. Typically, this class of contractor employer pays the equivalent of union fringes as part of the base pay rate directly to the employee. When this is the case, no additional money for employee fringes need be included in the labor column. However, these contractors usually provide fringe benefits such as health and welfare, pension, life insurance, and/or some form of profit-sharing benefit to their salaried employees. Also, contractors operating under collective bargaining agreements who provide fringe payments to their hourly paid labor generally provide health and welfare, pension, life insurance, and profit-sharing benefits to their salaried employees as well. Salaried employee benefits, which commonly amount to 20 to 25% of base pay, are carried in the labor column.

Payroll Taxes

A number of payroll taxes result in costs to construction employers. The largest of these is the Federal Insurance Contribution Act tax (FICA). In addition, the employer must pay Federal Unemployment Insurance (FUI) tax and, in most states, a state unemployment tax (SUI). These federal and state taxes are established by statute and, like most state and federal taxes, are subject to periodic increases. At any particular point in time, the percentage rates can be obtained by contacting the federal and state tax authorities. The sum of FICA, FUI, and SUI taxes typically amounts to 12 to 15 percent of base pay. Construction employer costs resulting from these taxes are carried in the estimate labor column.

Payroll Insurance

There are two principal types of labor-related insurance that result in costs to construction employers: worker's compensation and employer's liability insurance and third-party liability insurance. The premium for worker's compensation and employer's liability insurance is generally stated as a percentage of base pay, although one state (Washington) bases the premium on labor-hours of employment. Premiums for worker's compensation and employer's liability insurance are carried in the labor column. Contractors obtain the rates they will be required to pay from in-house insurance specialists or from their insurance and bond brokers.

Premiums for worker's compensation insurance depend on the type of work involved and the particular state of the union in which the work is performed. For underground work, premiums as high as 30% of base pay are not uncommon. In Hawaii, premiums for even routine work such as general carpentry approach 50% of base pay.

Some contractor employers also pay premiums for third-party liability insurance as a percentage of base labor cost. In this event, these premiums would also be carried in the labor column. However, many contractors pay third-party liability insurance premiums as a percentage of the total contract price, and under this method, the third-party liability insurance premium is included separately as part of the indirect cost and is usually carried in the expendable materials column rather than in the labor column.

Labor Column Estimate Mechanics

As stated previously, you should not be concerned at this point with how the various elements of labor cost find their way into the estimate labor column. This is accomplished in various ways, as will be made clear in later chapters. You should note that the computations made in this chapter for cost elements such as shift differential, overtime premium, portal-to-portal pay, and working-through-lunch pay are not intended to illustrate how these costs are calculated for entry in the labor column but rather, simply illustrate the magnitude of such costs relative to base labor pay.

PERMANENT MATERIALS

The permanent materials category of total project cost represents the cost of the construction materials to be incorporated in the work and remain as a permanent part of the project. More common permanent materials include structural steel, reinforcing steel, concrete, piping, miscellaneous metal work, anything embedded in concrete, and installed equipment. Permanent materials also include concrete materials such as cement, admixtures, various gradations of concrete aggregates, processed sand and gravel materials placed in backfills on the project, or any other type of purchased material permanently incorporated into the project.

The usual practice is to consider permanent materials as including only those that are specifically detailed or shown on the construction drawings. In the case of project elements such as cast-in-place concrete or materials placed in backfills, only that portion of such materials designated within "neat lines" on the drawings would be carried in the permanent material column. When placed in actual construction, part of these materials will fall outside of the neat lines shown on the drawings. This portion falling outside the neat lines, and/or materials that are wasted as cutoffs or trim wastage, usually are excluded from material included in the permanent material column even though the unit cost of such material is exactly the same. Cost of such waste material is usually carried in the expendable material column as explained below.

EXPENDABLE MATERIALS

By definition, expendable materials are building materials required to construct the project that are not shown on the drawings or otherwise specified to remain as a permanent part of the work. Typical examples are materials required for concrete formwork, steel sheet piling that will be extracted when its function is completed, explosives and detonators required for rock excavation, and small tools and supplies such as slings and shackles, oxygen and acetylene, and air and water hose and fittings.

As noted above, permanent materials falling outside the neat lines specified on the drawings or waste permanent materials are also generally carried in the expendable material column. A particularly important example is "overbreak" concrete, that is, concrete placed outside the neat lines shown on the drawings due to the structural concrete element being placed directly against an irregular soil or rock foundation.

Another important category of expendable material cost is the cost of materials utilized for temporary structures, such as steel and timber used for ground support systems, access trestles, scaffolding, or vertical shoring systems.

SUBCONTRACTS

The category of subcontracts is the cost of on-site work performed by forces other than the prime contractor's under a subcontract agreement between the prime contractor and a subcontractor. It is important to understand that a subcontract cost paid by the prime contractor conceptually represents all costs incurred by the subcontractor in performing the subcontract work plus the subcontractor's overhead and profit. Thus, all elements of the subcontractor's costs, plus overhead and profit, are included in the subcontract column of the prime contractor's cost estimate, although they are not broken out separately.

CONSTRUCTION EQUIPMENT OPERATING EXPENSE

In the particular horizontal format discussed here, the costs for construction equipment consist of three major divisions, the first of which is **equipment operating expense.** A number of different costs incurred to keep construction equipment operating are included here.

Fuel, Oil, and Grease

The **fuel, oil, and grease** (FOG) category includes in part the cost of fuel (diesel oil, gasoline, and/or the cost of electric power for equipment driven by electric motors). Fuel costs can be immense for certain types of equipment such as large earth movers or large hydraulic dredges. Also included in this category is the cost of lubricating oil and various types of hydraulic fluid typically required for hydraulically controlled and/or hydraulically driven modern-day equipment. FOG costs also include the costs of heavy greases and similar lubricants required for construction equipment.

Repair Parts

All types of construction equipment require **repair parts** ranging from fan belts and spark plugs to parts required for major overhaul such as caterpillar tracks and final drives for heavy earth-moving equipment. In some equipment accounting systems, repair parts are subdivided into running supplies (meaning small parts that are typically replaced in the field) and major overhaul parts (required when the equipment is rebuilt). Another parts category that is sometimes separately accounted for includes ground contact parts, which consist of ripper teeth, bulldozer and motor grader cutting edges, end bits, and so on.

Note that costs for drill steels and bits for soil and rock drilling equipment are normally not carried as an equipment operating expense. Rather, costs for these replacement items are usually carried separately as an expendable material in the expendable material column.

Tires

Many types of construction equipment are mounted on **pneumatic tires.** Today, earth and rock excavation haul units have reached tremendous sizes, requiring extremely large tires. Tread widths can easily approach 3 to 4 ft and the height of tires for these units often exceeds the height of a tall man. The cost of an individual tire of this size runs in the thousands of dollars.

Tire expense is usually a major consideration in large earth and rock excavation projects. The magnitude of the cost obviously depends on the quantities of excavation to be hauled. Also, costs will depend on the abrasiveness of the material and the daytime ambient temperatures that will prevail during the haul operation.[1] The tire costs for tough rock excavation projects or projects where common earth materials are being hauled over long distances in high daytime ambient temperatures can be very large.

Third-Party Service Costs

Third-party service costs represent charges for mechanic and tire service organizations engaged by the contractor to assist in the on-site repair of equipment or to rebuild the equipment. These charges would include many of the costs discussed above, as well as overhead and profit for the organization providing the service.

---

[1]High ambient temperatures cause tires to fail due to heat buildup within the tire during high-speed hauling operations.

Equipment Operating Expense Breakdown

Because the costs of all the individual items discussed above are carried in a single column, the equipment operating expense total derived at the end of an estimate will consist of a mix of costs for different commodity types and/or services. It is often helpful to have a breakdown of these costs. Today, when heavy construction costs estimates are computer aided, the computer program is often organized in a manner that produces a breakdown of the equipment operating expense column into the above individual categories (or certain commonly understood combinations of these categories) on a total job basis.

CONSTRUCTION EQUIPMENT REPAIR AND SERVICE LABOR

The cost of construction equipment **repair and service** labor is the labor cost for the mechanics and service personnel. An example is the tire and lubrication crew, which repairs, fuels, lubricates, and cleans the construction equipment. On projects involving extensive heavy equipment, particularly underground work where tunnel-boring machines and similar equipment are needed, it is often useful to identify the cost of mechanics and service personnel separately and distinctly from other direct labor costs, hence, the need for the equipment repair and service labor column.

This specialized labor is subject to the same fringes, taxes, and insurance, and to all of the same premium pay rules, as other direct labor. Although carried in a separate column, this specialized labor cost is handled in the same manner as other labor costs.

CONSTRUCTION EQUIPMENT RENTAL (DEPRECIATION OR WRITE-OFF)

There are at least three separate sources from which construction equipment can be furnished to a jobsite:

1. Equipment may be rented by the contractor from a third-party entity.
2. Equipment may be owned by the contractor.
3. Equipment may be rented from a separate business entity on an "all found" or operated-and-maintained basis (O&M rental).

Equipment Rented from a Third-Party Entity

Equipment obtained from a **third-party entity** is paid for according to the terms of a rental agreement between the contractor and that entity (usually an equipment rental organization) or sometimes is obtained from an equipment manufacturer on a long-term lease basis. In either case, the cost is usually incurred as a monthly payment made to the supplier of the equipment. The rental or lease agreement will usually specify that the contractor is responsible for fuel, oil, grease, and running repairs, whereas the equipment supplier is responsible for major overhaul-type expenses that the equipment may require.

Contractor-Owned Equipment

If the contracting organization performing the contract is a joint venture,[2] it is not uncommon for that joint venture to acquire the required equipment outright at the beginning of the project and dispose of it by sale to an individual partner or to third parties at the end of the project. In this case, the money charged to the estimate in the equipment rental column would represent the decline in value of the equipment during the performance of the job, that is, the **depreciation** or **write-off,** of the equipment. Even when it is not a joint venture, the contractor may obtain construction equipment from a separate in-house equipment rental company and pay a monthly rental charge for the use of the company's equipment on the project. In this event, the money would be charged to the estimate in much the same way as if the equipment were being rented from third parties, even though the rental transaction is an internal paper transaction within the same parent company.

Operated and Maintained Equipment

Usually, some of the equipment used on construction projects is obtained on an hourly basis from separate business entities that furnish the equipment on an "all found," or **operated and maintained (O&M), basis.** Intermittent equipment needs are often met in this way. The third party furnishes not only the equipment, but fuels and maintains it and also furnishes the operating personnel. Common examples of this type of equipment rental include crane service, concrete pumping service, and on-highway haulage services for supply of concrete aggregates, sand, or similar products to a jobsite. Another example is rental of on-highway trucking services for excavated materials that must be hauled and disposed of some distance from the site.

Depending on how the equipment is furnished to the project, the business and/or economic consequences to the contractor vary considerably. For instance, equipment rented from third parties represents a cash expense depleting the contractor's cash reserves and improving the cash position of the rental company, whereas depreciation on equipment owned by the contractor is not a cash expense.

Equipment Rental Breakdown

Like the equipment operating expense column, the equipment rental column in the completed cost estimate will include a mix of the types of equipment rental expenses discussed above. It is often useful to have a breakdown of the separate kinds of equipment rental costs; when the estimate is computer aided, the computer program will provide a summary that breaks down the several kinds of costs on a total project basis.

THE SIGNIFICANCE OF THE VARIOUS KINDS OF COSTS IN THE HORIZONTAL COST FORMAT

Labor Costs

Labor costs (L) are the most **volatile** of all construction costs; that is, they are the most difficult to estimate and control during project performance. Hourly paid labor is much more volatile than salaried payroll costs.

---

[2]A separate business entity composed of two or more separate contractors operating under a joint-venture agreement for that particular project.

Field labor costs for construction work put in place by the hourly paid construction trades are highly **productivity related.** The higher the work crew productivity achieved, the lower the labor cost, and vice versa. Uncertainty in forecasting labor productivity therefore results in commensurate uncertainty in estimated labor costs.

Permanent Materials

Unlike labor, permanent material (PM) costs are not productivity related. Although they often constitute a large portion of total project costs, they usually can be accurately determined by careful examination of the bidding plans and specifications and therefore are not volatile. The possibility of error in determining the correct permanent material costs for a project represents little risk to the bidding contractor.

Expendable Materials

Normally, expendable materials (EM) costs are not particularly productivity related. However, they can be very difficult to determine accurately and thus are always volatile to some extent. In some projects, the cost of expendable materials can be highly volatile. For example, consider the sketch shown in Figure 2-1 for the cross section of a railway tunnel constructed in British Columbia in the early 1980s.

The tunnel was specified to be constructed by drill-and-blast methods and the specifications required that no rock "tights" project inside the A line shown. When tunnels are constructed by this method, the rock can be expected to break in an irregular pattern outside of the A line in a manner similar to that indicated on the sketch. The extent of such "overbreak" depends partly on the care exercised by the heading crew and partly on the pattern of joints and other discontinuities existing in the rock through which the tunnel is to be driven. Twelve inches of overbreak is common, and in extreme cases, overbreak can

FIGURE 2-1 Example of the potential importance of the expendable material item of overbreak concrete.

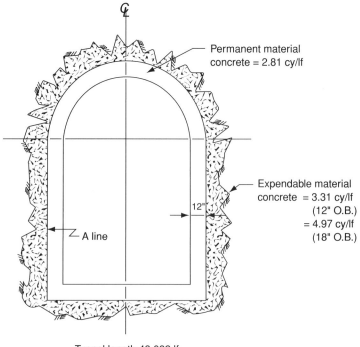

Permanent material
concrete = 2.81 cy/lf

Expendable material
concrete = 3.31 cy/lf
(12" O.B.)
= 4.97 cy/lf
(18" O.B.)

12"

A line

Tunnel length 48,082 lf
Approximate cost of concrete = $80/cy (FOB forms)

extend 3 ft or more beyond the A line. The contractor's takeoff figure for permanent material concrete (lying between the inside form surface of the tunnel lining and the A line) was 2.81 cy/lf. The takeoff figure for expendable material concrete (falling outside of the A line) was 3.31 cy/lf, based on 12 in. of overbreak, and 4.97 cy/lf, based on 18 in. of overbreak. The tunnel was 48,082 lf long and the estimated cost of batched and mixed concrete delivered to the tunnel forms was approximately $80.00 per cy.

Regardless of which overbreak assumption is made, the cost of permanent material concrete is the same. However, if overbreak is estimated at an average of 12 in., the expendable materials cost would equal

$$EM = (3.31)(48,082)(80.00) = \$12,732,114$$

If overbreak is estimated at an average 18 in., the expendable material costs would be

$$EM = (4.97)(48,082)(80.00) = \$19,117,403$$

The difference, depending on the overbreak assumption, equals $19,117,403 − $12,732,114, or $6,385,289, a very large number. This example is an extreme case, but it illustrates admirably the volatility of some expendable material costs.

When temporary structures are involved, such as ground support systems consisting of braced sheet piling or braced soldier pile and lagging systems for cut-and-cover construction, the determination of the expendable material cost for the materials used involves considerable layout and engineering work that must be performed by the bidding contractor. This contrasts sharply with the determination of permanent material costs, which do not require contractor layout and engineering except in special cases where the contract requires permanent project design work to be performed by the contractor.

Subcontracts

Although highly productivity related from the subcontractor's standpoint, subcontract costs (S) entered in the prime contractor's bid estimate are not productivity related from the standpoint of the prime contractor. Subcontract costs can vary from zero for some heavy engineered projects to 85 to 90% of the total job costs for commercial building projects.

When subcontract costs are in the form of quotations "per plans and specifications," secured by a 100% performance bond, they are not volatile and represent little risk to the prime contractor. However, when a subcontract quotation is in the form of a unit price to be applied against actual quantities of work to be done (estimated by the prime contractor), the money recorded in the subcontract column of the prime contractor's bid estimate can be very volatile, depending on the accuracy of the prime contractor's takeoff for the quantity of work to be performed. Also, when the subcontract quotation is from an unknown or nonbondable subcontractor, it represents considerable risk to the prime contractor and should be considered highly volatile. The use or nonuse of such quotations is a matter of business judgment, but it must be recognized by the bidding contractor that when such quotations are utilized, they represent considerable risk.

Equipment Operating Expense

Equipment operating expense (EOE) will always be highly related to the kind of work being performed in the construction contract. It will be a large cost in projects involving substantial numbers of heavy equipment units and will be a small cost in projects involving minimal construction equipment. It can be fairly volatile on equipment-intensive jobs under adverse physical conditions, such as excavation jobs in tough abrasive rock, jobs involving high-speed haulage operations in hot climates, or work in which the equipment is subjected to the action of salt water. An extreme example of an adverse operating condition is underground work where the equipment is continually inundated by groundwater inflows in a saltwater environment.

The estimator is confronted by a dual problem when estimating equipment operating expense. There are two major variables. One is the magnitude of the costs for the various EOE elements per hour of equipment operation. These costs are usually determined by reference to records of similar usage of equipment on previous projects. The second variable is the number of hours of equipment usage that will be required for the given amount of construction work to be accomplished. Successfully determining this variable depends on correctly estimating the productivity expected to be achieved by the equipment under the given job conditions.

Repair and Service Labor

Determining repair and service labor (RL) costs involves the same dual problem encountered in estimating equipment operating expense. The first part of the problem is determining how many labor-hours of repair and service labor will be required per equipment operating hour, and the second part is determining how many equipment operating hours will be required to accomplish the given amount of physical construction work. The first is generally determined by referring to records of past performance on similar work in which the total hours of repair and service labor expended for the various equipment units are recorded along with the number of hours of equipment usage. This establishes a labor-hour per equipment-hour relationship that generally will be applicable to future work of a similar nature where the skill and experience of the personnel involved is expected to be comparable.

Repair and service labor has all the attributes of other direct labor but normally will not be as large a cost.

Equipment Rental

The equipment rental (R) cost element is not only highly dependent on the type of construction project but will also vary greatly depending on the contractor's business philosophy and equipment policy, which in turn is often highly influenced by competitive conditions. This element is not particularly volatile, although, obviously, if the equipment usage estimate is grossly in error, the resulting estimate of rental expense will also be grossly in error.

A unique aspect of monies recorded in the equipment rental column is that they may not represent direct cash outlays. For instance, depreciation on equipment owned by a joint venture during the life of the project under contract is not a month-by-month cash outlay. The joint-venture partners invested the funds for equipment acquisition costs at the start of the job. Similarly, rental from the in-house equipment division of a company or from a subsidiary company is not a month-by-month cash outlay. The acquisition costs for the equipment have already been incurred. On the other hand, costs for any type of rental equipment from an outside entity *is* a month-by-month cash outlay. With the exception of O&M rental, this outlay is not as volatile as direct labor. However, O&M rental is highly productivity related and is just as volatile as direct labor.

Heavy construction estimates	L	+	PM	+	EM	+	S	+	EOE	+	RL	+	R	=	T
A simpler form	L	+	M			+	S	+			E			=	T
A simpler form yet	L		M	+	M	+					E			=	T
And yet simpler	L		+		M									=	T
And the ultimate (single value)			T											=	T

FIGURE 2-2 Forms of estimate cost expression.

POSSIBLE VARIATIONS IN HORIZONTAL FORMAT OF COST EXPRESSION

All discussion up to this point has focused on a particular horizontal format for cost expression that has proven effective for heavy engineered construction estimates. Many present-day companies operating in this field use a similar format. However, many variations are possible, depending on the degree of cost detail desired. Figure 2-2 indicates the more common format variations (the letters are abbreviations for the column headings).

The top line on Figure 2-2 is the format that fits the needs of heavy engineering construction contractors, as discussed in preceding sections of this chapter. By combining permanent and expendable materials into a single materials (M) column and the three separate equipment cost columns into a single equipment (E) column, the simpler form of cost expression in the second line of Figure 2-2 results. This format is used by many industrial contractors and some heavy engineering contractors.

A yet simpler form shown on the third line of Figure 2-2 results by further combining the already combined materials (M) column with the subcontracts (S) column and forming a single column, also entitled materials (M). This format is used by many specialized civil work contractors, such as pile-driving contractors and light-grading and paving contractors, and is also used in most published industry construction cost manuals in common use.

The fourth line of Figure 2-2 further combines the previously consolidated materials (M) column and the consolidated equipment (E) column into a now highly consolidated column also under the heading of materials (M), which now contains many other cost types in addition to strictly material costs. This format is widely used by electrical and piping contractors and contractors performing specialty trades work, such as painting, tile setting, and roofing.

A final consolidation results in the single value estimate shown on the last line of Figure 2-2, commonly used for conceptual estimates and the other previously discussed estimates made by architect/engineers (A/Es) during the development and design phase of a project.

RELATIONSHIP OF RISK TO ESTIMATE HORIZONTAL FORMAT COST PROFILE

As the previous discussion might indicate, one important aspect of the various types of costs in the estimate horizontal cost format is the relationship that the distribution, or profile, of these costs has to the degree of **risk of performance** for the project at hand.

Project	Labor	PM	EM	S	EOE	RL	R	Total
High-rise building	5.0	2.0	12.0	80.0	0.5	—	0.5	100.00
Sewerage treatment plant	14.9	37.9	5.0	38.7	1.0	0.4	2.1	100.00
Bridge river crossing substructure	32.5	11.1	24.8	1.9	4.0	1.4	24.3	100.00
Rapid transit tunnels in rock by TBM	29.2	5.4	17.8	20.1	6.5	1.8	19.2	100.00
Large earth-filled dam for BUREC	33.5	21.5	7.4	4.0	20.0	5.3	8.3	100.00
Hand-mined drainage tunnels	53.5	1.9	31.3	0.3	2.6	1.3	9.1	100.00

FIGURE 2-3   Variation in horizontal cost profile for various types of construction projects.

When the cost estimate for the project has been completed (including the direct and indirect costs), a key management decision is to determine the monetary value of **contingency allowances and profit** to be included in the bid or cost proposal. The amount of contingency allowances clearly relates to the amount of project risk, and a prudent entrepreneurial philosophy dictates that profit to be gained should correlate to the risk of loss to be assumed. Therefore, contractors pursuing new construction work by means of competitive bidding or by negotiation will look to the profile of the types of estimated cost they face in performing the work of the project and will consider this in evaluating the risk in the project. The contingency and markup sections of the next chapter in this text discuss the particular mechanics involved in this consideration.

To understand the great variation that can occur in the profile of construction costs elements, refer to Figure 2-3. This figure tabulates a profile of the total costs for six different kinds of construction projects ranging from high-rise building construction to several types of high-risk heavy engineered projects. In each case, the figures shown for each of the various cost elements are the percentages of the **total project cost** that each particular element represents. The table was derived from cost data taken from actual projects, either from competitive bids for the project or from the contractor's final cost ledgers when the project had been completed.

High-Rise Building: For Comparison to Heavy Construction

By considering the high-rise building project first, you will immediately note the extremely small percentage of total cost represented by the labor column, permanent material column, and equipment cost columns. On the other hand, subcontract costs comprise 80% of the project cost. The remaining 12% for costs recorded in the expendable material column are a mix of the cost of "general conditions" materials and services provided by the prime contractor for the benefit of the subcontractors who were performing the actual work. These costs include utility bills for construction water and electric power, performance and payment bond premiums, premiums for third-party and other insurances provided by the prime contractor, and a number of other items of a minor nature. The prime contractor furnished minimal permanent materials because the bulk of the permanent materials were furnished by the subcontractors. The money recorded in the subcontract column represents subcontract quotations secured by 100% performance bonds.

The profile of costs shown for this typical high-rise building project represents very little risk for the prime contractor. Accordingly, the contingency and profit additions to the cost estimate for this project were minimal.

Sewerage Treatment Plant Construction

For the project represented by the cost profile shown, the general contractor subcontracted all site excavation and yard piping and all electrical, mechanical, and architectural finish trades work. The prime contractor's own forces performed all structural concrete work and interior piping work. Accordingly, the recorded cost elements for equipment usage are small, but labor costs (including both indirect and direct labor) were 3 times those for the high-rise building, whereas the permanent material costs (representing purchase of ready-mix concrete, reinforcing steel, and the interior piping materials) represent nearly 40% of the total cost of the project. In this particular example, the expendable materials expense was small.

Although only moderate, the potential risk of loss for this project was 3 to 4 times that for the high-rise building example.

Bridge River Crossing Substructure Construction

This next example was a project involving construction of large bridge piers in deep water across a major river. Much of the work was constructed from floating equipment. The bridge piers were founded on reinforced concrete circular caissons 5 and 6 ft in diameter. The caissons were sunk through 30 ft of river sediments overlying a limestone foundation into which they were socketed.

One is immediately struck by the magnitude of the costs for the required equipment, which in this case totals nearly 30% of the total project cost. Most of this money is in the rental column. The bridge contractor owned most of the heavy equipment required with the exception of the fleet of work barges and the several tugboats required to move them from pier to pier during the work. A large part of the rental expense represented monies paid to third parties for the rental of these tugs and barges.

The general contractor's own forces performed practically all the work, with the exception of a small amount of temporary and permanent electrical work performed by an electrical subcontractor. The work involved deep sheet pile cofferdams constructed in the riverbanks and in the river proper, which required a large expenditure for sheet pile and bracing material for these temporary structures. The bridge pier design lent itself to the use of custom-built steel concrete formwork. All these project features are reflected in the nearly 25% of total project cost shown in the expendable material column. The only significant permanent material required for the project was transit-mix concrete and reinforcing steel for the bridge piers.

The balance of the work, nearly 33% of the total project cost, consisted of the prime contractor's direct and indirect labor, the majority of which was direct labor.

This type of heavy engineered construction project, as reflected by the horizontal format cost profile, represents great risk of financial loss to the contractor. The monies estimated for direct labor, expendable materials, and equipment expense (all volatile costs) total nearly 85% of the total project cost.

Rock Tunnels Excavated by Tunnel Boring Machines

By its very nature, underground construction is a high-risk undertaking. Today, contractors performing underground work under fixed-priced contracts are usually protected from the risk of encountering unexpected adverse underground conditions by a differing site conditions clause in the contract. Nonetheless, contractors still face great risk in evaluating probable costs, even when the encountered geotechnical conditions

PHOTO 2-1 These photographs were all taken during the installation of the 5- and 6-ft-diameter concrete caissons described in the text. They illustrate the employment of heavy floating equipment that is typical for substructures of major river crossings.

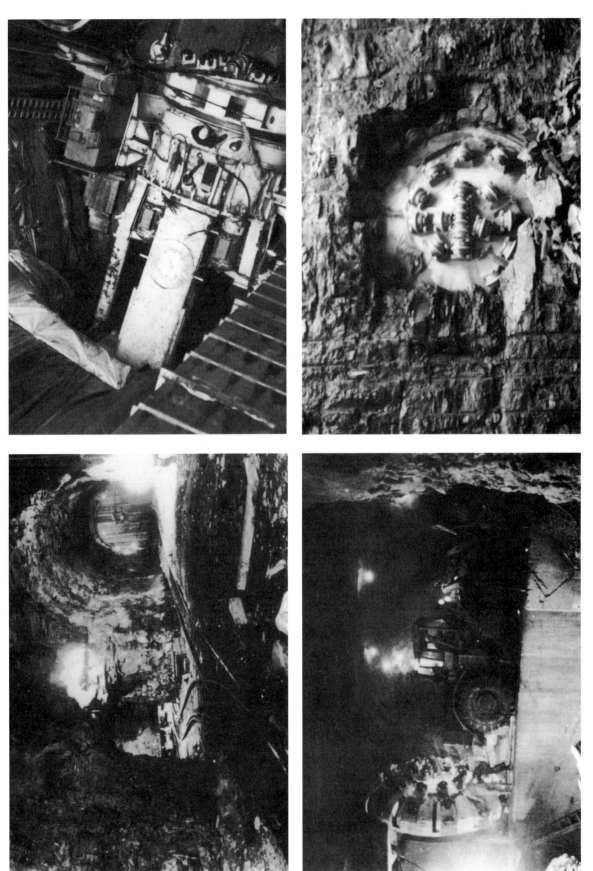

PHOTO 2-2 These photos were all taken on the TBM excavated rock tunneling project discussed in the text. This type of project is very labor and equipment intensive. Over 70% of the total project cost consisted of labor, equipment, and expendable materials.

turn out to be as expected. The cost profile displayed in Figure 2-3 was derived from a contract where two 11,000 ft parallel tunnels were excavated in hard limestone by the use of tunnel boring machines (TBMs) and lined with concrete for a rapid transit project. The contract required that the tunnels be excavated simultaneously, thus requiring two TBMs, each with all of the supporting underground equipment. All told, the required investment to procure the TBMs and supporting underground equipment (in 1979) was $10 million. The general contractor's own forces performed all TBM excavation, ground support, and the concrete lining of the tunnels and work shafts.

The subcontract money shown reflects the required services of an electrical subcontractor to install the power drops and the surface and underground electrical distribution systems to supply electric power to the TBMs, each of which involved a connected load of 1,200 hp. The subcontractor also supplied all service and maintenance electricians for the full duration of the project. Also included in the subcontract column was the cost for a haulage subcontractor to remove the spoil generated by the TBMs from the project site and dispose of it at a site some distance from the jobsite.

Most of the equipment rental expense, approximately 20% of the total cost, represented a write-off on the TBMs and other underground equipment. The money in the expendable material column was required primarily for small tools and supplies, underground support materials, indirect cost expendable material items, and bond and nonpayroll insurance costs. A number of temporary surface facilities required to support the underground operations also required expendable materials.

The permanent materials, approximately 5% of the total cost, represented the cost of transit-mix concrete for the lining of the tunnels.

The nearly 30% of total cost that was expended for direct and indirect labor is typical for this type of project. All told, the combination of labor, expendable materials, and equipment costs exceeded 70% of the total project cost. All these costs are highly volatile and represented a large financial risk for the contractor.

Earth-Fill Dam Construction

The fifth line in Figure 2-3 shows the cost profile of a bid estimate for an earth-fill dam for the U.S. Bureau of Reclamation. This 40 million cubic yard dam was to be constructed primarily from borrow excavation obtained about 3 mi upstream from the dam site. The project also contained a small grouting and drainage tunnel, concrete inlet and discharge works, and an extensive reinforced concrete emergency spillway. The cost profile shown reflects the characteristics of this type of project.

As is typical in the other heavy engineered projects discussed, the equipment costs for this estimate total well over 30% of the total project cost. The distribution of costs within the equipment group is interesting. Equipment-operating expense and repair labor are many times higher than for the other heavy engineered projects in Figure 2-3. Similarly, the monies in the rental column were roughly one-half to one-third of the rental expense for the other two heavy engineered projects. The majority of the equipment expense reflects the usage of a fleet of earth-moving equipment consisting of large loaders, 100 t bottom dump haul units, track-mounted bulldozers, motor graders, and assorted minor units. Although the acquisition cost of this type of equipment is high, its useful life is long enough that a considerable salvage value could be expected at the end of the project. Under these circumstances, a lower equipment write-off cost was reasonable. On the other hand, the 3 mi haul over several years of operation in high ambient temperatures resulted in relatively higher estimated costs for equipment spare parts and tires and for the associated service and repair labor.

The permanent material cost, equal to 2.5% of total project cost, primarily reflected the cost of concrete materials, reinforcing steel, embedded items, and permanent

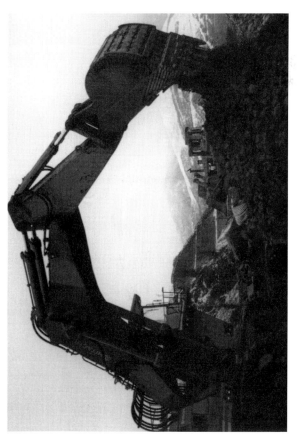

PHOTO 2-3   Large earth- or rockfill dams typically employ the type of heavy equipment shown and generate the type of cost profile described in the text.

33

PHOTO 2-4 These photographs show some of the construction details for the hand-mined drainage tunnel project discussed in the text. The major costs for this project were labor (53.5% of total) and expendable material (31.3%), with comparatively little equipment (13%).

dam-operating equipment. Expendable material costs totaling 7.4% of total project cost included form materials for the concrete work and small tools and supplies. Subcontract costs were small (4% of the total) and were mainly required for temporary power and lighting distribution systems and for permanent equipment power and controls.

In common with the other heavy engineered projects discussed, a comparatively large percentage of total costs was required for direct and indirect labor, in this case comprising 33.5% of the total estimated project cost. This labor cost compares to 32.5 and 29.2% for the bridge substructure and TBM tunnel projects, respectively.

Considerable risk is eliminated when a contractor is protected from the consequences of unexpected adverse site conditions by virtue of a differing site conditions clause in the contract, as was the case here. Nonetheless, projects of this type still involve risk, although not as much as for either the bridge substructure project or the TBM excavated rock tunnels. Contractors experienced in this type of work would regard the risk of the earth-fill dam project as moderate.

Hand-Mined Drainage Tunnels through Railroad Embankment

The final project depicted in Figure 2-3 consisted of eight side-by-side, closely spaced, 20-ft-diameter, 300-ft-long drainage tunnels driven by hand-mining methods and supported by bolted unitary steel-liner plate furnished by the owner. About the only equipment required was a track-mounted loader used for removing muck from the tunnel face after it was loosened by hand spading, a large air compressor, and portable pumps, welding machines, and so on. The total equipment expense was only 13% of total project cost, subcontracts were virtually zero, and, because the tunnel-liner plate was furnished by the owner, permanent material expense was also very low.

The major expense items consisted of expendable materials (31.3% of total project cost) and direct and indirect labor (53.5% of total project cost). The expendable materials consisted mainly of temporary supplementary support materials for the tunnels, small tools and supplies, bond and nonlabor insurance premiums included in the indirect cost, and other similar expenses.

Labor costs were a far larger portion of total cost than for any of the other projects discussed. This fact indicates the very high risk inherent in this project, particularly when the type of labor involved is unusual and not often performed, as was the case here. With the possible exceptions of encountering unexpected adverse site conditions without a differing site conditions clause in the contract, or absorbing catastrophic losses due to the vagaries of adverse weather or similar forces beyond the contractor's control, the risk inherent in this project typifies the highest category of risk to which a construction contractor can be exposed. The contingency allowance and markup for profit included in the contractor's bid for such a project should be much larger than for any of the other examples shown in Figure 2-3.

CONCLUSION

This chapter has explained the nature of the various elements of cost as they are displayed horizontally across the estimate page for any line entry made anywhere in the estimate. The same horizontal format is used for both the direct and indirect cost parts of the cost estimate.

The significance of each of the various cost elements was discussed from the standpoint of the impact of the cost element on the risk profile of the project. It was further explained that the project risk profile will affect the magnitude of the contingency allowance and markup percentage for profit that should be included in the contractor's competitive bid or cost proposal.

Also, the risk profiles for five different heavy engineered projects were compared to each other and to that for a typical high-rise building construction project, illustrating the great differences between heavy construction and building construction.

The following chapter discusses the vertical organization, or format, of construction cost estimates including the direct cost, the indirect cost, and other vertical format elements.

QUESTIONS AND PROBLEMS

1. What are the seven elements of the horizontal cost format discussed in this chapter?

2. What are the ten separate ways that labor costs originate that are explained in the chapter?

3. What are the usual hours paid and worked for day, swing, and graveyard shift, respectively?

4. What two types of projects were discussed that involve the possibility of portal-to-portal pay? What type of project is likely to involve working-thorough-lunch pay?

5. In the case of union fringes, to whom is the money paid? In the case of open shop or merit shop projects, to whom is the equivalent of union fringes paid?

6. Who receives payroll tax payments from the contractor?

7. What two types of contractor-paid insurance premiums were discussed in this chapter?

8. If third-party liability insurance premiums are based on a percentage of payroll cost, in which horizontal format column is the cost of the required premiums carried? If the premium is based on a percentage of the total construction contract price, in what part of the contractor's estimate and in which horizontal format column is the premium cost carried?

9. What is the feature of use in the project that distinguishes permanent materials from expendable materials?

10. For work expected to be performed by subcontractors, where in the prime contractor's horizontal cost format are the costs to be incurred by the subcontractors carried? How about the subcontractor's overhead costs and profit?

11. What are the three separate kinds of equipment cost incurred by the prime contractor that were discussed in this chapter?

12. What four separate kinds of costs are included in the equipment operating expense column?

13. Aside from being incurred for different purposes, are there any other distinguishing differences in the nature of the costs carried in the labor and repair labor columns?

14. What are the three separate sources from which construction equipment can be furnished to a construction jobsite that are discussed in this chapter?

15. Of the seven cost elements in the horizontal format, which are highly productivity related (from the prime contractor's standpoint), somewhat productivity related, or not productivity related at all?

16. From the standpoint of volatility, which cost element in the horizontal format is the most volatile, and which two cost elements are the least volatile?

17. Based on the discussion in this chapter, rate the six projects shown in Figure 2-3 in terms of risk of financial loss to the contractor, highest risk first, next highest next, and so on.

3

The Vertical Estimate Format

Key Words and Concepts

Direct cost

Lump sum bid

UCI format

Schedule-of-bid-items bid

Bid item

Unit price

Lump sum price

Bid quantity

Bid-item extension

Indirect cost

Salaried payroll

Time-related overhead expense

Non-time-related overhead expense

Insurance and taxes

General plant in-and-out

Escalation

Interest (the cost of money)

Cash-flow analysis

Cash balance

Accumulative cash balance

Identified contingencies

Unidentified contingencies

Markup (margin)

Relation of markup to annual construction volume

Bond premiums

Whereas the previous chapter focused on the horizontal organization of construction cost estimates, this chapter is concerned with vertical format, that is, what is included in the estimate, or the estimate's scope. The distinct parts of a contractor's bid estimate or cost proposal for a typical heavy engineered construction project usually consist of the following:

1. Direct cost
2. Indirect cost
3. Escalations
4. Interest (that is, the cost of money)
5. Contingency allowances
6. Markup or margin
7. Contract bond premiums

The horizontal format of cost expression discussed in the previous chapter applies to the first five of the above listed elements of vertical format. Dollar amounts for the last two, markup and contract bond premiums, are carried only in the far right-hand, or total, column.

THE DIRECT COST

Definition of Direct Cost

Direct cost is usually thought of as that part of the contractor's costs that can be directly associated with discrete physical parts of the project construction work. On a hydroelectric project, for instance, powerhouse rock excavation would be such a discrete part of the project. So would other physical parts of the project such as powerhouse substructure concrete, the turbines, the generators, dam foundation excavation, and dam embankment. The number of distinct line items in the direct cost obviously will depend on the complexity of the project and on the proclivities of the engineer who drafts the bid documents.

Direct Cost Format

Usually, the vertical format is a matter that the drafter of the bid documents controls and will depend on whether the owner desires to obtain the contractor's bid or proposal in the form of a single lump sum price or, alternately, in the form of unit prices and/or lump sum prices to be bid by the contractor against a series of bid item quantities established by the owner's engineer. This latter form of bid would be referred to as a schedule-of-bid-items bid and is more common in heavy construction.

Lump Sum Bids

If a lump sum bid *is* called for, bidding contractors are free to use any vertical format they wish when preparing the direct cost estimate portion of the bid. Generally speaking, the most convenient format will be one that best parallels the format of the technical specifications section of the bidding documents. In the United States, the technical specifications usually conform to the format of the Uniform Construction Index (UCI). The major divisions of the UCI are as follows:

Division 1	General requirements
Division 2	Site work
Division 3	Concrete
Division 4	Masonry
Division 5	Metal
Division 6	Wood and plastics
Division 7	Moisture-thermal control
Division 8	Doors, windows, and glass
Division 9	Finishes
Division 10	Specialties
Division 11	Equipment
Division 12	Furnishings
Division 13	Special construction
Division 14	Conveying systems
Division 15	Mechanical
Division 16	Electrical

Each of the above major divisions contains a listed number of subdivisions. These subdivisions furnish a convenient format by which to cap up direct costs pertaining to a particular subdivision; if desired, each subdivision may then be broken down into further line items for ease in estimating according to the choice of the individual bidding contractor.

Use of the UCI or some other consistent, commonly understood format is advantageous because, typically, the horizontal cost breakdown for each division and subdivision will contain subcontract and permanent material components, items for which prime contractors will seek subbids and material quotations from subcontractors and material suppliers. For a major bid, this fact results in extensive informal oral and written communications between material suppliers, subcontractors, and bidding prime contractors concerning the various items that are being quoted to the prime contractors for inclusion in their bids to the owner. Normal industry practice is to stick to a consistent format to avoid misunderstandings and to be sure that nothing is inadvertently duplicated or omitted. Communicating by reference to major divisions and subdivisions of the technical specifications is easier when suppliers, subcontractors, and the prime contractors are using the same format.

Format for Schedule-of-Bid-Items Bids

Bidding contractors usually utilize a vertical format for the direct cost for schedule-of-bid-items bids that conforms to the particular schedule of bid items in the bid form determined by the owner's engineer who drafted the bidding documents.

A **bid item** is a single line item, numbered sequentially, containing a description of the work included in the item and providing either a unit price to be bid for a stated number of units of the work described or a price to be bid for a single lump sum to be paid to the contractor for performing all of the work described. In the case of a lump sum bid item, the single **lump sum price** to be bid must include the total payment the contractor expects to receive for performing every item of work required by the particular specification sections and construction drawing references that are described and/or clearly implied to be included by the description of the lump sum bid item. In the case of a unit price bid item, the bid form will indicate for each bid item a **bid quantity**—that is, a quantity of individual units of work that represents the owner's engineer's estimate of the number of units of that type of work included in the contract. The bidding contractor will write in the intended **bid unit price** in a blank on the bid form and also write in the product of the bid unit price multiplied by the stated quantity of units in the bid quantity. This will yield the total amount of money to be paid to the contractor for performing the stated number of work units. The arithmetic result of multiplying the unit price written in by the bidding contractor and the owner's stated quantity is called the **bid-item extension.**

The bidder will then total the bid item extensions for unit price bid items and the lump sum figures for lump sum bid items and enter this total at the bottom of the bid form on a line designated as the total bid price.

An example of a lump sum bid item taken from the bidding schedule for one of the federal contracts for Lock and Dam 26 on the Mississippi River, which was bid in the mid-1980s, reads as follows:

Bid item no.	Description	Quantity	Unit	Unit price	Estimated amount
19	Steel sheet piling (overflow dike cutoff walls), drive, (government-furnished, including fabricated piling)	Sum	Job	—	$_____

An example of a unit price bid item in the Lock and Dam 26 bid form read as follows:

Bid item no.	Description	Quantity	Unit	Unit price	Estimated amount
21	Steel sheet piling, remove, clean and sort (first stage cofferdam, cells Nos. 15–27 and 40–45, including fabricated piling)	417,000	lf	$_____	$_____

Sometimes, a unit price bid item will be separated into two parts, as the following example from the Lock and Dam 26 bid schedule illustrates:

Bid item no.	Description	Quantity	Unit	Unit price	Estimated amount
17	Foundation fill:				
	(a) First 100,000 cy	100,000	cy	$_____	$_____
	(b) All over 100,000 cy	42,800	cy	$_____	$_____

All told, the Lock and Dam 26 bid form contained 181 separate bid items, and many of the unit price bid items contained two separate parts. A bid form of this length and complexity is typical for large heavy engineered public works projects. The bid form for smaller projects would be less extensive.

It should be emphasized here that both the lump sum and unit prices that bidding contractors write into the bid form contain any applicable cost contributions from *all* of the elements of the project vertical format enumerated at the beginning of this chapter. The significance of the bid form to the foregoing discussion of the **direct cost portion** of these unit price and lump sum prices is that the bid form will determine the vertical format of the contractor's **direct cost estimate.**

THE INDIRECT COST

Definition of Indirect Cost

Indirect costs are those project costs that cannot be easily associated with discrete physical parts of the project. Examples are general and administrative expenses, such as the salaries of project management, engineering, and craft supervisory personnel including project managers, engineers, and craft superintendents. There are many other kinds of such general and administrative expenses as well—for instance the costs of furnishing, maintaining, and removing project offices, shops, and other facilities. All these types of expenses are almost impossible to link or associate with individual physical parts of the construction project.

Relationship of Indirect Cost to Direct Cost

The above-described items of indirect cost are easily distinguishable from the cost items typically included as part of a contractor's direct cost estimate. However, some costs arguably could be included in either the direct or indirect cost portion of the estimate depending on the inclination of the contractor's estimators. An example of such an item would be the cost of furnishing and operating tugboats required on a bridge-substructure project on a major waterway, where substantial parts of the work must be constructed from barge-mounted floating equipment that is not self-powered. This type of equipment must be moved from location to location by one or more heavy-duty tugboats that must be continually present on the project from start to finish.

In this situation, determining the number of tugboats that will be required and the time duration over which they must operate during the performance of the work is always a problem. Some estimators prefer to include the tugboats as part of the equipment component of the individual work crews that are costed out in the direct cost portion of the estimate. Under this approach, the tugboat would generally be required for only a portion of the total required working hours of the individual work crews, a time duration that is usually difficult to estimate. For this reason, some estimators prefer not to include the tugboats in the direct cost work crews. Instead, they make a global estimate of the number of tugboats required to service the project for the number of shifts that work crews are expected to work anywhere on the project from a fixed beginning date to a fixed ending date (both dates dictated by the time-completion requirements of the project).

In the first case, the cost of furnishing and operating the tugboats would be included in the direct cost and should therefore be excluded from the indirect cost estimate; in the second case, the tugboat cost should be excluded from the direct cost estimate because it is included in the indirect cost estimate. If the tugboat usage is carefully estimated in each separate approach to the problem, a comparable estimate for the tugboat cost should result.

Clearly, the direct and indirect costs are closely related in matters of this kind. It is not uncommon for different estimating personnel to prepare the direct and the indirect cost portions of the contractor's estimate. Obviously, if the direct cost estimator had made the decision to carry items such as the aforementioned tugboat cost in the direct cost, the estimator preparing the indirect cost estimate must be aware of that fact so that the cost will not be doubled up. On the other hand, if the direct cost estimator had made the decision to omit this cost in the direct cost, the indirect cost estimator also must be aware of that fact so that the tugboat cost will not be omitted from the estimate entirely. Such errors as doubling up or leaving out important costs can be costly and can be avoided only by carefully coordinating the direct and indirect cost portions of the estimate. This is more easily accomplished if the direct cost portion of the estimate is completed, or nearly completed, before undertaking the preparation of the indirect cost estimate.

Another reason why it is generally preferable to defer preparation of the indirect cost until the direct cost estimate has been completed is that much of the information developed during the direct cost estimate is extremely important to the competent preparation of the indirect cost. By the time the estimating team has completed the direct cost estimate, it will be very familiar with the details of the project at hand with respect to, for instance, the kind and numbers of management and supervisory personnel likely to be required, the kind and size of temporary job shops and facilities that will be needed, and, most importantly, the time spans over which the different classifications of salaried personnel must be present on the jobsite and over which other forms of time-related expenses will be incurred. Such information is not at all clear at the beginning of estimate preparation. Only after the direct cost estimate has been largely completed does the estimator understand these and other similar matters well enough to formulate a comprehensive and competent indirect cost estimate.

INDIRECT COST VERTICAL FORMAT

Unlike the direct cost, indirect cost vertical formats do not depend on whether the bid is a lump sum bid or a schedule-of-bid-items bid. One format is equally applicable to either. Formats vary, but one good one for heavy engineered project indirect cost estimates contains the following major divisions:

- Salaried payroll
- Time-related overhead expense
- Non-time-related overhead expense
- Insurance and taxes
- General plant in-and-out

Salaried payroll includes the salaries, insurance, taxes, and company benefits paid to all employees other than hourly paid personnel. The major classifications included in this cost category are management, supervisory, engineering, administrative, and safety employees.

Time-related overhead expense refers to those types of recurring expenses that are time-related (other than salaried payroll), such as recurring expense for rentals, outside provided services, utility bills, maintenance expense on jobsite facilities, office consumables such as stationery, stamps, and similar recurring expenses.

Non-time-related overhead expense consists of overhead expenses that occur only once during the life of the project (usually near the beginning of the project) or at most, two or three times only during the life of the project. Typical examples include the purchase cost for things like office furniture; office, engineering, and safety equipment; outside engineering and outside surveying expense; employer association dues; medical and legal fees; and salaried personnel expense items such as relocation and per diem living expenses.

Insurance and taxes expense includes non-labor-related insurance premiums and various taxes. Insurance premiums normally carried in this section of the indirect cost include automotive and construction equipment floater insurance, builder's risk, and third-party liability insurance.[1] Typical taxes include personal property taxes on construction equipment and salvageable construction materials.

General plant in-and-out expense includes all costs required to furnish temporary facilities, utilities, roads, and so on, at the project site preparatory to undertaking permanent construction work. Also included are the costs for freight of the construction equipment to the jobsite, unloading and erection expense for the equipment, and the breakdown and load-out expense for the equipment at the end of the project. Freight-out costs normally would be paid by the entity receiving the equipment once it leaves the jobsite, with the exception of equipment rented from third parties, in which case freight expense for the return of the equipment to the third party would also be included.

More complete treatment of the subject of indirect cost is included in Chapter 11.

ESCALATIONS

Estimates for heavy engineered construction projects (which usually have durations of several years) are normally prepared on the basis of labor rates and costs of permanent and temporary materials that exist at the time the estimate is prepared. Because of the extended duration of such projects, these costs usually will increase during the performance of the project. If this fact is not recognized in some manner, the estimate will not contain sufficient monies to pay the costs that will be incurred when the actual work is performed. The normal method for handling this problem is to separately calculate the **costs of escalation** and include them as an additional line item in the bid estimate.

Kinds of Costs Normally Subject to Escalation

During the life of a typical heavy construction project, the unit cost rates for the following kinds of items will usually increase:

- Labor (both hourly and salaried payroll)
- Expendable materials and services
- Equipment operating expense

Permanent material and subcontract costs also will be subject to escalation and bidding prime contractors must consider it when assumed material and subcontract prices (that is, "plug" prices) are replaced with actual quoted prices received on the day of the bid or shortly before the day of the bid.[2] However, material and subcontract quo-

---

[1]As discussed in Chapter 2, third-party liability insurance premiums are sometimes based on labor and included in the estimate as a component of the loaded labor rates used to develop the monies carried in the labor column of the direct and indirect cost. In this case, premiums for third-party liability insurance would not be carried in the insurance and taxes section of the indirect cost.

[2]This adjustment process is not included within the scope of this book.

tations made to prime contractors are often quoted firm for the life of the project and thus would not be escalated. If such quoted prices are intended to be subject to escalation, the quotation should make this fact clear and furnish a basis for the escalation so that the prime contractor may calculate and include it in the prime bid or price proposal as part of the plug price adjustment.

Method of Determining Escalation Amount

The anticipated time schedule for performing the various parts of physical construction work in the project is generally known prior to the completion of the direct cost estimate. Once the estimating team has completed both the direct and indirect cost estimates, it is usually possible to allocate the various elements of these estimates to discrete time periods for the performance of the project by carefully relating the cost estimates to the schedule. Also, estimates can be made for the percentage amount that each various kind of expense is expected to escalate at specific points in time during the expected performance of the project. This type of forecasting is far from an exact science, but experienced estimators can make reasonable forecasts based on the prevailing economic climate at the time the estimate is prepared and by examining past cost trends. The following simplified example illustrates this general procedure.

Assume that an estimate had been prepared for a 3-year project by using the labor rates that were being paid when estimate preparation began. Further, these rates were expected to remain the same for the first 6 months of the project performance period and then were estimated to increase by 5%. Thereafter, they were expected to increase by 6% at the end of each succeeding 12-month period. In all cases, the estimated percentage increase was intended to apply to the labor rate levels that the job labor force was being paid immediately prior to the increase. Also, the estimating team's forecast of how the direct labor component of the estimate was expected to be spent was as follows:

Description	Total, $	Year 1, $		Year 2, $		Year 3, $	
		First 6 months	Second 6 months	First 6 months	Second 6 months	First 6 months	Second 6 months
Total direct labor	13,250,000	250,000	1,000,000	3,500,000	4,500,000	2,500,000	1,500,000

The amount of direct labor escalation that should be added to the amount of labor dollars already included in the estimate, in this case, $1,285,700, can be calculated as follows:

Description	Total, $	Year 1, $		Year 2, $		Year 3, $	
		First 6 months	Second 6 months	First 6 months	Second 6 months	First 6 months	Second 6 months
Direct labor		250,000	1,000,000	3,500,000	4,500,000	2,500,000	1,500,000
Escalation factor[1]			0.0500	0.0500	0.1130[2]	0.1130	0.1798[3]
Escalation, $	1,285,700		50,000	175,000	508,500	282,500	269,700

[1]This factor is a number, such that when multiplied by the unescalated estimate labor dollars expected to be expended during a particular 6-month period, will yield the escalation dollars that must be added to the estimate for labor expended during that 6-month period.
[2](1.0500)(1.06) − 1.0000 = 0.1130.
[3](1.1130)(1.06) − 1.0000 = 0.1798.

Required escalation on other estimated costs can be calculated in the same general manner as the above example.

INTEREST (THE COST OF MONEY)

This cost represents the **time value of the contractor's working capital** that is invested in the project. Heavy construction projects often require large capital investments at the beginning of the project that are not recovered until some point during project performance. In extreme cases, this does not occur until the contractor has completed the work and has received all job revenues, including the retained percentage, from the owner.

This situation is illustrated by Figure 3-1, which is a hypothetical plot of dollars received versus project time duration commencing with notice to proceed (NTP) and ending with project completion.

Three curves are shown in the figure: **accumulative job costs** (cash out), **accumulative project revenue** before deduction of **retained percentage,**[3] and **accumulative net revenue** after deduction of retained percentage. At any particular point in time, if the accumulative job costs exceed accumulative net revenue, the contractor is in an **out-of-pocket cash position** (the project is in the "red"). Once accumulative net revenue exceeds accumulative job costs, the project moves into the "black," resulting (in the case of the scenario represented in Figure 3-1) in the contractor realizing some measure of profit before the return of the retained percentage at project completion. Once the retained percentage has been returned to the contractor at project completion,

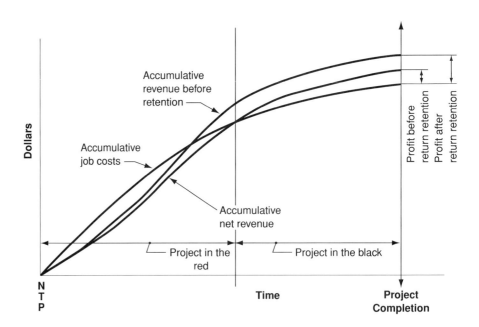

FIGURE 3-1   Typical project cost and revenue relationships.

[3]Retained percentage is an amount of the contractor's project earnings revenue (often 10%) that is deducted and held by the owner, often until all work has been completed and accepted, to form a fund to assure money will be available to remedy any defects in the work that the contractor does not satisfactorily remedy. Any retained funds not used to remedy defects are returned to the contractor when all work has been accepted by the owner at the completion of the project.

the profit will increase to the final figure shown. This case illustrates a profitable project that remains in the red for approximately one-half of the project duration, moving into the black only in the latter period of the project.

The **funds invested in the project** (that is, the difference at any point in time between accumulative job costs and accumulative net revenue while the project is in the red) have a **time value** that is just as much a cost to the contractor as any other cost item. It is unrealistic not to include this time value of money (that is, **interest expense**) as a cost element of the contractor's bid or cost proposal.

Interest expense is highly influenced by the type of work undertaken, the duration of the project, and the contract payment/retention provisions. It can be a very large cost. Projects requiring the purchase of large amounts of construction equipment, particularly highly specialized equipment, generally require very large cash investments at the inception of the project, greatly increasing the magnitude of the interest expense that should be included as part of the total estimated project cost. This is particularly true for projects that do not include a **mobilization allowance.**[4]

Calculation of Interest Expense

After the estimating team has completed direct and indirect project cost estimates, and determined escalations and contingencies, the interest expense may be calculated by making a **cash-flow analysis.** This should be done by estimators familiar with the project cost estimate and the project time schedule. First, cash outlays that the contractor must periodically make are carefully forecast for successive time increments of the project for direct labor, purchase of permanent and expendable materials, equipment operation costs, subcontract costs, and any other broad items of contract cost requiring cash outlays. The total of the various costs forecast for each time period of project performance must equal the total estimated project cost. Next, the monies expected to be received from the project owner in the form of progress payments as the work progresses (less the retained percentage deduction specified by the contract terms) are forecast for the various successive time periods for the project. The validity of both forecasts requires a clear understanding of the relationship between the expenditure of project costs and the receipt of progress payments to the project time schedule. For cash-flow purposes, the total project duration is usually divided into 3-month, or alternately, 6-month time periods.

For each time period, the net **cash balance** can be calculated as the difference between the cash-out figure for the period and the revenue received (net, after deduction of retained percentage). The **accumulative cash balance** is then determined for the end of each time period comprising the total project duration. The accumulative cash balance figure at the end of the project duration equals the anticipated profit to be gained from the project. Obviously, rough estimates for contingency allowances and the desired project profit must be included in the total project revenue figure that was used when forecasting progress payments for the various periods of the project.

Once the accumulative cash balance figure is determined for the end of each time period comprising total project duration, the **time value** of the **average accumulative cash balance** (which equals the *average* total investment of contractor funds during that particular period) can be calculated by multiplying the average accumulative cash balance during the period by the current interest rate that could be obtained by an alternate available investment during that time period. Because interest rates for future periods of time are not known, they must be estimated based on the value of money at the time the calculation is made and on expected general economic trends. Some contractors make

---

[4]A monetary allowance, sometimes a figure of millions of dollars, that is part of the contract price but is paid to the contractor at the beginning of the project before performance of any project work to provide funding for the heavy expenses incurred at the beginning of the project.

this calculation on the basis of compound interest, but in view of the approximate nature of much of the input data for the cash-flow analysis, a simple interest calculation is more generally used. The sum of the calculated interest figures for all periods of the job when the job is in a negative average accumulative cash balance position equals the **total interest cost** that should be added to the bid or proposed contract price. The following simplified example illustrates this general procedure.

Assume that the estimated cash outlays and project revenues (before retention) are as follows for the eight successive 6-month periods of a 4-year project.

Description	Total $ in estimate[1]	Year 1, $		Year 2, $		Year 3, $		Year 4, $	
		First 6 months	Second 6 months	First 6 months	Second 6 months	First 6 months	Second 6 months	First 6 months	Second 6 months
Cash outlays	123,000	12,000	20,000	22,000	26,000	26,000	12,000	3,000	2,000
Revenue (before retention)	141,000	10,000	17,000	24,500	31,500	30,000	13,000	10,000	5,000

[1]All dollar amounts in $1,000s.

Assume further that the contract was a standard federal government contract where 10% was to be retained by the government from all progress payments until the contract had been 50% completed (based on revenue earned). Thereafter, progress payments were to be made in full. All retained funds held by the government at the completion of the job can be assumed, for purposes of this analysis, to be returned to the contractor at the end of the final 6-month period of project performance. Assume that the value of money is 12.5% per annum. The calculation of the interest cost (simple interest), totaling $1,402,000, for the facts stated is as follows:

Description	Total $ in estimate	Year 1, $		Year 2, $		Year 3, $		Year 4, $	
		First 6 months	Second 6 months	First 6 months	Second 6 months	First 6 months	Second 6 months	First 6 months	Second 6 months
Revenue (before retention)	141,000	10,000	17,000	24,500	31,500	30,000	13,000	10,000	5,000
Retention @10%		<1,000>	<1,700>	<2,450>	<1,900>[2]				7,050
Net revenue	141,000	9,000	15,300	22,050	29,600	30,000	13,000	10,000	12,050
Cash outlays	123,000	12,000	20,000	22,000	26,000	26,000	12,000	3,000	2,000
Cash balance	18,000	<3,000>	<4,700>	50	3,600	4,000	1,000	7,000	10,050
Accumulative cash balance		<3,000>	<7,700>	<7,650>	<4,050>	<50>	950	7,950	18,000
Interest	1,402	94[3]	334[4]	480	366	128			

[1]All dollar amounts in $1,000s.

[2]Because the 10% retention from progress payments will cease at the 50% completion point, total retention will be limited to (10%)(50%)($141,000) = $7,050. Additionally, $7,050 − $1,000 − $1,700 − $2,450 = $1,900.

[3]Average investment during the period equals (0 + $3,000)/2 = $1,500. Interest for the 6-month period = ($1,500)(0.125)(0.50) = $94 (rounded to nearest $1,000).

[4]Average investment during the period equals ($3,000 + $7,700)/2 = $5,350. Interest for 6 months equals ($5,350)(0.125)(0.50) = $334 (rounded to nearest $1,000).

Note that, in the foregoing example, the contractor was in an out-of-pocket situation until the end of the first 6 months of year 3 of the 4-year project.

CONTINGENCY ALLOWANCES

Contingency allowances are included in project cost estimates to cover special risks for which the potential costs are not included in the estimate, things that simply may go wrong, or cost items that possibly have been overlooked when the estimate was prepared. Several different kinds of contingency allowances are illustrated by Figure 3-2.

As indicated, contingency allowances can be broadly subdivided into **identified** and **unidentified contingencies.**

Identified Contingencies

Productivity hedges are allowances to cover the risk inherent in aggressively priced work operations—that is, work operations for which management, upon review, feels the estimator assumed a level of crew productivity that may be better than actually obtainable. This situation is illustrated by the pile-driving example discussed in Chapter 1. Additional money is simply added to the estimate as a "hedge" against the possibility that the estimated productivity will not be achieved.

An **identified special risk** is a situation in which it is recognized that some adverse event could or might occur, but the estimator did not include money in the estimate to cover such an eventuality. In review, management decides to provide a sum of money to cover costs if the event in fact does occur and the money is then added to the estimate.

A **"soft" subcontract price** is a subcontract quotation for an important part of the contract work where the subcontractor cannot furnish a performance bond and the bidding prime contractor believes the subcontractor may have difficulty performing at the quoted price. Instead of discarding such a subcontract quotation and using the next higher quotation, the prime contractor bidder may choose to soften the potential impact of the subcontractor failing to fully perform by adding an amount of money to the estimate midway between the low subcontract quotation and the next higher one.

FIGURE 3-2 Kinds of bid contingency allowances.

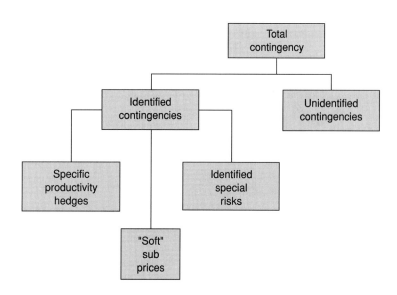

Unidentified Contingencies

A firm price bid imposes a level of risk to the bidding contractor that depends on the type of construction work involved, the potential physical risks such as encountering unfavorable site conditions, and the level of aggressiveness reflected in pricing the various required work operations. Contractors allow for these inherent uncertainties by including a **risk coverage** allowance as protection from monetary loss. This type of contingency can be thought of as a general allowance to make the bidder comfortable and is generally included as part of the markup consideration as discussed below.

Generally speaking, the bidding contractor's overall contingency analysis will be made along with the markup consideration shortly before finalizing the bid or cost proposal.

MARKUP (MARGIN)

Markup, or **margin,** is the amount of money added to the bid price in addition to all estimated project site costs and identified contingency allowances discussed above. It is usually not finalized until shortly prior to the submission of the bid or cost proposal. As shown in Figure 3-3, markup is composed of three elements.

Risk Coverage

The risk component of markup, discussed earlier under Unidentified Contingency, is simply a general allowance to make the bidding contractor comfortable with the bid.

Home Office General and Administrative Expense

All contracting organizations, with the exception of extremely small firms, operate over a particular geographical area from a home office facility. Usually, this facility houses the company's top management organization as well as the administrative, engineering, accounting, personnel, safety, and equipment management organizations. Home office expenses (Home Office G & A) incurred by the company typically include salaries, office rental or real property depreciation expense, office supplies, telephone and utility bills, transportation expenses, and so on. In the case of large companies, these expenses total many millions of dollars per year. They must be met by revenue received from construction projects performed by the company's field organizations, prior to the realization of a profit that contributes to the company's net worth. Therefore, an allowance must be included in bids on other contract price proposals for construction projects performed by the company to generate money to meet these expenses.

FIGURE 3-3   Bid markup elements.

Profit

Profit may be thought of as that portion of the total annual revenues received by the company after all project costs and home office general administrative expenses have been met. Corporate income taxes will be computed upon this figure. After payment of taxes, the net balance of earned profit increases the company's net worth.

Industry Methods of Markup Determination

The risk coverage element of markup for a particular construction project is closely related to the **horizontal format cost distribution** of the total project cost. The profit element is highly influenced by **competitive business conditions.** Contractor bidders may, if they wish, temporize with the profit element, but they must not fail to provide adequate monetary allowances for risk coverage and home office general and administrative expense in the total markup figure if the company expects to remain in business. These monies are required to meet potential costs, and no business will survive unless project revenues generate at least as much money as necessary to meet all the costs.

Contractors use at least **three general methods** to determine the magnitude of markup to be added to bids or cost proposals:

1. Percentage-of-total-cost method
2. Percentage-of-labor-exposure method
3. "Broad-based" method (considers total horizontal format cost distribution)

These three methods are illustrated by Figure 3-4.

The markup computed by any of these three methods includes all components discussed above: risk coverage, home office general and administrative expense, and profit. In the percentage-of-total-cost method, the markup is simply taken as a percentage of total estimated costs (including contingency allowances). The particular percentage utilized depends on the degree of risk believed to be inherent in the project work, ranging from 2 to $2\frac{1}{2}$% for commercial building construction up to 12 to 20% for most heavy engineered construction projects.

(a) Percentage of total cost method	(b) Percentage of labor method	(c) Broad-based method
x% of total estimated cost	x% of total labor	Σ of 25 to 30% of total labor
		+
x depends on risk	x = 25 to 30% for sewer treatment plants, etc.	15% (EM + EOE + R)
		+
$x \approx$ 2 to 2.5% for commercial high-rise construction	x = 40 to 60% for bridge substructure or tunnel work	2% (PM + Subs)
$x \approx$ 12 to 20% for most heavy engineering projects	x = 30 to 40% for earth-filled dams and similar projects	
	x = 60 to 100% for extremely labor-intensive, high-risk projects	

FIGURE 3-4   Common methods of bid markup determination.

By the **percentage-of-labor method,** markup is computed as a percentage of labor in the total project costs, including contingency allowances (both direct and repair labor in the direct cost and indirect cost are considered). The percentage utilized varies from 25 to 60 to 100%, as indicated in Figure 3-4, depending on the type of construction work and perceived risks involved.

The **broad-based method** determines markup by application of a formula whereby total markup equals 25 to 30% of total labor in the project cost (including direct and repair labor in both the direct cost and indirect cost) plus 15% of the sum of expendable materials, equipment operating expense, and equipment rental, plus 2% of the sum of permanent materials and subcontracts. This method considers all of the cost elements in the horizontal format of cost expression.

The importance of the distribution of costs in the horizontal cost format is clearly illustrated in the determination of markup by the percentage-of-labor and the broad-based methods. Only the percentage-of-total-cost method does not consider the cost distribution. Most heavy engineered construction contractors determine markup by the percentage-of-labor method, the broad-based method, or some similar approach that considers the distribution of costs in the horizontal format.

Relationship of Markup Elements to Contractor's Annual Construction Volume

Assume that a hypothetical construction company's business plan is to put in place a $115,000,000 annual volume of high-risk heavy engineered construction. Further, this company projects annual home office general and administrative (G & A) expense of $9,000,000 and hopes to realize a pretax profit of $5,000,000. To better assure this pretax profit on high-risk work, it would be reasonable for such a company to include a total additional $5,000,000 in bids or other contract cost proposals as an unidentified contingency for general risk coverage. On this basis,

Annual revenue from projects	=	$115,000,000
Less home office G & A	=	<9,000,000>
Less risk coverage	=	<5,000,000>
Less residual pretax profit	=	<5,000,000>
Project costs (at estimate level)	=	$96,000,000

Home office G & A	(9,000,000/96,000,000)(100)	=	9.38%
Risks coverage	(5,000,000/96,000,000)(100)	=	5.21%
Profit	(5,000,000/96,000,000)(100)	=	5.21%
Required markup on project costs		=	19.80%

Typically, applying the percentage-of-labor and broad-based methods of determining markup (both of which consider the distribution of costs in the horizontal cost format) will generate a total markup percentage in a range consistent with the hypothetical calculation shown above.

For instance, the total bid figure[5] for the tunnel boring machine (TBM) excavated rock tunnel project described in Chapter 2 consisted of the following components (including direct and indirect cost estimates, all escalations, interest expense, identified contingencies, bond costs, and final material and subcontract quotation adjustments):

---

[5]In 1979 dollars.

Equipment operating expense (EOE)	$ 1,523,900
Repair labor (RL)	628,100
Rental (R)	5,969,800
Labor (L)	10,231,800
Expendable materials (EM)	7,272,300
Permanent materials (PM)	3,265,900
Subcontracts (Subs)	3,558,200
Estimated project cost	$ 32,450,000
Markup	6,500,000
Total bid	$ 38,950,000

The actual markup used, in percentage of labor, was [6,500,000/(628,100 + 10,231,800)](100) = 59.8% of total labor. This figure is at the upper limit of the range presented previously for determination of markup by the **percentage-of-labor-exposure** method.

Application of the **broad-based** method produces a somewhat smaller markup figure:

(628,100 + 10,231,800)(0.30)	=	3,258,000
(1,523,900 + 5,969,800 + 7,272,300)(0.15)	=	2,214,900
(3,265,900 + 3,558,200)(0.02)	=	136,500
Total markup	=	$5,609,400

The actual markup of $6,500,000 amounts to (6,500,000/32,450,000)(100) = 20.0% of estimated project cost (the upper end of the range for markups computed on **percentage-of-job-cost** basis), whereas the lesser markup computed by the broad-based method is (5,609,400/32,450,000)(100) = 17.3% of estimated project cost.

The following actual operating results of a large successful present-day heavy engineering contractor for 1995 illustrate a final interesting point on this subject. All figures are reported in percentage of total project revenue received:

Total project revenues	100.00%
Total project costs	87.00%
Home office G & A expense	9.50%
Pretax profit	3.50%

Putting these results in context with the previous discussion,

Home office G & A	=	(9.5/87.0)(100) = 10.92% of project costs
Pretax profit	=	(3.5/87.0)(100) = 4.02% of project costs
Home office G & A + pretax profit		= 14.94% of project costs

To produce an overall 4.02% pretax profit, an unidentified contingency allowance (at least to some extent) for risk coverage would have to have also been included in bids and other project cost proposals. Taking this unidentified contingency allowance into account, the general range of markup on project costs would have been higher than the 14.94% figure calculated above for this company, which agrees well with the 12 to 20% markup range of Figure 3-4 for heavy engineered construction projects.

BOND PREMIUMS

All public construction projects and many private projects require the contractor to furnish project bonds consisting of a **bid bond,** a **performance bond,** and a **labor and material payment bond.** The premiums charged by sureties for furnishing these bonds are normally combined into a single premium quotation. The quotation is based on contract price and is usually furnished by the contractor's insurance bond broker in the following typical form:[6]

$15.00/$100 of the first $100,000 of bid price
$ 7.50/$100 of the next $100,000 of bid price
$ 4.50/$100 of the next $1,000,000 of bid price
$ 2.50/$100 of the next $5,000,000 of bid price
$ 1.75/$100 of the balance of the bid price

Thus, final determination of the bond premium is not possible until all other elements of the contractor's bid are known, including the markup. Therefore, computation of the bond premium is the last step in finalizing the contractor's bid or cost proposal.

DISTRIBUTION OF COST COMPONENTS TO FINAL BID-ITEM PRICE

Figure 3-5 illustrates the structure of a typical lump sum bid. In this case, the only step necessary to complete the bid form is to add all the vertical components of the estimate to arrive at the total bid figure, which appears in the lower right-hand corner of Figure 3-5. Nothing else is required.

On the other hand, a schedule-of-bid-items bid is considerably more complicated (see Figure 3-6). For schedule-of-bid-item estimates in which the bid figure is determined by adding all the vertical components of the estimate, the estimators must do considerable additional work before they can fill out and submit the bid form. Lump sum bid prices for lump sum bid items and unit prices and extensions for unit price bid items must be individually determined for each bid item in the estimate for entry into the required blanks on the bid form. The direct cost portion of the total bid amount of each bid item has already been allocated to individual bid items via the format by which the estimate was made. The balance of the bid price for each bid item must be distributed individually to the bid items from a sum of money consisting of the indirect costs, escalations, interest cost, contingencies, markup, and bond premiums. Figure 3-6 illustrates diagramatically how this sum of money (which equals the amount designated on Figure 3-6 as Dt) is then distributed to individual bid items in amounts of D1, D2, D3, D4, and so on, the sum of these distributions equaling Dt. The total bid amount for each bid item then will be the total direct cost portion plus the distributed amount. The sum of the total bid figures thus determined for each bid item will total the intended bid figure, which is the sum of all the vertical components. The involved process by which this distribution is made and unit bid prices determined is explained in Chapter 13.

---

[6]The specific rates shown are for example only.

Description	Labor	PM	EM	Subs	EOE	RL	Rent	Total
Direct costs								
General requirements	x		x	x	x	x	x	x
Site work								
Excavation	x		x	x	x	x	x	x
Pile driving	x	x	x	x	x	x	x	x
Utilities	x	x	x	x	x	x	x	x
Concrete work, etc.	x	x	x	x	x	x	x	x
Masonry				x				x
Metals, etc.				x				x
Mechanical				x				x
Electrical				x				x
Total direct costs	xx	xx	xx	xx	xx	xx	xx	xx
Indirect costs								
Salaried payroll	x							x
Time-related overhead	x		x	x	x	x	x	x
Non-time-related overhead			x	x				x
Insurance and taxes			x					x
Construction plant in & out	x		x	x	x	x	x	x
Total indirect costs	xx	xx	xx	xx	xx	xx	xx	xx
Labor escalation	x					x		x
Material escalation		x	x					x
Equipment escalation					x		x	x
Interest			x					x
Contingency	x		x		x	x	x	x
Total cost w/o bond	xx	xx	xx	xx	xx	xx	xx	xx
			Markup					xx
			Total bid w/o bond					xx
			Bond					xx
			Total bid					xx

FIGURE 3-5   Structure of a typical lump sum bid.

Description	Labor	PM	EM	Subs	EOE	RL	Rent	Total cost	Distribution	Total bid
Direct costs and distribution										
BI #1	x	x	x	x	x	x	x	x	D1	xx
#2	x	x	x	x	x	x	x	x	D2	xx
#3	x	x	x	x	x	x	x	x	D3	xx
#4	x	x	x	x	x	x	x	x	D4	xx
Etc.									Etc.	Etc.
Total direct costs & bid	xx	xx	xx	xx	xx	xx	xx	xx	Dt	xxx
Indirect costs										
Salaried payroll	x							x		
Time-related overhead	x		x	x	x	x	x	x		
Non-time-related overhead			x	x				x		
Insurance and taxes			x					x		
Construction plant in and out	x		x	x	x	x	x	x		
Total indirect costs	xx	xx	xx	xx	xx	xx	xx	xx		
Labor escalation	x					x		x		
Material escalation		x	x					x		
Equipment escalation					x		x	x		
Interest			x					x		
Contingency	x		x		x	x	x	x		
Total cost w/o bond	xx	xx	xx	xx	xx	xx	xx	xx		
			Markup					xx		
			Total bid w/o bond					xx		
			Bond					x		
			Total bid					xxx		

FIGURE 3-6   Structure of a typical schedule-of-bid-items bid.

CONCLUSION

The first three chapters of this book have discussed general estimate considerations and both the horizontal and vertical formats of construction cost estimates. Chapter 4 will be devoted to the quantity takeoff, that is, the process by which the various physical work and material quantities that are required to be priced out are determined.

QUESTIONS AND PROBLEMS

1. What are the seven major elements or divisions of the vertical estimate format discussed in this chapter?

2. What is the difference between a lump sum bid and a schedule-of-bid-items bid? Which is more common for heavy construction projects? For which form of bid is the bidding contractor free to determine the vertical format of the direct cost estimate? For which form of bid is the bidding contractor free to determine the format of the indirect cost estimate?

3. If a lump sum bid is called for, why is it important that a prime contractor establish a consistent, commonly understood, vertical format for the cost estimate?

4. Explain the following terms: bid item, unit price bid item, lump sum bid item, unit price, bid item extension.

5. In what way is the relationship of the indirect cost to the direct cost important as explained in this chapter? Why is this relationship one reason why the direct cost portion of the project cost estimate should be completed or nearly completed before the indirect cost is undertaken? What is another reason why the direct cost should be completed first?

6. What are the five major divisions in the vertical format of the indirect cost as presented in the chapter?

7. What are the three elements of the horizontal cost format that usually will increase during project performance over the level at which the costs were estimated and therefore are included in the escalation element of the vertical estimate format? What are the two additional elements of the horizontal cost format for which cost escalation is usually considered as part of the final adjustment process for changes in price just prior to bid?

8. What are some of the factors about the project that determine the magnitude of the interest expense that should be included as a line item in the vertical format of the contractor's cost estimate?

9. What is a mobilization allowance? What is it's effect on the magnitude of interest expense? On what type of project is its use generally appropriate?

10. In regard to a cash flow analysis, what is net cash balance and how is it determined? Answer the same questions with regard to accumulative cash balance. What is retained percentage and what effect does it have on net cash balance and accumulative cash balance?

11. What are the two broad classifications or types of contingency allowances discussed in this chapter? What three subtypes of identified contingencies were mentioned?

12. What are the three elements comprising a contractor's bid markup? Which element is related to the horizontal cost distribution of the contractor's bid estimate? Which

is related to competitive business conditions? Which elements must always be adequately covered in the markup figure if the contractor is to survive in business?

13. Name three common industry methods of bid markup determination.

14. What is the final step discussed in this chapter for finalizing the contractor's bid or cost proposal and upon what is it based?

15. The direct and indirect cost estimates for a contractor's bid to the U.S. Army Corps of Engineers for a large lock and dam project contain the following summary figures:

Cost category	Direct cost, $	Indirect cost, $
EOE	4,847,800	344,200
RL	877,200	68,800
R	4,255,200	253,000
L	11,926,400	4,331,700
EM	4,452,300	3,447,800
PM	54,456,700	—
S	30,328,600	667,800
T	111,144,200	9,113,300

Additionally, the estimate team has prepared the following forecasts over the life of the project. All figures are in thousands of dollars:

	Total $ in estimate	Year 1, $		Year 2, $		Year 3, $		Year 4, $	
		First 6 months	Second 6 months	First 6 months	Second 6 months	First 6 months	Second 6 months	First 6 months	Second 6 months
Direct labor & RL	12,803	1,380	1,820	2,020	2,460	2,020	1,180	1,380	543
Indirect labor & RL	4,400	600	500	600	500	650	650	450	450
Direct EOE	4,848	585	627	727	970	827	385	485	242
Direct EM	4,452	545	768	668	790	568	545	345	223
Indirect time-related EM	773	100	100	100	97	96	93	93	94
Cash outlays	126,000	20,000	20,000	22,000	26,000	26,000	12,000	3,000	2,000	<5,000>
Revenue before retention	136,500	10,000	15,000	32,000	29,500	25,000	10,000	10,000	5,000

Develop (either manually or by computer spreadsheet) a full display of the project horizontal and vertical estimate detail summary in the following format:

Description	EOE	RL	R	L	EM	PM	S	T
Direct cost	X	X	X	X	X	X	X	X
Indirect cost	X	X	X	X	X		X	X
Total cost	X	X	X	X	X	X	X	X
Escalation	X	X		X	X			X
Interest					X			X
Contingency	—	—	—	X	X	—	—	X
Total cost w/o bond	X	X	X	X	X	X	X	X
Markup								X
Total bid w/o bond								X
Bond								X
Total bid								X

Develop separate line totals for escalation, interest, and contingency, utilizing the data forecast by the estimate team based on the following assumptions for top-management conclusions about the project:

Escalation

Direct labor rates are expected to increase 4.5% at the end of the first year, an additional 5.5% at the end of the second year, and an additional 6.2% at the end of the third year. Indirect labor rates are expected to increase 6.0% at the end of the first year and an additional 5.0% at the end of each year thereafter. EOE is expected to increase 6.5% at the end of the first year and an additional 5.5% at the end of each year thereafter. EM (both direct cost EM and indirect time-related EM) are expected to increase 6.5% at the end of the first year and 4.5% at the end of each year thereafter.

Interest

Ten percent will be retained from all progress payments until the contract is 50% completed (based on earned revenue). Thereafter, progress payments will be made in full. All retained funds will be returned to the contractor at the end of the second 6 months of year 4. Interest is at 12.5% per annum, simple.

Contingencies

It is believed that in spite of the contractor's best efforts, one of the contractually specified milestone completion dates might not be possible to meet until 70 calendar days later than the date specified in the contract. The government has declined to alter the contractually specified date and the specified liquidated damages (LD) are $12,500 per day for every day the milestone achievement date is late. Contingent LD payments should be carried in the EM column.

Markup

Management has directed that markup be determined by the broad-based method using the following percentages:

30% on labor (including RL)
15% on EOE, R, and EM
 2% on Subs and PM

Bond

The contractor's bond broker's quoted rates are:

$15.00 per $100 on the first $100,000 of bid price
$ 7.50 per $100 on the next $100,000 of bid price
$ 4.50 per $100 on the next $1,000,000 of bid price
$ 2.50 per $100 on the next $5,000,000 of bid price
$ 1.75 per $100 on the balance of bid price

For each line item in the project estimate summary, prepare a backup calculation sheet (either manually or by computer spreadsheet) that supports the line item total figures entered on the summary sheet.

4

The Quantity Takeoff

Key Words and Concepts

Quantity takeoff defined

Five major categories of quantity takeoff

Relation of takeoff quantities to bid-item structure

Relation of takeoff quantities to measurement and payment provisions

Takeoff quantities for items not explicitly shown on contract drawings

Importance of reasonable assumptions when developing takeoff quantities

Importance of establishing meaningful units of measurement for work operation takeoff quantities

Importance of sequence in which takeoff computations are made

Takeoff practice where numerous small items are involved

Use of tabular computational approach

Chapter 1 pointed out that the cost-generating mechanism in construction cost estimates depends on associating, or connecting, the quantities of units to some form of known monetary and time impact for accomplishing a single quantity of that unit. The starting point is the determination of these "quantities." This chapter is devoted to this process, broadly referred to as the **quantity takeoff.**

GENERAL CONSIDERATIONS

Admittedly, the effort required to take off the quantities for a major heavy engineered construction estimate is a Herculean chore, but the major problem is not in performing this detailed work. Rather, the larger problem lies in determining **what quantities** are necessary to take off in a given case. The takeoff problem is much broader than simply determining quantities of materials that remain in place in the permanent project. Takeoff is typically required for

- Quantities of materials originally in place on the project site that must be removed (soil and rock excavation, for example)
- Quantities of things that must be temporarily installed or erected and then later removed (concrete formwork, for example)
- Quantities of things that are not installed or removed but are required to be treated or altered in place in some manner—for instance, quantities of insitu foundation soil or rock to be treated by grouting[1] to improve their strength or to make them more watertight.

Although the variety of things required to be taken off is virtually unlimited, as suggested above, takeoff quantities logically fall into the following **five major takeoff categories:**

1. Quantities of work to be performed by subcontractors (S)
2. Quantities of permanent materials (PM) that the prime contractor must furnish
3. Quantities of expendable materials (EM) that the prime contractor must furnish
4. Quantities of work that will be measured for payment to the prime contractor by the owner (the pay quantities)
5. Quantities of work, by distinct significant work operation to be performed by the prime contractor's own forces

Quantities of Work to Be Performed by Subcontractors

Although subcontractors will carefully take off these work quantities as a preliminary step in furnishing their cost quotations to the prime contractor, the prime contractor should, to the extent practical, make an independent takeoff of this work for at least two reasons. First, it is necessary to have a rough idea of the quantities of subcontract work to be performed to establish a preliminary range of the total value of subcontract work that will be included in the prime bid. Prime contractors expecting to submit bids for large heavy engineered projects usually will discuss the proposed bid in a preliminary way with their bonding company to confirm that the bonding company is willing to

---

[1]The forced intrusion of cementations or chemical materials into the voids in soil or into joints or other discontinuities in rock.

issue the required performance bond. Bonding companies usually want to know the general monetary level of the expected bid, which is difficult to determine without knowing the likely dollar range of subcontract work to be included. A second reason is to assist the prime contractor in evaluating the subcontract bids. Final subcontract quotations are usually not available until shortly before the time of the prime bid. Prime contractors need to quickly determine that the subcontract bids received are realistic. To make this type of judgment, the prime contractor must have an approximate idea of the quantities of work the subcontractor is bidding upon.

Permanent Materials to Be Furnished by Prime Contractor

Prime contractors must furnish all permanent materials required for work intended to be performed by their own forces. It is therefore necessary to take off accurate permanent material quantities against which to apply unit price quotations received from material suppliers. This category also includes taking off, or determining in some manner, the amount of permanent material waste likely to be incurred, even though the cost of this waste material is usually carried in the expendable material column.

Expendable Material Quantities to Be Furnished by Prime Contractor

All expendable materials required for work to be performed by the prime contractor's own forces must also be furnished by the prime contractor, and quantities of these materials against which to apply unit price quotations from suppliers must be determined. It was mentioned earlier that the determination of expendable material quantities for heavy engineered projects that involve ground support systems, trestles, decking structures, and such often requires extensive engineering analysis and design work. Further, where reuse of expendable materials is possible (for formwork, scaffolding, sheet piling, and similar uses), the determination of quantities will require careful consideration of the anticipated construction schedule.

The Pay Quantities

The pay quantities consideration applies to schedule-of-bid-items bids only. It is critical that bidding contractors make an independent assessment of the amount of work that will actually be measured for payment in accordance with the measurement and payment provisions of the contract documents if the work is performed as shown on the bid drawings. The bid quantities stated by the owner on the bid form are sometimes overstated, creating the potential for serious monetary loss, even when the work is performed at or below the costs estimated. Problems associated with inaccurate bid quantities are discussed in Chapter 13 of this text.

Work Quantities for Operations to Be Performed by the Prime Contractor's Own Forces

Prime contractors bidding heavy engineered construction projects perform the majority of the work with their own forces, sometimes virtually all of the work. Proper analysis and pricing of such work requires careful definition of each significant work operation required to build the job. This definition in turn requires detailed knowledge of construction processes. Once these work operations are properly defined, quantities of work required for each operation can be taken off the bid drawings, method study sketches, and so on. The particular work operations required will obviously be closely related to the particular construction scheme or method the contractor intends to employ.

To complete a construction scenario on paper, which through the pricing process, can later be converted to dollars and cents of cost and time required, the estimator must conceptualize the required work into a reasonable number of individual work operations. For simplicity's sake, minor work operations, which occur in almost unlimited numbers and descriptions, must be consolidated into larger work packages that are manageable in the estimate. One of the most important skills that a heavy construction estimator must possess is the ability to conceptualize complicated construction operations into a series of reasonable work operation packages that may be analyzed and priced.

To some, the above five categories of takeoff quantities may seem too limited. However, I have never encountered a takeoff quantity that did not logically fit into one or the other of the above-listed categories.

The application of takeoff calculation principles is illustrated by the three following hypothetical projects. Each has been greatly simplified for purposes of discussion. In the explanations that follow, the intended emphasis concerns applications of the takeoff principles involved rather than an exercise in arithmetic computations.

EXAMPLE TAKEOFF FOR HYPOTHETICAL ROAD PROJECT

Figure 4-1 shows a greatly simplified profile and cross sections for a small road project. This hypothetical project is modeled after the type of secondary road commonly constructed in the western United States. Assume that a contractor who does not perform paving work wishes to take off the quantities involved for the direct cost portion of a bid cost estimate. The project involves both soil and rock excavation in hilly terrain. Assume further that the project has been designed so that the earth work quantities balance[2] and that the specifications permit the mixed placement of soil and rock materials from the cuts into the fills. Typically, there would be a number of separate bid items for the measurement and payment of the project work, and **each bid item would have its own hierarchy of required takeoff quantities.**

Conceptually, the required takeoff quantities, organized by takeoff category, might consist of the following items.

Work to Be Performed by Subcontractors

Because the prime contractor does not perform paving work, all paving quantities fall into this category. In the simplified example shown, the following quantities would have to be determined for work that would be subcontracted:

- Subbase material to be furnished and placed, probably measured for payment by the ton (t)
- Asphalt paving to be furnished and placed, probably measured for payment by the ton, or alternately, by the square yard (sy) of a designated thickness
- Quantities of minor items such as roadway signs, guardrails, and pavement striping

---

[2]The quantities of soil and rock excavation produced in the roadway cuts will be totally consumed by, and furnish sufficient material for, the construction of the fills.

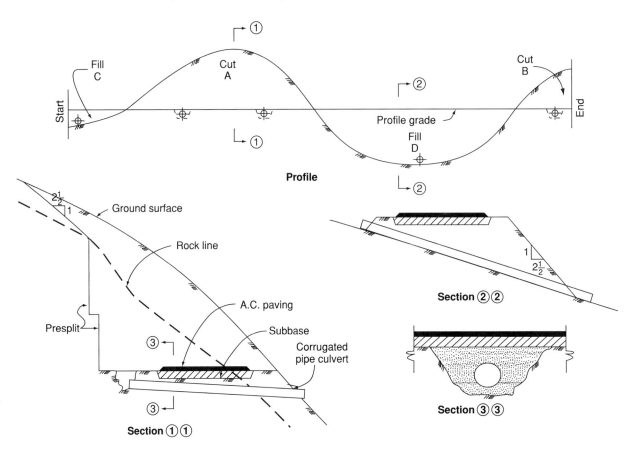

FIGURE 4-1   Hypothetical road project.

Permanent Materials

The only permanent materials required in the example shown would be the culvert pipe material and the material for the compacted backfill around this pipe, which probably would be a processed material obtained from a supplier's crushing and screening plant. The conceptual takeoff quantities required include the following:

- Linear feet of culvert pipe for each pipe size required, measured for payment by the linear foot (lf)
- Processed granular material for culvert bedding and backfill, measured for payment by the cubic yard (cy), or alternately, by the ton

Expendable Materials

In this example, required expendable materials would include waste quantities of permanent materials, explosives, and drill bits, steels, and other drill expendables for rock excavation operations. No permanent material waste quantities should be required for the culvert pipe, but considerable wastage typically occurs in handling processed bedding materials. This material must be delivered from a plant some distance from the jobsite, and it may not be possible to coordinate the delivery of the material with the placing operation around the culverts. It would therefore be necessary to take delivery of the material in advance and stockpile it at several points along the alignment. This would result in a high stockpile loss in addition to the amount of material that, when placed, necessarily would fall outside of the neat ditch lines shown on the drawings. The total waste and overbreak allowance in these circumstances could easily amount to 50 to 100% of the theoretical quantity shown on the drawings.

The computations of the pounds of explosives required and the number of drill bits, steels, shanks, and couplers required to drill the required presplit and blast holes would be based on the takeoff figure for rock excavation and the square footage of presplit surface determined from the contract drawings. Computations for these types of drill and blast expendables are explained Chapter 7.

The Pay Quantities

The number and description of bid items to establish the unit prices to be paid to the prime contractor for various required items of work vary, depending upon the inclination of the owner's engineer who drafts the contract documents. Assume, for purposes of this example, that the bid item structure for the major bid items for this hypothetical project consisted of the following:

> *Bid item 1.* Clear and grub all original ground surfaces, measured in acres, to 5 ft beyond the limits of all required excavations and embankments.
>
> *Bid item 2.* Perform common excavation, to be measured in bank cubic yards (Bcy) within the slope lines shown on the drawings from the natural surface down to the rock line.
>
> *Bid item 3.* Excavate rock measured in bank cubic yards below the rock line within the slope lines and subgrade lines shown on the drawings.
>
> *Bid item 4.* Place and compact roadway embankment measured in embankment cubic yards (Ecy) above the natural surface lines and within neat slope lines and subgrade lines shown on the drawings.
>
> *Bid item 5.* Place compact roadway subbase to the specified thickness measured in square yards between the limits shown on the drawings.
>
> *Bid item 6.* Furnish, roll, and compact asphalt paving to the specified thickness measured in sy between the pay limits shown on the drawings.
>
> *Bid item 7.* Place culvert pipe including inlets and exits, including trench excavation and compacted bedding and backfill material measured by the linear foot. A separate bid item would be required for each separate diameter of culvert pipe.

Additional bid items would also have been included for such minor items as signs, guardrails, and pavement striping.

After reading the measurement and payment provisions of the contract documents carefully, the contractor should accurately compute the quantity for payment that will be received if the work is constructed precisely as indicated in the drawings and specifications and measured for payment according to the rules stated in the bid documents. If the bid quantities stated on the bid form have been carefully determined by the owner's engineer, the contractor's takeoff quantities should compare with them closely. However, discrepancies often occur, sometimes very large discrepancies, and the contractor's takeoff identifies these discrepancies for proper handling by the contractor when pricing out the bid items and when pricing the bid form. Procedures for both of these latter activities are discussed in Chapters 6 and 13, respectively.

Work Quantities for Operations to Be Performed by the Prime Contractor's Own Forces

Certain of the major bid items will be largely or entirely performed by the contractor's own forces. The first step is to define the particular work operations involved. This can be done in a number of different ways, depending upon the experience and inclination of the estimator organizing the takeoff. For example, one particular arrangement for the contractor's work operations could be as follows:

Operation 1: *Clear and Grub All Excavations and Embankments*

The work quantity, measured in acres, can be calculated directly from the contract drawings.

Operation 2: *Load and Haul Common Excavation from Cut A to Fill C*

The particular length of cut A from which the common material lying above the rock will be excavated starting at the beginning of the cut would be determined to a certain following station by use of mass diagram analysis, explained in Chapter 8. The work would involve loading this material, hauling it to fill C, and dumping and spreading the material in layers for compaction. The work quantity would be measured in bank cubic yards. The quantity would be computed by use of earthwork quantity takeoff computer software or by manual computation by the average end area method.[3]

Operation 3: *Load and Haul Common Excavation Cut A to Fill D*

Operation 3 is the same as operation 2 except the destination of the excavated material is fill D rather than fill C.

Operation 4: *Load and Haul Common Excavation Cut B to Fill D*

Operation 4 is the same as operation 2 except the material is excavated from cut B and hauled to fill D, where it is dumped and spread for compaction.

Operation 5: *Drill Presplit Holes for Cuts A and B*

Cross-section 1-1 on Figure 4-1 indicates that the presplit holes would be drilled in two stages, the first stage down to the horizontal berm and the second stage from the berm to subgrade. The construction drawings would not show either the size of the holes or their center to center spacing. This is an example of the many instances where the estimator must develop **takeoff quantities for items not explicitly shown on the contract drawings.** Commonly, these holes would be 2 inches in diameter and spaced at 18 to 24 in. centers. Based on this information, the total quantity of holes to be drilled can be calculated from the drawings. The work quantity would be measured in linear feet.

Operation 6: *Drill Blast Holes for Cuts A and B*

Blast holes required for loading explosives to break the rock will not be shown on the drawings. However, once the quantity of rock excavation required for the job is determined (see operations 8, 9 , and 10, below), the linear feet of blast holes can be calculated by methods explained in Chapter 7. Typical blast holes for rock excavation of this type would be 3 in. in diameter.

---

[3]This earthwork computation method is described in any standard text on surveying and will not be further discussed here.

Operation 7: *Load Powder and Blast for Cuts A and B*

The work quantity in operation 7 is normally measured by the number of holes to be loaded. The work quantity number is developed as part of the calculation required to determine the total linear feet of blast holes required to be drilled for operation 6, above.

Operation 8: *Load and Haul Rock Excavation from Cut A to Fill C*

Operation 8 is similar to operation 2 except that the material being loaded and hauled is the rock excavation portion of cut A. The station to which rock excavation from cut A will be loaded and hauled for placement in fill C is determined by mass diagram analysis, explained in Chapter 8. The quantity is measured in bank cubic yards.

Operation 9: *Load and Haul Rock Excavation from Cut A to Fill D*

Operation 9 is the same as operation 8 except the material is deposited in fill D.

Operation 10: *Load and Haul Rock Excavation from Cut B to Fill D*

Operation 10 is similar to operation 8 except the material is excavated from cut B and deposited in fill D.

Operation 11: *Spread and Compact Embankment on All Fills*

Previous work operations delivered the soil and rock materials to fills C and D from cuts A and B. Separate work operations were required for the various load-and-haul operations because the haul distances varied for each of them. However, haul distance is immaterial as far as the spread and compact operation is concerned. Assuming that the spread and compact equipment will be of a type that can handle both soil and rock materials delivered from the cuts, one operation will suffice for all the embankment to be placed. If the nature of the materials involved and/or any of the requirements of the compaction specifications made necessary the use of separate compaction equipment for soil and rock, two separate operations would be required, one for soil and one for rock. In any case, the work quantity would be measured in embankment cubic yards. Once the quantities of soil and rock materials in bank cubic yards delivered to the fills are known (see operations 2, 3, 4, 8, 9, and 10, above), the corresponding quantities in embankment cubic yards can be computed by methods explained in Chapter 8.

Operation 12: *Drill and Blast Culvert and Drain Trench Excavation*

Operation 12 is a minor item, so the drilling and blasting work may be combined into one work operation. It has been kept separate from the mass drill-and-blast work because the nature of the operation will be considerably different. The blast holes will be shallow and small in diameter, and the general efficiency of the operation will be low because of the small quantities involved and the scattered locations. The work quantity will be measured in bank cubic yards and can be determined by computation from the construction drawings.

Operation 13: *Load and Haul Trench Excavation*

The quantity of material excavated from the culvert and drain trenches is small and will be handled intermittently during the project. If a convenient fill area is available to accept it at the time of excavating, the material would be directly placed there. Otherwise, it would be wasted. The quantities involved are not large enough to worry about potential differences in haul length or similar questions. The work quantity would be measured in bank cubic yards.

Operation 14: *Place and Compact Culvert Bedding and Backfill Material*

Operation 14 is a labor-intensive operation that would be assisted by a small front-end loader to obtain the processed granular bedding material from the on-site stockpiles and tram to the ditch locations. The work quantity is the volume of the ditches (including overbreak) less the volume occupied by the culvert pipe. It can be computed from the construction drawings and is measured in embankment cubic yards.

Operation 15: *Install Culverts Including Entrances and Exits*

Operation 15 is primarily a labor item. The work quantity is measured in linear feet. If the diameter of the culverts differ significantly, separate operations will be needed for the various sizes or alternately, one operation could be set up for the average size culvert.

Relation of Work Operations to Bid-Item Structure

Although no differentiation was made in the above discussion, each of the previously described work operations is required for only certain of the previously discussed bid items.

- Operation 1 pertains to bid item 1.
- Operations 2, 3, and 4 pertain to bid item 2.
- Operations 5, 6, 7, 8, 9, and 10 pertain to bid item 3.
- Operation 11 pertains to bid item 4.
- Operations 12, 13, 14, and 15 pertain to the culvert bid items (bid items 7 and above).

EXAMPLE TAKEOFF FOR A PILE-DRIVING PROJECT

An understanding of the measurement and payment provisions in the contract documents is critically important if correct takeoff quantities are to be determined. The following example for a hypothetical project modeled after pile-driving work common for lock and dam construction along the Mississippi River illustrates this point.

Assume that the bid form for such a project contains a bid item for pile-driving work stated as follows:

Bid item 17   Furnish and drive
                HP14 × 73 bearing   312,500 lf at $_____ per lf = $_____
                pile

Thus, the owner's bid quantity is 312,500 lf.

Further assume that the piles are shown on the plans extending from a specified cut-off elevation to a specified tip elevation and that

- There are 2,573 piles with a cutoff elevation = 375.0 and a tip elevation = 330.0.
- There are 2,764 piles with a cutoff elevation = 375.0 and a tip elevation = 315.0.
- The pile-driving work is described in the specifications as follows:

PHOTO 4-1   This aerial view of the first-stage construction of the Lock and Dam 26 project on the Mississippi River near St. Louis, Missouri, involved the type of pile driving work illustrated by the example bid item in the text. There are three separate pile-driving rigs visible in the photo.

Piles will be driven to the tip elevations shown on the drawings and to a driving resistance of at least 100 blows of a Vulcan OR Hammer for the last foot of penetration. If, upon reaching the specified tip elevation, the above specified driving resistance has not been reached, contractor shall weld an additional length of pile onto the driven pile and continue driving until the specified driving resistance is reached. If, when the specified driving resistance is reached, some of the added length remains above the cutoff elevation shown on the drawings, the contractor shall cut the pile off at the specified cutoff elevation.

- The measurement and payment section of the specifications reads as follows:

  Measurement: Piles will be measured for payment by the linear foot from the final tip of the pile as driven to the specified cutoff elevation. Any undriven length of pile remaining above specified cutoff elevation upon competition of driving will not be measured for payment.
  Payment: Payment will be made at the bid price per linear foot for all piling measured for payment for the complete work of furnishing and driving piling, including the work of making any required splice welds and the work of any required cutting off of pile.

To make the required quantity takeoff calculations, the estimator must make certain assumptions about this pile-driving work. **If the quantities are to be realistic, these assumptions must be reasonable.** In the following example, the estimator concluded

- To minimize the number of pile extension splices required (which would be for the contractor's account), all piling would be ordered from the mill 5 ft longer than the theoretical length required.

- Fifteen percent of the piling would not develop the required driving resistance when the mill-supplied length has been driven and would therefore require extensions.[4]
- A 10 ft. extension piece would be sufficient to assure achieving driving resistance for all piles required to be extended.[5]
- For the piles that reach the specific driving resistance within their ordered length (85% of the total pile), an average of 50% of the extra footage within the ordered length of the pile would actually be driven and would be measured for payment, and 50% would be cut off after driving and would not be measured for payment. All piles, both those not requiring extensions and those that do require extensions, would need to be cut off after driving.
- For the piles that are extended by welding on a 10 ft extension piece (15% of the total pile), an average of 50% of the 10 ft extension piece footage would be cut off after driving and thus would not be measured for payment. The 50% remaining in place would be measured for payment

The takeoff quantities for this hypothetical pile-driving project would be determined as follows based on the above information obtained from the drawings and specifications and the assumptions made by the estimator.

Work to Be Performed by Subcontractors

On lock and dam projects of this type, prime contractors usually perform large pile-driving operations with their own forces. Thus, no work quantities are to be performed by subcontractors.

Permanent Materials

Shown on drawings:	$(2,573)(375 - 330)(73 \text{ lb/ft})$	$= 8,452,300 \text{ lb}$[6]
	$(2,764)(375 - 315)(73 \text{lb/ft})$	$= 12,106,300 \text{ lb}$
Extra 5 ft:	$(2,573 + 2,764)(0.15)(5 \text{ ft})(73 \text{ lb/ft})$	$= 292,200 \text{ lb}$
	$(2,573 + 2,764)(0.85)(5 \text{ ft})(0.50)(73 \text{ lb/ft})$	$= 827,900 \text{ lb}$
Extensions:	$(2,573 + 2,764)(0.15)(10 \text{ ft})(0.50)(73 \text{ lb/ft})$	$= \underline{292,200 \text{ lb}}$
		$21,970,900 \text{ lb}$
Total permanent materials:		$= 219,700$ hundred weight (Cwt)

Expendable Materials

Except for small tools and supplies for which, as explained in Chapter 6, a takeoff quantity is not required, the only expendable material quantities would be the piling lengths that are purchased for which no payment will be received. These may be calculated as follows:

---

[4]This assumption is a judgment call, typically based on the estimator's study of the soil borings and consultation with a geotechnical engineer familiar with the soils at the project site.

[5]This is the same kind of judgment call described in footnote 4.

[6]Piling weight and linear foot quantities rounded to the nearest 100 lb or 100 lf, as the case may be.

Extra 5 ft:	$(2{,}573 + 2{,}764)(0.85)(5\text{ ft})(0.50)(73\text{ lb/ft})$	$=$	827,900 lb
Extensions:	$(2{,}573 + 2{,}764)(0.15)(10\text{ ft})(0.50)(73\text{ lb/ft})$	$=$	292,200 lb
			1,120,100 lb
Total expendable materials:		$=$	11,200 Cwt

Pay Quantities

Shown on drawings:	$(2{,}573)(45\text{ ft})$	$=$	115,800 ft
	$(2{,}764)(60\text{ ft})$	$=$	165,800 ft
Extra 5 ft:	$(2{,}573 + 2{,}764)(0.15)(5\text{ ft})$	$=$	4,000 ft
	$(2{,}573 + 2{,}764)(0.85)(5\text{ ft})(0.50)$	$=$	11,300 ft
Extensions:	$(2{,}573 + 2{,}764)(0.15)(10\text{ ft})(0.50)$	$=$	4,000 ft
Total pay quantities		$=$	300,900 ft

Work to Be Performed by Contractor's Own Forces

Although some estimators might conceptualize these required work operations differently, the following is a typical approach that would produce satisfactory results.

Operation 1: *Unload Pile*

The permanent and expendable material prices usually include the freight of the piling from the mill to the jobsite where it must be unloaded by the contractor. The material would be delivered to a job of this type either by barge or by on-highway tractor-trailer units. Assuming that the work operation of unloading the delivery vehicle would be essentially the same for the two modes of delivery, only one work quantity need be determined.

An important decision the estimator must make is to determine the unit of measurement for the work quantity. In this case, if the project engineer were to ask the work crew foreman how much material was unloaded the previous day, the foreman would likely reply that the crew unloaded some specific number of pieces. The foreman would not be likely to state that the crew had unloaded so many tons or so many linear feet. Each piece of piling must be handled by a crane using slings fitted with end hooks. It makes little difference if the individual piece is 45, 60, or, for that matter, 15 ft long. The same steps must be followed to unload a single piece regardless of its length. The end hooks must be slipped over the ends of the pile; the crane must hoist the pile, swing it over to the unloading area, and set it on the storage stockpile; and the hooks must be disengaged. On a project of this type, it is unlikely that pile deliveries from a mill some distance from the jobsite can be coordinated with pile-driving production, so the piling material must be ordered in advance and stockpiled in a lay-down area at the site. Based on the foregoing, the operation quantity would be calculated as follows:

$$\text{Work quantity} = (2{,}573 + 2{,}764)(1.15) = 6{,}138 \text{ pieces}$$

Operation 2: *Rehandle and Move to Pile Driver*

Prior to being driven, the piling must be rehandled from the lay-down area and transported within the jobsite to the location at which it will be driven. The work quantity is the same as for unloading the pile, that is, 6,138 pieces.

Operation 3: *Drive 50-Ft-Long Pile*

Operation 3 refers to driving the 50 ft pile sections as they were received from the mill without regard to whether or not some of them must be spliced onto and driven further. These piles are shown on the drawings as a 45-ft-long pile extending from elevation 330.0 to elevation 375.0. The known quantity is 2,573 piles.

Operation 4: *Drive 65-Ft-Long Pile*

Operation 4 refers to driving the 65 ft pile sections as they are received from the mill without regard to whether or not some of them are to be spliced onto and driven further. Again, the quantity is known (they show on the drawings as 60 ft piles extending from elevation 315.0 to elevation 375.0). The quantity is 2,764 piles.

Operation 5: *Pile Splices*

From the information stated and assumptions made, the work quantity in operation 5 is calculated as follows:

$$\text{Work quantity} = (2{,}573 + 2{,}764)(0.15) = 801 \text{ splices}$$

Operation 6: *Drive Extension Pieces after Splicing*

The work quantity for operation 6 is the same as the number of splices, that is, 801 extensions.

Operation 7: *Cutoffs*

The estimator's assumption was that all piles would have to be cut off after driving regardless of whether or not an extension piece was welded on. This assumption was based on the belief that the pile would be difficult to drive exactly to cutoff elevation.

$$\text{Work quantity} = (2{,}573 + 2{,}764) = 5{,}337 \text{ cutoffs}$$

Operation 8: *Remove Cutoffs*

Depending on their length, the cutoffs would be too heavy to handle by hand. A small crane or other unit of loading equipment plus a truck for collection would be required. The work quantity is the same as the cutoff work quantity, that is, 5,337 cutoffs.

In the previous two projects illustrating the quantity takeoff procedure, the discussion followed the order in which the five major categories of takeoff quantities were initially presented. **Although these five categories should always be kept in mind, the takeoff computations are often simplified and made more efficient by taking off the quantities in a different sequence,** as the following example illustrates.

EXAMPLE TAKEOFF FOR DRAINAGE CHANNEL PROJECT

The balance of this chapter is devoted to illustrating the takeoff procedure for a project involving excavation, structural concrete work (including reinforcing steel), and structural backfill. Although conceptually similar to actual projects of this type, this hypothetical project has been greatly simplified to facilitate the explanation of takeoff concepts. Actual

PHOTO 4-2 These four separate photos taken from several different projects illustrate the type of structural concrete work involved in the simplified drainage channel project in the text. Upper left, job built form panels. Upper right, transverse bulkhead joint with waterstop, keyway, and reinforcing steel penetrations. Lower left, backfill placement. Lower right, completed portal structure. Also shows horizontal construction joints.

73

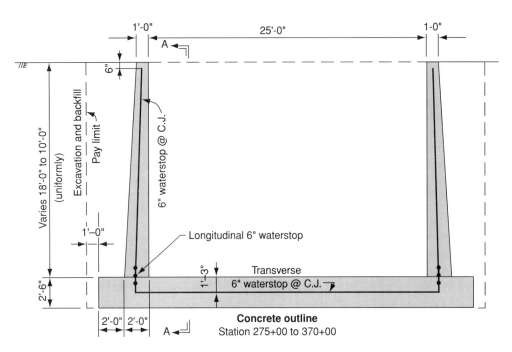

FIGURE 4-2   Cross section of drainage channel showing concrete outline.

projects will be more complex, but the principles illustrated here are equally applicable—the only difference is that more computational effort will be required to take off comparable quantities for an actual project.

General Description of the Project Work

A typical cross section of a reinforced concrete drainage structure is shown as Figure 4-2. Note that all horizontal dimensions and the thickness of the bottom slab are constant from one end of the structure to the other. The wall thicknesses at the top and bottom of the walls are also constant, but the height of the walls varies, ranging from 18 ft 0 in. at one end of the structure to 10 ft 0 in. at the other. The structure is 9,500 ft long, extending from station 275+00 to station 370+00. Vertical construction joints occur every 50 ft, so there will be a total of 9,500/50 = 190 individual sections between construction joints. A typical longitudinal section AA is shown as Figure 4-3. Reinforcing steel details are shown on Figures 4-4 and 4-5.

For simplicity's sake, assume that the project work is described and paid for by means of the following four bid items:

Bid item			
21	Structural excavation	210,000 Bcy at $_____ per Bcy	= $_____
Bid item			
22	Structural backfill	36,900 Ecy at $ _____ per Ecy	= $_____
Bid item			
23	Structural concrete	44,750 cy at $ _____ per cy	= $_____
Bid item			
24	Reinforcing steel	5,675,000 lb at $_____ per lb	= $_____

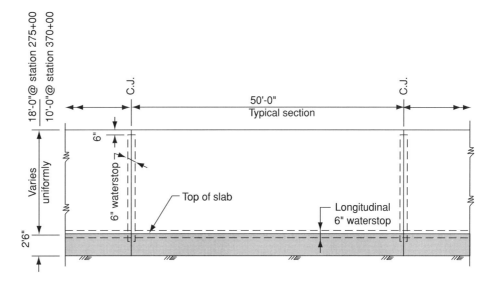

FIGURE 4-3   Longitudinal section of drainage channel showing concrete outline.

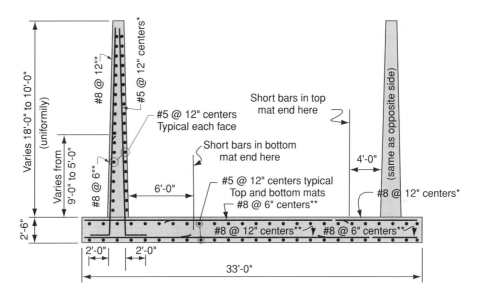

* See sheet 2 of 2 for detail of wall steel.
** See sheet 2 of 2 for detail of slab steel.

FIGURE 4-4   Reinforcing details: Sheet 1 of 2.

Takeoff Computations for the Structural Excavation Bid Item

Assume that the estimator has concluded that an open cut for construction of the concrete drainage channel can be excavated with $\frac{3}{4}$ horizontal to 1 vertical side slopes with toes of the slopes at a point 3 in. below the bottom of the structure and 5 ft outside the sides of the base slab on each side of the structure. Thus, the theoretical excavated depth would be approximately 3 in. below the bottom of the base slab.

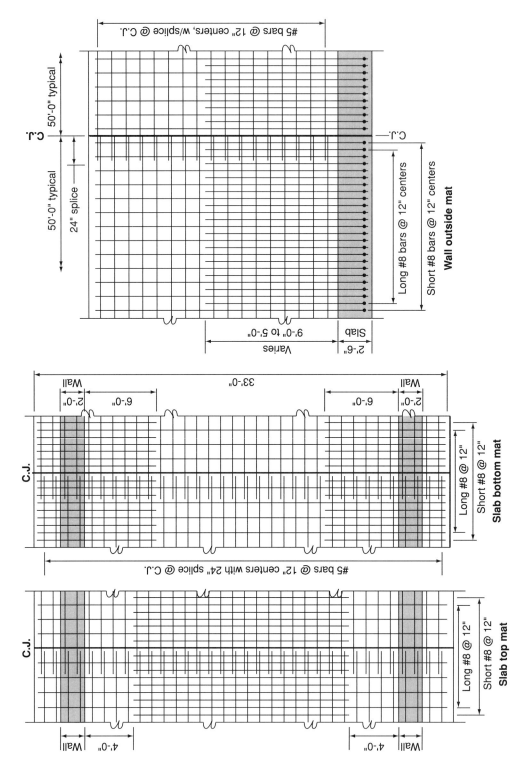

FIGURE 4-5   Reinforcing details: Sheet 2 of 2.

The measurement and payment provisions of the contract documents state that only material between the original ground and the bottom of the concrete structure and bounded by the vertical paylines shown on Figure 4-2 will be measured for payment. The only takeoff quantities to be computed are the pay quantity and quantities for work operations to be performed by the contractor's own forces, which in this case consists of one quantity only—the quantity of material that must be actually excavated.

The Pay Quantity

For measurement and payment purposes, the depth of the structure at one end is 18 ft 0 in. + 2 ft 6 in. = 20 ft 6 in. (20.50 ft), varying uniformly to a depth at the other end of 10 ft 0 in. + 2 ft 6 in. = 12 ft 6 in. (12.50 ft). The width between the vertical paylines is 25 ft 0 in. + (2)(5 ft 0 in.) = 35 ft 0 in. (35.00 ft).

$$\text{Pay quantity} = \frac{(20.5 \text{ ft} + 12.5 \text{ ft})(35.00 \text{ ft})(9500 \text{ ft})}{(2)(27)} = 203{,}194 \text{ Bcy}$$

Work Operation Quantity for Actual Excavation

The depth of actual excavation is 3 in. deeper than that measured for payment. The actual excavation depth at one end of the structure is 18 ft 0 in. + 2 ft 6 in. + 0 ft 3 in. = 20 ft 9 in. (20.75 ft), varying uniformly to a depth at the other end of 10 ft 0 in. + 2 ft 6 in. + 0 ft 3 in. = 12 ft 9 in. (12.75 ft). The average depth of the excavation therefore is (20.75 ft + 12.75 ft)/2 = 16.75 ft. The bottom width of the actual excavation is 25 ft 0 in. + (2)(4 ft 0 in. + 5 ft 0 in.) = 43 ft 0 in. (43.00 ft). The slope width for the 16.75 ft average depth is (0.75)(16.75) = 12.56 ft. The average top width of the excavation is therefore 43.00 ft + (2)(12.56 ft) = 68.12 ft.

$$\text{Actual excavation quantity} = \frac{(68.12 \text{ ft} + 43.00 \text{ ft})(16.75 \text{ ft})(9500 \text{ ft})}{(2)(27)} = 327{,}444 \text{ Bcy}$$

Takeoff Computations for the Structural Backfill Bid Item

After the concrete structure has been completed, compacted backfill must be placed up to the original ground surface from the outside surface of the concrete structure to the excavated slopes. As was the case for the excavation bid item, the only categories of takeoff required are the pay quantity and the work quantity to be performed by the contractor's own forces.

Pay Quantity

From computations made for the excavation bid item, the average depth of the backfill measured for payment falling outside the end of the foundation slab is 20 ft 6 in. (20.50 ft) at one end of the structure, varying uniformly to 12 ft 6 in. (12.50 ft) at the other end. The portion of the backfill falling on top of the foundation slab against the outside of the vertical wall is 18 ft 0 in. (18.00 ft) at one end of the structure and 10 ft 0 in. (10.00 ft) at the other end. Therefore, average depth of backfill placed outside the end of the foundation slab is (20.50 ft + 12.50 ft)/2 = 16.50 ft, whereas the average depth of backfill placed on top of the foundation slab is (18.0 ft + 10.00 ft)/2 = 14.00 ft. The width of the backfill measured for payment on each side of the structure falling outside the end of the foundation slab is 1.00 ft. The average width of the backfill falling on top of the foundation slab on each side of the structure is (3.00 ft + 2.00 ft)/2 = 2.50 ft.

$$\text{Backfill pay quantity} = \frac{(2)(16.50 \text{ ft})(1.00 \text{ ft})(9500 \text{ ft}) + (2)(14.00 \text{ ft})(2.50 \text{ ft})(9500 \text{ ft})}{27}$$

$$= 36,241 \text{ Ecy}$$

Work Operation Quantity

The actual backfill quantity the contractor must place is computed similarly. Utilizing computations made for the excavation bid quantity, the average depth of actual backfill placed outside the end of the foundation slab is 16.75 ft. The average depth of backfill placed on top of the foundation slab against the vertical wall of the structure is 14.00 ft. The backfill cross section on one side of the structure consists of two trapezoidal areas, one 16.75 ft deep and the other 14.00 ft deep. The top width of the first trapezoidal area is (12.56 ft + 5.00 ft) = 17.56 ft, whereas the bottom width is 5.00 ft. The top width of the second trapezoidal area is 3.00 ft. and the bottom width 2.00 ft.

Actual backfill quantity =

$$\frac{(2)(17.56 \text{ ft} + 5.00 \text{ ft})(16.75 \text{ ft})(9500 \text{ ft}) + (2)(3.00 \text{ ft} + 2.00 \text{ ft})(14.00 \text{ ft})(9500 \text{ ft})}{(2)(27)}$$

$$= 157,587 \text{ Ecy}$$

Structural Concrete Bid Item

Many takeoff quantities must be determined for structural concrete work—in this case a total of 25 separate quantity items. The nature of each will be explained and the appropriate quantity computed as follows.

Foundation Fine Grade (Work Operation 1 Quantity)

The structural excavation for the structure would be performed by heavy construction equipment to an accuracy of approximately ± 0.10 ft. Prior to commencing work for the concrete structure, a labor crew must smooth and compact the machine-excavated foundation to a reasonably level firm surface suitable for blocking up the lower mat of slab reinforcing steel. If the excavation had been performed by blasting in rock, the foundation would be very irregular and probably would contain some shattered and loosened rock left by the excavating equipment. Cleanup of this surface, including removal of loose and/or drummy areas of rock, is called *foundation preparation* and is a far more labor intensive and expensive operation than *fine grade*. The fine-grade operation should be carried out to at least 1 ft beyond the edges of the foundation slab.

$$\text{Fine grade quantity} = (25.00 \text{ ft} + 5.00 \text{ ft} + 5.00 \text{ ft})(9,500 \text{ ft}) = 332,500 \text{ sf}$$

Erect and Strip Foundation Slab Edge Forms (Work Operation 2 Quantity)

Concrete formwork quantities are usually broken down into the several different types of forms required for the structure, inasmuch as different formwork elements vary in complexity and in the labor required to erect and strip (E & S) them. For this simple example, only three classifications will be used, the first of which is the slab edge form extending, for takeoff computation purposes, the full length of the structure on each edge of the slab. Note that when the structure is built, these forms will be erected and set in 50 ft increments (the length of an individual section between construction joints).

The unit of measurement for formwork quantities shown on the drawings is square foot of contact area (sfca).

$$\text{E \& S quantity} = (2)(2.75 \text{ ft})(9,500 \text{ ft}) = 52,250 \text{ sfca}$$

E & S Wall Forms (Work Operation 3 Quantity)

These formed surfaces vary in height from 18.00 ft at one end of the structure to 10.00 ft at the other, averaging 14.00 ft. Even though the actual form sections will be erected and stripped in 50 ft lengths, the total contact area is computed as follows:

$$\text{E \& S wall forms} = (4)(14.00 \text{ ft})(9,500 \text{ ft}) = 532,000 \text{ sfca}$$

E & S Slab and Bulkhead Forms (Work Operation 4 Quantity)

Bulkhead forms are required to contain the ends of the concrete placement in each 50 ft. section of the structure. They usually contain a keyway and are normally penetrated by longitudinal reinforcing steel bars. They are an expensive form to erect and strip and are classified separately for this reason. Considering the forms at each end of the structure, the number of individual bulkhead formed surfaces will be 9,500 ft/50 ft + 1 = 191 separate surfaces.

E & S slab bulkheads = (191)(2.75 ft)(33.00 ft)	=	17,333 sfca
E & S wall bulkheads = (2)(191)(1.5 ft)(14.00 ft)	=	8,022 sfca
Total E & S bulkhead forms	=	25,355 sfca

Form Fabrication (Work Operation 5 Quantity)

The previous calculations were for the erect and strip quantities of form work measured in square feet of contact area. Because concrete forms are typically reused, the contractor will not need form panels to form the full square feet of contact area at one time. The number of form panels required depends on the reuse factor for each classification of forms, which in turn depends on the construction schedule, the configuration of the structure being constructed, and the quality of form surface required by the job specifications. The total number of form panels is a matter of judgment. For this example, assume that the estimator's judgment is that the foundation slab-on-grade (SOG) edge forms will be used 10 times, the wall forms, 20 times, and the slab and wall bulkhead forms, 3 times. On this basis, the total quantity of forms that must be fabricated is computed as follows:

$$\text{Form fabrication, SOG edge} = \frac{52,250}{10} = 5,225 \text{ sf}$$

$$\text{Form fabrication, walls} = \frac{532,000}{20} = 26,600 \text{ sf}$$

$$\text{Form fabrication, bulkheads} = \frac{25,356}{3} = 8,452 \text{ sf}$$

$$\text{Total form fabrication} = 40,277 \text{ sf}$$

SOG Concrete (Work Operation 6 Quantity)

If the quantity takeoff calculations are performed in the following sequence and manner, **earlier calculated quantities will facilitate the calculation of later required quantities.**

Generally speaking, there will always be some quantity of concrete waste, depending on the job circumstances. This quantity varies between 2 and 5% of the neat volume shown on the drawings. Also, concrete replacing the extra 3 in. of excavation (overbreak concrete) should be set out separately. In this example, the waste is assumed to run 2% of the neat volume:

$$\text{Shown on drawings} = \frac{(2.50 \text{ ft})(33.00 \text{ ft})(9,500 \text{ ft})}{27} = 29,028 \text{ cy}$$

$$\text{Overbreak concrete} = \frac{(0.25 \text{ ft})(33.00 \text{ ft})(9,500 \text{ ft})}{27} = 2,903 \text{ cy}$$

$$\text{Waste concrete} = (0.02)(29,028 \text{ cy}) = \underline{581 \text{ cy}}$$

$$\text{Total SOG concrete} = 32,512 \text{ cy}$$

Wall Concrete (Work Operation 7 Quantity)

The 2% waste figure also applies to the wall concrete, but there is no overbreak. The average wall height is 14 ft 0 in. (14.00 ft) and the average thickness is 1 ft 6 in. (1.50 ft.):

$$\text{Shown on drawings} = \frac{(1.50 \text{ ft})(14.00 \text{ ft})(9,500 \text{ ft})(2)}{27} = 14,778 \text{ cy}$$

$$\text{Waste concrete} = (0.02)(14,778 \text{ cy}) = \underline{296 \text{ cy}}$$

$$\text{Total wall concrete} = 15,074 \text{ cy}$$

Base Slab Wet Concrete Finish (Work Operation 8 Quantity)

After concrete is deposited in the forms, the top horizontal surface generally will receive some type of specified finish except for those parts of that surface where additional concrete is intended to be placed. The contract specifications define the quality of the finish required for the various horizontal finished areas. In the simple example shown, the top surface of the base slab between the walls would probably be specified to receive a high-quality wood-float finish, whereas the 2 ft projection beyond the outside of the wall that will be covered with backfill would require nothing further than a strike-off. Because this latter area is small, it may be combined with the high-quality finished area without introducing serious error.

$$\text{Base slab wet finish} = (25.00 \text{ ft} + 2.00 \text{ ft} + 2.00 \text{ ft})(9,500 \text{ ft}) = 275,500 \text{ sf}[7]$$

Top of Wall Wet Finish (Work Operation 9 Quantity)

The 1 ft-wide area on the top of each wall usually requires a high-quality finish. This requirement is extraordinarily expensive because when each 50 ft concrete section of walls is placed (which might occur on a one-section-per-working-shift schedule), there will be very little wet finish work to perform. Job supervision must find other tasks to gainfully employ the required concrete finisher for the balance of the shift, resulting in inherent inefficiencies. Because of the high cost of this type of wet finish, the quantity is usually taken off separately:

$$\text{Wall top wet finish} = (1.00 \text{ ft} + 1.00 \text{ ft})(9,500 \text{ ft}) = 19,000 \text{ sf}$$

---

[7]Note that the top surface of the base slab under the side walls does not receive a wet finish.

Concrete Cure (Work Operation 10 Quantity)

All concrete placed in the structure must receive some form of cure treatment,[8] either shortly after placement, or—in the case of formed surfaces—immediately after the forms are stripped. By examining the structure cross section, one can see that the cure area will be all horizontal areas, including the area within the walls on the foundation slab where additional concrete will be placed, plus all formed areas:

Top of slab	=	(33.00 ft)(9,500 ft)	=	313,500 sf
Add slab edge sfca			=	52,250 sf
Add all bulkhead sfca			=	25,355 sf
Add wall sfca			=	532,000 sf
Add wall top wet finish			=	19,000 sf
Total cure			=	942,105 sf

Sandblast Horizontal Construction Joints (Work Operation 11 Quantity)

Horizontal surfaces that are to receive additional concrete placement must be treated in some way to remove laitance from the joint, which if not removed, would prevent good bond of the following placement, weakening the structure. One way of removing this laitance is to "green cut" the surface approximately an hour after the concrete has taken its final set. This is done by directing a stream of high-pressure compressed air and water from a jet pipe onto the freshly placed concrete, which will cut away the laitance. Another method is to sandblast these surfaces after the concrete has hardened prior to setting up the formwork for the following placement. Assume that in this example, these areas, consisting of the 2-ft-0-in.-wide strips on top of the base slab where the base of the walls will rest, are to be sandblasted:

$$\text{Sandblast quantity} = (2)(2.00 \text{ ft})(9,500 \text{ ft}) = 38,000 \text{ sf}$$

Nonexposed Dry Finish (Work Operation 12 Quantity)

When formwork is stripped, additional work must be performed on the concrete surfaces to meet specification finish requirements. If the concrete surface will not be exposed to public view (that is, the surface is in a nonaccessible location or will be covered with backfill), the amount of work is minimal. Form tie holes must be filled and any honeycomb[9] repaired. In this example, the work quantity is computed as follows:

Base slab edges	=	52,250 sf
Outside wall surface = 532,000/2	=	266,000 sf
Total nonexposed finish	=	318,250 sf

Exposed Dry Finish (Work Operation 13 Quantity)

When formed concrete surfaces will be exposed to public view or to flowing water, a generally smooth surface finish will be required, depending on the precise requirements of the specifications. In addition to plugging tie holes, all irregularities must be

---

[8]Keeping the concrete moist for a specified period (usually 14 days) or, alternately, preventing the evaporation of moisture by use of a sprayed-on membrane curing compound.

[9]Honeycomb is an area on the exposed surface where the concrete paste has not completely filled the voids in the coarse aggregate. It is repaired by chipping out the affected area and replacing it with "dry pack" mortar.

ground off and air bubbles remaining in the concrete surface filled with cement mortar by a process known as *sack rubbing.* In this example, the only exposed surfaces requiring this treatment are the inside surfaces of the walls:

$$\text{Exposed dry finish} = 266,000 \text{ sf}$$

Waterstop (Work Operation 14 Quantity)

Waterstop is a polyvinyl extrusion placed in vertical and horizontal construction joints to prevent the passage of water through the joint. This kind of item is typical of a large number of similar items frequently embedded in the concrete that are not measured separately for payment and which must be taken off and priced as part of the concrete item. In this case, the waterstop runs the full length of vertical construction joints in the walls, the length of the vertical construction joints between walls in the base slab, and the full length of the structure in the construction joint at the base of each wall, all as shown on Figures 4-2 and 4-3. The required quantity is computed as follows:

Longitudinal runs = (2)(9,500 ft)	= 19,000 lf
Transverse construction joints = (25.00 ft + 1.00 ft + 1.00 ft + 1.25 ft + 1.25 ft + 14.00 ft + 14.00 ft)(191)	= 10,983 lf
Total waterstop quantity	= 29,983 lf

Quantity of Concrete to Be Purchased (PM)

The major portion of the quantity of concrete to be purchased falls in the permanent material category. The quantity can be determined from previous quantity computations for the concrete volumes shown within the neat lines on the drawings.[10]

SOG quantity shown on drawings	= 29,028 cy
Wall quantity shown on drawings	= 14,778 cy
Total concrete quantity (PM)	= 43,806 cy

Waste and Overbreak Concrete (EM)

Again, the waste and overbreak portion of concrete to be purchased is easily determined from previously computed quantities:

SOG overbreak	=	2,903 cy
SOG waste	=	581 cy
Wall waste	=	296 cy
Total waste and overbreak concrete	=	3,780 cy

Purchase Waterstop (PM)

Purchase waterstop quantity is the same as the quantity of waterstop that must be placed, = 29,983 lf.

---

[10]As previously mentioned, the order in which quantities are taken off greatly facilitates the determination of additional quantities.

Waterstop Waste (EM)

An allowance of 10% of the quantity shown on the drawings is reasonable:

$$\text{Waterstop waste} = (0.10)(29,983) = 2,998 \text{ lf}$$

Concrete Pay Quantity (Pay Quantity)

The concrete pay quantity is the quantity of concrete shown on the drawings measured in cubic yards that the engineer will measure for payment. It is the same as the purchased concrete (PM) item above, 43,806 cy.

Buy Dimension Lumber (EM)

Dimension lumber is required for the fabrication of the job-built panels used for concrete formwork. The quantity is usually estimated on the basis of an experience factor for the number of board feet[11] of dimension lumber required per square foot of fabricated form panel. Dimension lumber use factors are more fully discussed in Chapter 10, but for the moment assume that 3.0 FBM per square foot will be required.

$$\text{Dimension lumber required} = \frac{(40,277)(3.00)}{1,000} = 121 \text{ MBF[12]}$$

Purchase Plywood (EM)

The quantity of plywood required for the fabrication of job-built form panels is computed in a similar manner to that for dimension lumber. Assume a use factor of 1.1 sf of plywood per square foot of form panel fabricated:

$$\text{Plywood required} = (40,277)(1.1) = 44 \text{ MSF[13]}$$

Purchase Form Ties and Miscellaneous Form Supplies (EM)

These materials include form ties, form oil, screws, bolts, nails, and any other similar hardware item required in the fabrication or in the erect-and-strip operation of concrete formwork. **Quantities are not estimated for these items individually. Rather, money is generated in the cost estimate by the application of an experience factor against the total E & S form quantity measured in square feet of contact area.** The appropriate takeoff quantity is the total square feet of contact area required, which is readily available from previously calculated form quantities:

SOG	=	52,250 sfca
Walls	=	532,000 sfca
Bulkheads	=	25,355 sfca
Total square feet of contact area for miscellaneous form supplies	=	609,605 sfca

Purchase Sandblast Sand (EM)

This item is handled in a manner similar to that utilized for miscellaneous form supplies. To generate enough money in the estimate to purchase the sand required for construction

---

[11]A board foot, or foot board measure (FBM), means 1 square foot, 1 inch thick.

[12]1 MBF = 1,000 board feet.

[13]1 MSF = 1,000 square feet.

joint sandblasting, a quantity is taken off against which an experience use factor will be applied. The quantity is simply the area of construction joints to be sandblasted, in this case 38,000 sf.

Purchase Cure Material (EM)

To generate money in the estimate for the purchase of items such as membrane curing compound, curing mats, and similar materials by the application of an experience factor, the appropriate takeoff quantity is the area of concrete to be cured, which in this case equals 942,105 sf.

Purchase Dry Finish Materials (EM)

Money for dry finish materials such as Portland cement and sand is generated in the cost estimate by the application of an experience factor to the total area requiring dry finish, which in this case is the sum of the nonexposed and exposed dry finish areas:

$$\text{Dry finish area} = 318,250 + 266,000 = 584,250 \text{ sf}$$

As the above discussion illustrates, taking off quantities for structural concrete is tedious. Although the above quantity calculations were rather meticulously explained, and exact dimensions were used for purposes of illustration, absolute precision is not required in practice. A reasonable amount of rounding off greatly facilitates computations. Further, concrete takeoff computations lend themselves to a **tabular computational approach** in which all required takeoff quantities are determined element by element of the structure in a logical manner. An example of this type of tabular computation for a bridge pier project is shown in Figure 4-6.

Reinforcing Steel Bid Item

Prime contractors seldom take off reinforcing steel, which is usually purchased from a supplier cut and bent according to the requirements of the contract drawings. At bid time on most projects, the reinforcing-steel suppliers furnish a quantity takeoff that is relied upon for bidding purposes by prime contractors and reinforcing steel placing subcontractors alike. Nonetheless, construction estimators occasionally must take off reinforcing steel. Also, it is important to understand some of the details of this specialty to properly price the item and/or deal with suppliers and placing subcontractors on a sound business basis. For that reason, reinforcing steel takeoff computations are explained below for the example drainage channel project.

The reinforcing steel details shown in Figures 4-4 and 4-5 are similar to those that would be shown in the contract drawings. In this case, only two sizes of reinforcing steel bar are utilized, #5 bars weighing 1.043 lb/lf and #8 bars weighing 2.67 lb/lf.

Assume that the general contractor plans to purchase the fabricated reinforcing steel and engage a subcontractor to place it. The purchased fabricated steel must also include any nonpay splices that the general contractor wants, all templates for supporting the steel in place, and any other supports that are required. Under the terms of the subcontract, it is expected that the subcontractor will be paid a price per pound for the weight of steel installed including all splices, templates, and other supports, regardless of whether or not the general contractor will be paid by the owner on this basis.

Further assume that the measurement and payment provisions in the contract state that splices made for the contractor's convenience (that is, splices *other* than those shown on the drawings) and any steel used for templates and supports will not be measured for payment. Thus, under the commercial terms of the proposed placing subcon-

Example takeoff table (Figure 4-6). Columns: Pier | Element | L | W | H | Buy concrete* (Neat cy, Waste cy, Overbreak cy — each split 3000/4000 psi) | Pour concrete (Footing No./cy, Columns & webwall No./cy, Columns No./cy, Caps No./cy) | E & S forms (Footing sfca, Columns & webwall sfca, Columns sfca, Caps sfca) | Fdn. prep sf | C.J. clean sf | Cure sf | Wet finish (Footing sf, Columns & webwall sf, Caps sf) | Dry finish (Exposed sf, Nonexposed sf).

Pier	Element	L	W	H	Neat cy 3000	Neat cy 4000	Waste cy 3000	Waste cy 4000	Overbreak cy 3000	Overbreak cy 4000	Pour Footing No.	Pour Footing cy	Pour Col & WW No.	Pour Col & WW cy	Pour Columns No.	Pour Columns cy	Pour Caps No.	Pour Caps cy	E&S Footing sfca	E&S Col & WW sfca	E&S Columns sfca	E&S Caps sfca	Fdn. prep sf	C.J. clean sf	Cure sf	Wet Footing sf	Wet Col & WW sf	Wet Caps sf	Dry Exposed sf	Dry Nonexposed sf
#1	Footing (x1)	36.00	16.00	5.00	106.7		3.2		16		1	125.9							598				576		1174	576				598
	Column & WW (x1)	31.34	7.66	8.50	75.6		2.3						1	77.9						663				240	1021		118		663	
	Columns (x2)	8.00	7.66	16.00	72.6		2.2								2	74.6					1002			123	1125				1002	
	Columns (x2)	8.00	6.83	16.08	65.1		2.0								2	67.1					953			109	1062				953	
	Cap-middle	33.00	6.00	8.00		58.7		1.8									1	60.5				528 / 102		96	828			198	630	
	Cap - ends (x2)	10.00	6.00	7.00		31.1		1.0										32.1				280 / 120 / 72								
#2	Footing (x1)	36.00	18.00	5.00	120.0		3.6		18		1	141.6							621				648		592	648				621
	Column & WW (x1)	31.34	7.66	12.50	111.1		3.3						1	114.4						975				240	1269 / 1333		118		472 / 975	
	Columns (see above)				137.7		4.2								4	141.9					1955			232	2187				1955	
	Cap (see above)					89.8		2.8									1	92.6				1102		96	1420			318	1102	
	Etc.																													

*Takeoff on the basis of 4,000 psi concrete required for cap and 3,000 psi concrete for balance of structure

FIGURE 4-6   Example takeoff for concrete cost elements.

tract, the subcontractor will be paid for weights of material installed that the general contractor will not be paid for by the owner.

Utilizing exact dimensions when taking off reinforcing steel is not necessary. The clear distance between the end of a reinforcing steel bar and the form or horizontal surface against which it abuts can be disregarded. Takeoff computations follow:

- **Base Slabs**

	#8	#5
Bottom mat transverse bars, 50/#8/33 ft	1,650 ft	
Bottom mat transverse bars, 2 × 49/#8/10 ft	980 ft	
Top mat transverse bars, 50/#8/33 ft	1,650 ft	
Top mat transverse bars, 49/#8/17 ft	833 ft	
Bottom mat longitudinal bars, 33/5/54 ft		1,782 ft
Top mat longitudinal bars, 33/5/54 ft		1,782 ft
Totals: One base slab	5,113 ft	3,564 ft
Totals: 190 base slabs	971,470 ft	677,160 ft
Total Weight: 190 base slabs	2,593,825 lb	706,278 lb

- **Walls**

	#8	#5
Inside vertical bars, 50/#5/18.5 ft average		925 ft
Outside vertical bars, 50/#8/18.5 ft average	925 ft	
Outside vertical bars, 49/#8/11.5 ft average	564 ft	
Total longitudinal bars, 2 × 13 average /#5/54 ft		1,404 ft
Totals: One wall	1,489 ft	2,329 ft
Total: 380 walls	565,820 ft	885,020 ft
Total Weight: 380 walls	1,510,739 lb	923,076 lb

- **Total Weight Reinforcing Shown on Plans**

Slab #5 bars	706,278 lb
Slab #8 bars	2,593,825 lb
Wall #5 bars	923,076 lb
Wall #8 bars	1,510,739 lb
Total weight shown on plans	5,733,918 lb

- **Splices Not Shown on Plans:** After the base slab is poured, the wall steel would extend as high as 18 ft in the air with its lower ends embedded in the slab. Contractors often prefer to dowel these vertical bars out,[14] thus avoiding the long vertical projections. Assume that the prime contractor chooses to do this, requiring a 2.5 ft splice at the base of the wall. These splices will not be paid for by the owner, but the prime contractor must pay the subcontractor for this extra steel. Computations for the weight of steel involved follow:

---

[14]Putting a splice in the vertical bars at the base of the wall.

	#8	#5
#5 bars per 50 ft wall section, 50 × 2.5 ft		125 ft
#8 bars per 50 ft wall section, 49 × 2.5 ft	123 ft	
Total per 50 ft wall section	123 ft	125 ft
Total for 380 wall sections	46,740 ft	47,500 ft
Total weight for 380 wall sections	124,796 lb	49,543 lb

- **Templates and Supports** A reasonable allowance for the weight of this steel is 5% of the weight of the reinforcing steel in the base slab. From prior computations, the total weight of steel in the base slab is

$$706,278 \text{ lb (#5 bars)} + 2,593,825 \text{ lb (#8 bars)} = 3,300,103 \text{ lb}$$

Therefore, the total weight of templates and supports equals

$$(3,300,103 \text{ lb})(0.05) = 165,005 \text{ lb}$$

- **Takeoff Quantity: Purchase Reinforcing Steel (PM)** This quantity is the weight of reinforcing steel shown on the plans, previously computed as 5,733,918 lb.
- **Takeoff Quantity: Purchase Nonpay Reinforcing Steel (EM)** This quantity consists of the following:

#5 bar nonpay splices	=	49,543 lb
#8 bar nonpay splices	=	124,796 lb
Templates and supports	=	165,005 lb
Total nonpay rebar	=	339,344 lb

- **Quantity Measured for Payment to General Contractor (Pay Quantity)** This quantity is the same as the weight of reinforcing shown on the plans previously computed as 5,733,918 lb.
- **Subcontract Pay Quantity (S)** This quantity consists of the pay quantity to the general contractor plus the nonpay quantity, or 5,733,918 lb + 339,344 lb = 6,073,262 lb.

CONCLUSION

This chapter classified all takeoff quantities into five major categories and then discussed and explained the fundamental principles of the quantity takeoff operation. These principles were then demonstrated by application to three separate types of construction projects, a road construction project, a pile-driving project, and a cut-and-cover reinforced concrete drainage channel project. Countless other examples could also have been presented, but the purpose of this chapter was to emphasize basic principles rather than repetitive arithmetical computations. The examples shown should suffice to demonstrate the concepts and principles involved.

The next chapter moves on to the subject of the preparatory steps that should be taken prior to undertaking the detailed cost estimating work for any heavy engineered construction project.

QUESTIONS AND PROBLEMS

1. What are the five major takeoff categories discussed in this chapter?

2. What are the two reasons why prime contractor's make an independent takeoff of work expected to be subcontracted?

3. Explain why consideration of the construction schedule is important when determining the required quantities of certain expendable materials. What are some of the kinds of expendable materials for which consideration of the construction schedule is important?

4. Why is it important for bidding contractors on schedule-of-bid-item bids to make an independent takeoff of the pay quantities stated by the owner in the bid schedule?

5. Explain the relationship between the particular construction scheme or method the contractor intends to employ and the quantities of work expected to be performed by the contractor's own forces.

6. According to the hypothetical roadway example in this chapter, what are the units of measurement and how are the measurements for payment made for the following pay quantities?
   (a) Clear and grub
   (b) Common excavation
   (c) Rock excavation
   (d) Roadway embankment
   (e) Roadway subbase
   (f) Asphalt paving
   (g) Culvert pipe

7. According to the roadway project example, what units of measurement would be used and how would the quantities be determined for the following work operations to be performed by the contractors own forces?
   (a) Clear and grub work
   (b) Load and haul common excavation A to C, A to D, and B to D
   (c) Drill 2-in.-diameter presplit holes, cuts A and B
   (d) Drill 3-in.-diameter blast holes, cuts A and B
   (e) Load powder and blast, cuts A and B
   (f) Load and haul rock excavation A to C, A to D, and B to D
   (g) Spread and compact embankment
   (h) Drill and blast culvert and drain trench excavation
   (i) Load and haul trench excavation
   (j) Place and compact culvert bedding and backfill

8. Identify each work operation in question 7 with the pay quantity in question 6 to which the work operation applies.

9. According to the hypothetical pile-driving example in this chapter, what are the units of measurement for each of the following quantities that must be taken off?
   (a) Permanent material   (b) Expendable material
   (c) Pay quantity   (d) Unload pile
   (e) Rehandle and move to driver (f) Drive 50 and 65 ft piles
   (g) Pile splices   (h) Drive extension pieces
   (i) Cutoffs   (j) Remove cutoffs

10. For the hypothetical drainage channel example in this chapter, what are the units of measurement and how are the following bid-item pay quantities measured for payment?

   (a) Bid item 21   **(b)** Bid item 22

   (c) Bid item 23   **(d)** Bid item 24

11. Under which one of the five major takeoff categories does each of the following quantities fall and what is the unit of measurement for each?

   (a) Actual excavation

   (b) Actual backfill

   (c) Foundation fine grade

   (d) E & S forms

   (e) Fabricate forms

   (f) Place SOG and wall concrete

   (g) Concrete wet finish

   (h) Concrete cure

   (i) Sandblast construction joints

   (j) Dry finish

   (k) Install waterstop

   (l) Purchase pay concrete

   (m) Purchase waste and overbreak concrete

   (n) Purchase waterstop

   (o) Purchase waterstop waste

   (p) Purchase dimension lumber

   (q) Purchase plywood

   (r) Purchase form ties and miscellaneous form supplies

   (s) Purchase sandblast sand

   (t) Purchase cure materials

   (u) Purchase dry finish material

   (v) Purchase pay reinforcing steel

   (w) Purchase nonpay reinforcing steel

   (x) Reinforcing steel placement subcontract pay quantity

12. The work on a lock and dam consists of furnishing and driving HP14 $\times$ 89 steel bearing piling. The piles are shown on the plans as extending from a specified cutoff elevation to a specified tip elevation:

There are 2,875 piles with C.O. elevation of 375.00 and a tip elevation of 320.00.

There are 3,120 piles with C.O. elevation of 375.00 and a tip elevation of 305.00.

The pile driving work is described in the specifications as follows:

> Piles will be driven to the top elevations shown on the drawings and to a driving resistance of at least 100 blows of a Vulcan OR Hammer for the last foot of penetration. If, upon reaching the specified tip elevation, the above specified driving resistance has not been reached, contractor shall weld an additional length of pile onto the driven pile and continue driving until the specified driving resistance is reached. If, when the specified driving resistance is reached, some of the added length remains above the cutoff elevation shown on the drawings, contractor shall cut off the pile at the specified cut off elevation.

The measurement and payment section of the specifications reads as follows:

> *Measurement:* Piles will be measured for payment by the linear foot from the final tip of the pile as driven to the specified cutoff elevation. Any undriven length

of pile remaining above specified cutoff elevation on completion of driving will not be measured for payment:

Payment: Payment will be made at the bid price per linear foot for all piling measured for payment for the complete work of furnishing and driving piling including the work of making any required splice welds and the work of any required cutting off of pile.

The bid item in the bid form for the pile driving work reads as follows:

Bid item   Furnish and drive
  17   HP 14 × 89 bearing   405,000 lf at $_____ per lf = $_____
                  pile

Make calculations based on the following assumptions about this pile driving work.

(a) To minimize the number of pile extension splices required (which are for the contractor's account), all piling will be ordered from the mill 5 ft longer than the theoretical length required.

(b) Twenty percent of the piling will not develop the required driving resistance when the mill-supplied length has been driven and will therefore require extensions.

(c) A 12 ft extension piece will be sufficient to assure achieving driving resistance for all pile required to be extended.

(d) For the piles that make the driving resistance within their ordered length (80% of the total pile), an average of 50% of the extra 5 ft footage within the ordered length of the pile will actually be driven and will be measured for payment, and 50% will be cut off after driving and will not be measured for payment. All piles, both those not requiring extensions and those that do require extensions, will need to be cut off after driving.

(e) For the piles that are extended by welding on a 12 ft extension piece (20% of the total pile), an average of 50% of the 12 ft extension piece footage will be cut off after driving and thus will not be measured for payment.

Develop the following takeoff quantities:

(a) Quantity of piling in hundredweight to be purchased as permanent material (PM)

(b) Quantity of piling in hundredweight to be purchased as expendable material (EM)

(c) Quanity of piling in linear feet that the engineer is expected to measure for payment

(d) Number of pieces of piling material that must be unloaded at the job site

(e) Number of pieces of piling material that must be rehandled and moved to the pile driver

(f) The number of pile 60 ft long that must be driven (or mostly driven)

(g) The number of pile 75 ft long that must be driven (or mostly driven)

(h) The number of extended pile that must be driven

(i) The number of welded splices that must be made in the leads

(j) The number of pile that must be cut off at grade

Refer to the bridge pier foldout at the back of the book. This drawing was prepared by Alfred Benesch & Company, Consulting Engineers, Chicago, Illinois, as part of the contract drawings for the Jefferson Barracks Bridge Project across the Mississippi River. The bid form for the bridge substructure project of which the westbound roadway pier 4 shown on the foldout was a part (along with many other similar piers) contained separate bid items for structural concrete, seal coat (tremie) concrete, reinforcing steel, cofferdams, cofferdam excavation, and steel bearing piles.

13. Make the calculations for determining the following takeoff quantities for the pier 4 portion only of the total bearing piling bid item necessary for pricing the bid item (not including the test pile noted in the bill of material). Assume, that piling will be ordered from the mill 8 ft longer than the extra length shown on the drawing and that, when the piles are driven, 50% of the extra length will be measured for payment and that no additional weld-ons will be required. Also, note that, although the piling must be driven under water inside of a sheet-pile cofferdam, a follower will be used to accomplish this. Assume that when the required driving resistance is reached, the top of the driven pile will be near, but above, the cutoff elevation shown. The required quantities are
   (a)   The permanent material to be purchased in hundredweight
   (b)   The expendable material to be purchased in hundredweight
   (c)   The pay quantity in linear feet
   (d)   The unload quantity in pieces
   (e)   The rehandle and move-to-driver quantity in pieces
   (f)   The drive pile quantity in pieces
   (g)   The cutoff quantity in pieces
   (h)   The cutoff remove quantity in pieces

14. Make the calculations necessary to determine the takeoff quantities for pier 4 only necessary to price the structural concrete bid item. Use a tabular format similar to that shown in Figure 4-6. (The foot-total of each column will yield the necessary takeoff quantities for the pier 4 portion of the total.) Note the delineation at elevation 375.0 between structural concrete and seal coat (tremie) concrete. Also, the seal coat (tremie) concrete below elevation 375.0 is placed under water directly against the inside surface of the sheet-pile cofferdam, which extends to elevation 400.0. After cofferdam dewatering, this permits the structural concrete above elevation 375.0 to be formed and placed in the dry. Assume waste at 4%.

15. Calculate the following takeoff quantities for pier 4 only necessary to price the seal coat (tremie) concrete bid item:
   (a)   Purchase pay concrete in cubic yards
   (b)   Purchase waste and overbreak concrete in cubic yards
   (c)   Place underwater concrete in cubic yards
   (d)   Underwater foundation preparation in square feet

Base your takeoff calculations on the assumptions that the inside vertical face of the sheet piling is set 2 ft outside the ends of the structural concrete footing (note that the drawings show this surface to be 3 ft outside of the sides of the structural concrete footing). Due to the presence of the interior wells formed by the sheet piling, the average vertical boundary of the seal coat concrete placement will be 6 in outside of the inside vertical face of the sheet piles. Also, assume that the underwater cofferdam excavation carried out prior to driving the steel bearing piling will be taken to an average elevation 1 ft below the bottom of the seal coat shown on the drawings. Assume waste at 4%.

16. Calculate the following takeoff quantities for pier 4 only necessary to price the cofferdam bid item (excluding furnishing, installing, and removing the interior cofferdam bracing, which must be determined separately). In addition to the cofferdam information given in questions 14 and 15, assume that the cofferdam will be formed with vertically driven interlocking PZ-27 sheet piling (each interlocking sheet produces a 1.5 ft length of cofferdam wall and weighs 40.5 lb per vertical foot). Also, assume that the sheet pile extends from the cofferdam top at elevation 400.0 to an elevation 7 ft below the bottom elevation of the seal coat concrete shown on the drawings. The required takeoff quantities are
   (a)   Weight of PZ-27 sheet pile to purchase in tons

     (b)   Quantity to unload in pieces

     (c)   Quantity to rehandle to driver in pieces

     (d)   Quantity to drive in square feet

     (e)   Quantity to pull in square feet

17. Calculate the following takeoff quantities for pier 4 only necessary to price the cofferdam excavation bid item:

     (a)   The actual quantity to be excavated in bank cubic yards

     (b)   The pay quantity in bank cubic yards

     (c)   The number of interior wells to blow clean of foundation material prior to placing seal-coat concrete, each (one interior well is formed for each two sheet-piles used)

The measurement and payment provision states that excavation will be measured for payment from the existing river bottom to the elevation of the bottom of the seal coat shown on the drawings bounded by vertical projections of the outside surfaces of the seal coat shown on the drawings (see question 15 for the relationship of the outside surfaces of the seal coat to the outside surfaces of the structural concrete footing). Also, as stated in question 15, the actual excavation should be carried 1 ft below the bottom elevation of the seal coat shown on the drawings. Assume existing river bottom at elevation 392.0.

18. Calculate the following quantities for pier 4 only necessary to price the reinforcing steel bid item. Assume that extra steel for templates and supports will run $2\frac{1}{2}\%$ of the quantities shown on the drawings, that payment to the prime contractor will be made only for the quantity shown on the drawings, that the prime contractor will purchase all steel f.o.b. jobsite, and that placement will be made by a subcontractor, who will be paid by the prime contractor for all reinforcing steel placed, including templates and supports. The required quantities are

     (a)   Permanent material quantity in hundredweight

     (b)   Expendable material quantity in hundredweight

     (c)   Pay quantity to prime contractor in pounds

     (d)   Pay quantity to subcontractor in pounds

Refer to foldouts at the back of the book for the service spillway for Alan Henry Dam. These drawings were prepared by Freese and Nichols, Inc., Consulgint Engineers, Fort Worth–Arlington–Austin, Texas, as part of the contract drawings for Alan Henry Dam. The bid form contained separate bid items for the mud slab (3 in. lean protective concrete), the structural concrete, and the reinforcing steel.

19. Calculate the following takeoff quantities necessary to price the mud slab bid item:

     (a)   Permanent material concrete in cubic yards

     (b)   Expendable material concrete in cubic yards

     (c)   The pay quantity in cubic yards

     (d)   E & S edge forms in linear feet

     (e)   Fine grade in square feet

     (f)   Wet finish in square feet

     (g)   Cure in square feet

Assume that the average overexcavation is 2 in. and that waste runs 5%. Only concrete shown within the neat lines on the drawings is measured for payment.

20. Calculate the following quantities necessary to price out the structural concrete bid item:

     (a)   Permanent material concrete in cubic yards

(b) Expendable material concrete in cubic yards

(c) Deneff Swellseal Plus quantity in linear feet

(d) E & S slab-on-grade forms in square feet of contact area

(e) E & S slab edge forms in square feet of contact area

(f) E & S slab bulkheads with key in square feet of contact area

(g) E & S wall forms in square feet of contact area

(h) E & S wall bulkheads with key in square feet of contact area

(i) E & S horizontal keyway in linear feet

(j) Fabricate forms in square feet

(k) Fabricate horizontal keyway in linear feet

(l) Sandblast horizontal construction joints in square feet

(m) Place and vibrate slab-on-grade concrete in cubic yards

(n) Place and vibrate wall concrete in cubic yards

(o) Wet finish slab in square feet

(p) Wet finish wall tops in square feet

(q) Cure concrete in square feet

(r) Nonexposed dry finish in square feet

(s) Exposed dry finish in square feet

(t) Buy dimension lumber in thousands of board feet

(u) Buy form plywood in thousands of square feet

(v) Buy ties and miscellaneous form materials (allowance) in square feet of contact area

(w) Buy sandblast sand (allowance) in square feet

(x) Buy cure materials (allowance) in square feet

(y) Buy dry finish materials (allowance) in square feet

(z) The pay quantity in cubic yards

Assume the following form reuse and material use factors:

SOG edge forms, use four times

SOG bulkheads, use twice

Wall forms, use four times

Wall bulkheads, use once

Horizontal keyways, use once

Dimension lumber, 3.5 FBM/sf of flat forms fabricated and 3.5 FBM/lf of horizontal key

Plywood, 1.2 sf/sf of flatforms fabricated

Only concrete within the neat lines shown on the drawings will be measured for payment.

5

Setting Up the Estimate

Key Words and Concepts

Estimate sponsor
Chief estimator
Labor rate library
Administrative manager
Labor director
Outside insurance broker
Fully loaded labor rate
Labor rate with base pay and union fringes only
Variation of labor rate depending on location
  within project
Equipment rental library
Company equipment division

In-house rental rates
New purchase/write-off
Rental from third parties
Equipment operating expense library
Equipment repair labor library
Cost records from completed projects
Equipment rate guidebooks
Permanent material library
Subcontract library
Expendable material library
Manual estimates
Computer-aided estimates

The previous chapter described the quantity takeoff process. Following or simultaneously with the quantity takeoff, estimators should give thought to the organization of the estimate as well as to other preparatory steps they must take prior to starting actual estimating work. This process can be thought of as "setting up the estimate," the subject of this chapter.

Cost estimates for large heavy construction projects are so laborious and time consuming that they are usually prepared by a team of several estimators working together, led by a senior estimator (often called the **estimate sponsor**). Imagine the confusion that would result if the estimating team did not follow a coherent, consistent set of rules, including use of common labor and equipment hourly costs, unit costs of materials, subcontract costs, and so forth. The estimate sponsor must therefore ensure that the estimate is properly set up prior to any attempt by anyone to estimate the cost of anything. For large companies, this setup function is often the responsibility of the company's **chief estimator** (the person designated to be responsible for all cost estimating, commonly an officer of the company).

The required estimate setup tasks include the following:

- Establish the estimate direct cost format
- Create the labor rate library
- Create the construction equipment rental, operating expense, and repair labor rate libraries
- Create the permanent material and subcontract assumed price libraries
- Create the expendable material library

THE LABOR RATE LIBRARY

For estimates prepared by manual methods, the **labor rate library** is nothing more than a list of the labor costs per hour for each hourly paid craft worker expected to be required to perform work on the project. For computer-aided estimates, these labor rates are preloaded into a computer program so that the labor rate library becomes part of the computer database.

At a minimum, these rates are set up to include the base rate and the union or employer-provided fringe benefits (both of which were described more fully in Chapter 2). As explained below, the rates may also include payroll taxes, payroll insurance, and travel and/or subsistence pay (all of which were also described in Chapter 2). In the estimating system presented in this book, however, components of labor cost for such things as shift differential premium, overtime premium, portal-to-portal pay, and working-through-lunch pay (also described in Chapter 2) are not included as components of the labor rates in the labor rate library. If these latter labor cost elements are required, money to cover them is provided by use of the labor premium procedure described in Chapter 6.

Source Data

The source data for labor rate determination is normally provided in a large construction company by the company's **administrative manager,** assisted by the **labor director** and by the company's **outside insurance broker.** The labor director generally "keeps up" on all labor matters, including maintaining a file of all of the active labor agreements with the various craft unions, and is the best source of interpretation of the payment terms of these agreements. With the labor director's guidance, the estimate sponsor obtains the applicable labor rates, fringe benefits, shift work premium, overtime premium, and other costs applicable to the project at hand. The administrative manager furnishes current information on applicable payroll taxes and applicable workmen's compensation premiums, assisted when necessary by the company's outside insurance broker for workmen's compensation premium issues. The estimate sponsor's next task is to compile all of this source data into a convenient form for use by estimators preparing manual estimates or, in the case of computer-aided estimates, into a form appropriate for loading into the computer program's database. In either case, the estimate sponsor must determine the particular components of labor cost to be included in the hourly rates and which components are to be included elsewhere in the cost estimate. Two optional methods of setting up the labor hourly rates are discussed below.

Option 1: Fully Loaded Labor Rate

If the project work is not expected to involve extensive overtime or shift work, portal-to-portal pay, work-through-lunch pay, and so forth, all the components of labor cost may be included in the hourly rate that the individual estimators will use directly. In this case, this "loaded" rate would include base pay, union or company-provided fringe benefits, payroll taxes, worker's compensation insurance premiums, and any applicable travel and/or subsistence payments. In addition, if the contractor's third-party liability insurance premium payments are based on labor dollars expended, the premium cost for this insurance would also be loaded into the labor rate. Otherwise, these premium payments would be included in the indirect cost estimate, as explained in Chapter 2.

For example, assume the following basic data had been obtained for a journeyman carpenter:

Base pay	$22.93/hr
Union fringes:	
Health and welfare	$0.75/hr
Pension	$1.41/hr
Industry fund	$0.03/hr
Payroll taxes	
FICA	6.95% of base pay
FUI	0.85% of base pay
SUI	5.75% of base pay
Worker's compensation insurance premium	18.41% of base pay
Third-party liability insurance premium	4.75% of base pay

The fully loaded labor rate would then be

Base pay:	$22.93/hr
Union fringes:	$0.75 + $1.41 + $0.03 = $2.19/hr
Payroll taxes:	(0.0695 + 0.0085 + 0.0575)(22.93) = $3.11/hr
Worker's compensation insurance:	(0.1814)(22.93) = $4.16/hr
Third-party liability insurance:	(0.0475)(22.93) = $1.09/hr
Loaded labor rate	$33.48/hr

If the third-party liability insurance premium was not to be paid on the basis of labor dollars expended, but instead was to be paid on the basis of the contract price, the fully loaded labor rate would be decreased by $1.09/hr to $32.39/hr. Note that in all cases, payroll taxes, worker's compensation, and third-party liability insurance premiums are calculated on payroll base pay only.

Option 2: Labor Rate Including Base Pay and Union Fringes Only

In this case, only the base pay and the union fringes are included in the labor library rate. For the payroll data in the example above, the labor rate would be $22.93/hr + $2.19/hr, or $25.12/hr. The balance of the labor cost components would be included under the taxes and insurance section of the indirect cost estimate as explained in Chapter 11.

Variations in Worker's Compensation Premiums for Different Work Areas of Project

Many heavy construction projects require craft workers to perform work operations in discrete areas of the project that are subject to different worker's compensation insurance premium rates that depend on the degree of risk associated with the work performed. An example would be a bridge substructure project over a navigable river resulting in one worker's compensation insurance premium rate for work over land on the approaches to the bridge and a higher rate when the same worker is performing work over the water. Another example would be a combination surface and underground project where a particular craft would be subject to one rate for work on the surface and a much higher rate when working underground. Therefore, estimators computing the labor rates for a large project involving a number of trade classifications, some or all of which are subject to different worker's compensation insurance premium rates depending on the location where the work is performed on the project, should make computations in tabular form reflecting the different work situations for each craft employee. Computer spreadsheet applications are particularly suited for this purpose.

Example Labor Rate Determination for Hypothetical Bridge Substructure Project

Figures 5-1 and 5-2, prepared for a hypothetical bridge substructure project with variable workmen's compensation insurance premium rates depending upon whether the work is over land or water, illustrate the application of the foregoing principle. Figure 5-1 was prepared under the option 1 method explained above, whereas Figure 5-2 was prepared under option 2. Note that for Figure 5-2, only one loaded rate is developed because the workmen's compensation insurance consideration is excluded and will be included as a component of the indirect cost estimate. On the other hand, Figure 5-1 shows four separate rates computed for those particular crafts that will perform work on both land and water. With a little study of Figures 5-1 and 5-2, the methodology should be readily apparent.

Classification	Base pay, $/hr	Union fringes, $/hr	PR tax, 13.55%, $/hr	Subtotal, $/hr	Concrete work, land		Concrete work, water		Pile driver work, land		Pile driver work, water	
					WC Ins. 18.14%, $/hr	Total, $/hr	WC Ins. 25.64%, $/hr	Total, $/hr	WC Ins. 38.52%, $/hr	Total, $/hr	WC Ins. 46.02%, $/hr	Total, $/hr
Carpenter foreman	22.93	2.19	3.11	28.23	4.16	32.39	5.88	34.11	—	—	—	—
Carpenter	22.30	2.19	3.02	27.51	4.05	31.56	5.72	33.23	—	—	—	—
Pile driver foreman	22.93	2.19	3.11	28.23	—	—	—	—	8.83	37.06	10.55	38.78
Pile driver	22.30	2.19	3.02	27.51	—	—	—	—	8.59	36.10	10.26	37.77
Cement mason	21.15	2.10	2.87	26.12	3.84	29.95	5.42	31.54	—	—	—	—
Iron worker foreman	22.98	2.48	3.11	28.57	4.17	32.74	5.89	34.47	—	—	—	—
Iron worker	21.56	2.48	2.92	26.96	3.91	30.87	5.53	32.49	—	—	—	—
Labor foreman	20.25	1.15	2.74	24.14	3.67	27.82	5.19	29.34	—	—	—	—
Laborer	19.79	1.15	2.68	23.62	3.59	27.21	5.07	28.70	—	—	—	—
Crane operator (>150 ft)[1]	21.51	3.77	2.91	28.19	3.90	32.10	5.52	33.71	8.29	36.48	9.90	38.09
Crane operator (<150 ft)[2]	20.58	3.77	2.79	27.14	3.73	30.87	5.28	32.42	7.93	35.07	9.47	36.61
Pump operator	20.58	3.77	2.79	27.14	3.73	30.87	5.28	32.42	—	—	—	—
Air compressor operator	20.58	3.77	2.79	27.14	3.73	30.87	5.28	32.42	7.93	35.07	9.47	36.61
Concrete conveyor operator	20.58	3.77	2.79	27.14	3.73	30.87	5.28	32.42	—	—	—	—
Oiler	17.61	3.77	2.39	23.77	3.19	26.96	4.52	28.28	6.78	30.55	8.10	31.87
Teamster	19.69	1.16	2.67	23.52	3.57	27.09	—	—	—	—	—	—

[1] Boom over 150 ft in length.

[2] Boom under 150 ft in length.

FIGURE 5-1 Loaded labor rate calculations (option 1).

Classification	Base pay, $/hr	Union fringes, $/hr	Total, $/hr
Carpenter foreman	22.93	2.19	25.12
Carpenter	22.30	2.19	24.49
Pile driver foreman	22.93	2.19	25.12
Pile driver	22.30	2.19	24.49
Cement mason	21.15	2.10	23.25
Iron worker foreman	22.98	2.48	25.46
Iron worker	21.56	2.48	24.04
Labor foreman	20.25	1.15	21.40
Laborer	19.79	1.15	20.94
Crane operator (>150 ft)	21.51	3.77	25.28
Crane operator (<150 ft)	20.58	3.77	24.35
Pump operator	20.58	3.77	24.35
Air compressor operator	20.58	3.77	24.35
Concrete conveyor operator	20.58	3.77	24.35
Oiler	17.61	3.77	21.38
Teamster	19.69	1.16	20.85

FIGURE 5-2   Loaded labor rate calculations (option 2).

Use of Separate Identifying Codes When Variation of Labor Rate with Location of Work on Project Occurs

In the situation reflected by Figure 5-1 where the rate paid the worker depends on **where the work is performed on the project,** the labor rate library should be coded in a manner that makes it easy for the estimators to select the correct labor rate when setting up work crews. For example, the labor rate library for the rates computed in Figure 5-1 should be arranged as shown in Figure 5-3.

CONSTRUCTION EQUIPMENT RENTAL, OPERATING EXPENSE, AND REPAIR LABOR RATE LIBRARIES

The following are the general tasks required to establish the **equipment rental library:**

- Review work operations to be performed by the contractor's own forces to determine major equipment needs
- Determine the planned acquisition source for the equipment
- Depending on the planned acquisition source, determine an hourly rate for equipment rental (or write-off) for each class and size of equipment required

| Code | Description | | Rate, $/hr |
	Trade classification	Work location	
CARP 1	Carpenter foreman	Land	32.39
CARP 2	Carpenter	Land	31.56
CARP 3	Carpenter foreman	Water	34.11
CARP 4	Carpenter	Water	33.23
PD 1	Pile driver foreman	Land	37.06
PD 2	Pile driver	Land	36.10
PD 3	Pile driver foreman	Water	38.78
PD 4	Pile driver	Water	37.77
CM 1	Cement mason	Land	29.25
CM 2	Cement mason	Water	31.54
IW 1	Iron worker foreman	Land	32.74
IW 2	Iron worker	Land	30.87
IW 3	Iron worker foreman	Water	34.47
IW 4	Iron worker	Water	32.49
LAB 1	Labor foreman	Land	27.82
LAB 2	Laborer	Land	27.21
LAB 3	Labor foreman	Water	29.34
LAB 4	Laborer	Water	28.70
OPER 1	Crane operator (>150 ft)	Concrete/land	32.10
OPER 2	Crane operator (>150 ft)	Concrete/land	33.71
OPER 3	Crane operator (>150 ft)	Pile/land	36.48
OPER 4	Crane operator (>150 ft)	Pile/water	38.09
OPER 5	Crane operator (<150 ft)	Concrete/land	30.87
OPER 6	Crane operator (<150 ft)	Concrete/water	32.42
OPER 7	Crane operator (<150 ft)	Pile/land	35.07
OPER 8	Crane operator (<150 ft)	Pile/water	36.61
OPER 9	Pump operator	Concrete/land	30.87
OPER 10	Pump operator	Concrete/water	32.42
OPER 11	Air compressor operator	Concrete/land	30.87
OPER 12	Air compressor operator	Concrete/water	32.42
OPER 13	Air compressor operator	Pile/land	35.07
OPER 14	Air compressor operator	Pile/water	36.61
OPER 15	Concrete conveyor operator	Concrete/land	30.87
OPER 16	Concrete conveyor operator	Concrete/water	32.42
TEAM 1	Teamster	Concrete/land	27.09

FIGURE 5-3   Labor library with identifying work location codes.

Review of Work Operations to Be Performed by the Contractor's Own Forces to Determine Major Equipment Needs

Based on a study of the plans and specifications, the project sponsor must first determine in a general way the particular classes of construction equipment that will be required for the work operations intended to be performed by the contractor's own forces.

Determining the required number of individual units of each major class of equipment is not necessary or even possible at this time. Only the equipment classes themselves need be determined. This step requires considerable construction experience and is a matter that the estimate sponsor would probably discuss with the chief estimator and others expected to participate in the estimate.

Determination of the Acquisition Source

Generally speaking, a contractor could expect to obtain the necessary construction equipment in one of the following ways:

- From the contractor's presently owned fleet of equipment
- By purchase from third parties, either new or used
- Rental from third parties
- Rental from third parties on an operated and maintained (O&M) basis

Equipment from Presently Owned Fleet

Most large construction companies own substantial fleets of construction equipment. They often manage the ownership and maintenance of this equipment by placing it in an **equipment division,** operating it as a separate profit center within the company's corporate organization. When this method is used, various projects utilizing the equipment are charged an **in-house monthly rental rate** for each month the equipment is assigned to a particular project. If a single bidding contractor with an in-house equipment division contemplated performing the project, and the project was of such a size as to justify adding new equipment to the company's equipment fleet, that option could be exercised if the bid was successful. In that case, a rate would be established for use in the estimate on the understanding that if the bid is successful, the company will purchase the required equipment and rent it to the job at that monthly in-house rental rate. In any case, the equipment would be considered to come from the company's owned fleet of construction equipment administered by the equipment division.

New Purchase

New purchase is an option frequently utilized in **joint-venture bids.** The estimate sponsor delineates particular items of equipment that, in the event the bid is successful, would be purchased either new or used by the joint-venture partners at the beginning of the project and sold upon the completion of the work. In this event, the estimate would be charged the assumed loss of value during the performance of the project (**the write-off**) equal to the difference between the purchase price and estimated salvage value at the end of the project. It is not unusual for joint-venture bids to be predicated on the basis that the joint-venture partners will purchase many million dollars worth of construction equipment at the onset of the project if the joint-venture bid is successful.

Rental from Third Parties

Ordinarily, contractors are reluctant to execute large heavy construction projects by relying on equipment rented from third parties because this method is usually not competitive. They much prefer to use equipment they already own or equipment purchased especially for the job, either new or used, that can be realistically written off, or largely written off, during the performance of the project work. Therefore, contractors usually restrict rental from third-party sources to equipment items that they expect to need for either a very short time period or for a series of intermittent short time periods.

Operated and Maintained Rental

Some units of equipment are typically obtained from **owner-operators** who furnish the equipment on a fully operated and maintained basis at a quoted hourly rate for each hour's use. Examples include on-highway trucking equipment, concrete pumping services, crane services, and grade-all and small backhoe services. This source of construction equipment is often attractive for limited portions of the work.

Determination of an Hourly Rate for Rental

The method of determining an appropriate hourly rate for use in the equipment rental rate library depends on the source of the equipment. For equipment to be furnished from the contractor's presently owned fleet, it is necessary to estimate only the probable average hours of usage per month on the job and to divide the monthly in-house rental rate by that monthly average. For example, if an 18 t center-mount crane could be obtained from the company's equipment division for an in-house rate of $1,800 per month and the average usage on the project per month was expected to be 125 hr, the appropriate rate for the equipment rental rate library would be $1,800/125 hr, or $14.40/hr.

On the other hand, if a joint-venture bids a project with the expectation (if the bid is successful) of purchasing a 160 t crawler crane for $675,000 at the inception of the project and selling it for $375,000[1] upon completion of the work and, further, expects that the crane would be needed on the project for 48 months at an average of 150 hr usage per month, the appropriate rate for the equipment rental rate library would be

$$\frac{\$675,000 - \$375,000}{(48 \text{ mo})(150 \text{ hr/month})} = \$41.67/\text{month}$$

If the equipment was expected to be obtained through a third-party rental, the appropriate hourly rental rate would be determined in the same manner as if the equipment were coming from the company's in-house equipment division.

If the equipment was intended to be furnished on an O & M basis, the appropriate hourly rental rate would be determined by canvassing the local O & M equipment market.

Determination of Hourly Rates for Equipment Operating Expense and for Repair Labor

The hourly rates for equipment operating expense (EOE) and repair labor (RL) are typically developed by referring to the contractor's **cost records from previously completed projects** of a similar nature (when such records are available) or from **equipment rate guidebooks**[2] in the event that no in-house job records are available. In either case, the estimator usually must make adjustments to update the rates for cost increases due to inflation and to reflect site-specific conditions should the condition of service not be the same as conditions for previously completed projects. For instance, assume that job records indicated the following information for use of a 165 t crawler crane on a previous project:

Repair labor (RL)	0.25 labor hours per hour
Fuel, Oil, and Grease (FOG)	$ 6.75/hr
Spare parts	$11.48/hr

---

[1]Determination of estimated or resale values for construction equipment depends on many factors and is beyond the scope of this book.

[2]An excellent guidebook is the *Cost Reference Guide* currently published by Primedia Information, Inc. (formerly K 111 Directory Corporation and Dataquest).

Further, assume that at the time of the previous project, diesel fuel cost $0.65/gal and spare parts cost 45% of what they cost at the time of preparing the new estimate. Also at the time of the new estimate, the loaded mechanics wage rate was $30.25/hr, and diesel fuel cost $1.45/gal. On the basis of the above facts,

FOG would cost ($6.75)($1.45/$0.65) = $15.06/hr
Spare parts would cost ($11.48)(1/0.45) = $25.51/hr
Therefore, the EOE cost for the new project would be $15.06 + $25.51 = $40.57/hr
RL for the new job would cost (0.25)($30.25) = $7.56/hr

If there were significant changes in the expected service conditions that could cause any of the above cost components to run higher or lower than those calculated, those changes should be reflected in the update. For instance, if the service conditions for the crane were expected to be 25% more severe than for the previous job, the updated factors for EOE and RL would be (1.25)($40.57) = $50.71/hr and (1.25)($7.56) = $9.45/hr, respectively.

To illustrate the above-described procedure more completely, assume that the estimate sponsor had determined the equipment needs for a bridge substructure project and that the various classes of equipment could be acquired in the manner and at the costs shown in the following table. Bear in mind that the information in this table would be only the project sponsor's best determination prior to performance of any detailed estimating work on the project. Although the estimate will be prepared using these hourly rates, the total dollar figure for equipment rental in the project total direct cost will be analyzed on a global basis when the estimate has been completed. Any necessary adjustments will be made at that time. The procedure for making this global analysis and any necessary equipment adjustment is discussed in Chapter 12 of this book.

Planned Major Equipment Needs and Acquisition Information

Equipment class description	Months on job	Average hours per month	Rental, $/Month	O&M Rent, $/hour	Buy, $	Salvage, $
165 ton crawler crane	30	150			690,000	384,000
Crane barge 90 ft × 50 ft	30	150			510,000	330,000
Material barge 100 ft × 50 ft	30	150			234,000	60,000
600 hp tugboat	30	150			474,000	120,000
ICE 812 vibratory hammer	30	75			750,000	426,000
45,000 ft-lb pile hammer	10	75	4,500			
Pile leads and follower	10	75	2,625			
1600 cfm diesel air compressor	10	75	5,250			
185 cfm diesel air compressor	30	50	1,219			
400 amp diesel welder	30	50	406			
150 hp jet pump	9	75	12,000			
10 in. electric submersible pump	6	75	7,688			
3 in. electric submersible pump	30	730[1]	844			
Concrete belt conveyor delivery system	24	75			594,000	180,000
Truck-mounted concrete pump	24	75	8,438			
High-cycle generator with concrete vibrators	24	75	1,031			
Three-axle boom truck	30	173[2]	4,688			
Sedan	30	173[2]	406			
Pick-up	30	173[2]	494			
Clam shell bucket	6	75	1,188			
30 ton high-bed truck-trailer				54.00		

[1]These pumps were assumed to run 24 hr per day, 7 days per week. Thus, (365 days/12 months)(24 hr/day) = 730 hr.

[2]Automotive equipment usually assumed to be in service 5 days per week, 8 hr per day. Thus, (52 weeks/12 months)(5 days)(8 hr) = 173 hr.

Further, assume that the following information was available from the company's job records from a bridge substructure project performed a number of years earlier:

Job Records Project #77302

Equipment description	RL labor hours per hour	FOG, $/hr	Power, $/hr	Parts, $/hr
165 ton crawler crane	0.25	6.75		11.48
Crane barge 90 ft × 50 ft	0.10			2.02
Material barge 100 ft × 50 ft	0.05			0.68
600 hp tugboat	0.15	3.00		4.72
ICE 812 vibratory hammer	0.60	4.88		4.39
45,000 ft-lb pile hammer	0.45	0.75		7.76
Pile leads and follower	0.30			3.38
1600 cfm diesel air compressor	0.20	9.00		3.04
185 cfm diesel air compressor	0.15	3.15		1.15
400 amp diesel welder	0.05	2.63		1.01
150 hp jet pump	0.35	5.25		4.39
10 in. electric submersible pump	0.10		4.50	0.81
3 in. eectric submersible pump	0.10		2.25	0.68
Concrete belt conveyor delivery system	1.50		14.25	6.08
Truck-mounted concrete pump	1.50	6.75		6.41
High-cycle generator with concrete vibrators	0.25	1.43		4.05
Three-axle boom truck	0.25	2.62		2.02
Sedan	0.15	0.75		0.47
Pick-up	0.15	1.13		1.01
Clam shell bucket	0.15			3.38
30 ton high-bed truck-trailer	0.25	3.75		4.72

At the time of the previous job:

Diesel cost	$0.65/gal
Power cost	$0.065/kilowatt hour (kwh)
Parts cost	45% of today's cost

For the present job:

Mechanic wage rate	$30.25/hr (includes insurance and taxes)
Diesel	$ 1.45/gal
Power	$ 0.115/kwh

Once again, a spreadsheet application is ideal for developing the **rental rate library,** the **equipment operating expense library,** and the **repair labor library,** as illustrated by Figures 5-4 and 5-5.

PERMANENT MATERIAL AND SUBCONTRACT ASSUMED PRICE LIBRARY

By surveying the plans and specifications, the estimate sponsor should be able to compile the various classifications of important permanent materials required for the project

Description	Source	Months on job	Average hrs/month	New Buy, $	Purchase Salvage, $	Only Write-off, $	Rental only, $/month	Use in estimate Rental, $	Use in estimate O&M rental, $
165 ton crawler crane	Buy New	30	150	690,000	384,000	306,000	—	68.00	—
Crane barge 90 ft × 50 ft	Buy New	30	150	510,000	330,000	180,000	—	40.00	—
Material barge 100 ft × 50 ft	Buy New	30	150	234,000	60,000	60,000	—	38.67	—
600 hp tugboat	Buy New	30	150	474,000	120,000	174,000	—	78.67	—
ICE 812 vibratory hammer	Buy New	30	75	750,000	426,000	324,000	—	144.00	—
45,000 ft-lb pile hammer	Co. Rent	10	75	—	—	—	4,500	60.00	—
Pile leads & follower	Co. Rent	10	75	—	—	—	2,625	35.00	—
1,600 cfm diesel air compressor	Co. Rent	10	75	—	—	—	5,250	70.00	—
185 cfm diesel air compressor	Co. Rent	30	50	—	—	—	1,219	24.38	—
400 amp diesel welder	Co. Rent	30	50	—	—	—	406	8.12	—
150 hp jet pump	Co. Rent	9	75	—	—	—	12,000	160.00	—
10 in. electrical submersible pump	T.P. Rent[1]	6	75	—	—	—	7,688	102.51	—
3 in. electrical submersible pump	T.P. Rent[1]	30	730[3]	—	—	—	844	1.16	—
Belt conveyor concrete delivery system	Buy New	24	75	594,000	180,000	414,000	—	230.00	—
Truck-mounted concrete pump	Co. Rent	24	75	—	—	—	8,438	112.51	—
High-cycle generator with concrete vibrators	Co. Rent	24	75	—	—	—	1,031	13.75	—
Three-axle boom truck	Co. Rent	30	173[4]	—	—	—	4,688	27.10	—
Sedan	Co. Rent	30	173	—	—	—	406	2.35	—
Pick-up	Co. Rent	30	173	—	—	—	494	2.86	—
Clam shell bucket	Co. Rent	6	75	—	—	—	1,188	15.84	—
30 ton high-bed truck-trailer	O&M Rent[2]	30	75	—	—	—	—	—	54.00

FIGURE 5-4 Determination of rental (R) hourly rates for equipment rate library.

[1] Third-party rental.
[2] Operated and maintained rental.
[3] (365 d/yr ÷ 12)(24 hr/d) = 730 hr.
[4] (52 wks/yr ÷ 12)(5d)(8 hr/d) = 173 hr.

	Experience					Modifiers			Use for Estimate				
Description	RL, hr/hr	FOG, $/hr	Power, $/hr	Parts, $/hr	Experience source[1]	FOG[2]	Power[3]	Parts[4]	FOG, $/hr	Power, $/hr	Parts, $/hr	Total EOE $/hr	RL $/hr
165 ton crawler crane	0.25	6.75	—	11.48	Project #77302	2.22	1.77	2.23	14.98	—	25.61	40.59	7.56
Crane barge 90 ft × 150 ft	0.10	—	—	2.02	"	"	"	"	—	—	4.51	4.51	3.03
Material barge 100 ft × 50 ft	0.05	—	—	0.68	"	"	"	"	—	—	1.52	1.52	1.51
600 hp tug	0.15	3.00	—	4.72	"	"	"	"	6.66	—	10.53	17.19	4.54
Ice 812 vib. hammer	0.60	4.88	—	4.39	"	"	"	"	10.83	—	9.79	20.62	18.15
45,000 ft-lb hammer	0.45	0.75	—	7.76	"	"	"	"	1.66	—	17.31	18.97	13.61
Leads & follower	0.30	—	—	3.38	"	"	"	"	—	—	7.54	7.54	9.08
1,600 cfm comp.	0.20	9.00	—	3.04	"	"	"	"	19.98	—	6.78	26.76	6.05
185 cfm comp.	0.15	3.15	—	1.15	"	"	"	"	6.99	—	2.57	9.56	4.54
400 amp welder	0.05	2.63	—	1.01	"	"	"	"	5.84	—	2.25	8.09	1.51
150 hp jet pump	0.35	5.25	—	4.39	"	"	"	"	11.66	—	9.79	21.45	10.59
10 in. electric submersible pump	0.10	—	4.50	0.81	"	"	"	"	—	7.96	1.81	9.77	3.03
3 in. electric submersible pump	0.10	—	2.25	0.68	"	"	"	"	—	3.98	1.52	5.50	3.03
Belt conveyor concrete delivery system	1.50	—	14.25	6.08	"	"	"	"	—	25.21	13.56	38.77	45.38
Truck-mounted concrete pump	1.50	6.75	—	6.41	"	"	"	"	14.98	—	14.30	29.28	45.38
High-cycle generator with vibrators	0.25	1.43	—	4.05	"	"	"	"	3.17	—	9.03	12.20	7.56
Three-axle boom truck	0.25	2.62	—	2.02	"	"	"	"	5.82	—	4.51	10.33	7.56
Sedan	0.15	0.75	—	0.47	"	"	"	"	1.66	—	1.05	2.71	4.54
Pick-up	0.15	1.13	—	1.01	"	"	"	"	2.24	—	2.25	4.49	4.54
Clam shell bucket	0.15	—	—	3.38	"	"	"	"	—	—	7.54	7.54	4.54
10 ton high-bed trailer	0.25	3.75	—	4.72	"	"	"	"	O&M	O&M	O&M	O&M	O&M

[1]In the general case, the experience source will not necessarily be the same for all equipment required for the new project. If various sources were involved, they would be noted. In that case, the FOG, Power, and Parts modifiers would be different depending on the experience source and would be noted and applied, as applicable, to each class of equipment.

[2]FOG modifier = 1.45 / 0.65 = 2.23.

[3]Power modifier = 0.115 / 0.065 = 1.77.

[4]Parts modifier = 1.00 / 0.45 = 2.22.

FIGURE 5-5 Determination of EOE rates and RL rates for the equipment operating expense and repair labor libraries.

along with their units of measurement. Similarly, he or she should be able to determine major work operations from the estimate, sometimes comprising entire bid items (with appropriate units of measurement) that are expected to be partially or totally performed by subcontractors. Once these compilations have been completed, the estimate sponsor should establish an assumed unit price, or assumed lump sum price, for each separate permanent material and separate subcontract work item. An example format for **permanent material and subcontract-assumed price libraries** is shown in Figure 5-6.

The assumed prices, of course, will be adjusted just prior to bid in accordance with actual quotations received at that time from material suppliers and subcontractors.

Bid Item	Description	Unit	Quantity	Permanent material unit price, $	Subcontract unit price, $
–	Concrete 3,000 psi[1]	cy	–	35.00	–
–	Concrete 3,500 psi[1]	cy	–	37.50	–
–	Concrete 4,000 psi[1]	cy	–	39.50	–
–	Waterstop[1]	lf	–	4.00	–
–	Premolded joint filler[1]	sf	–	0.75	–
–	Anchor bolts[1]	lb	–	2.00	–
–	Weld plates[1]	lb	–	1.50	–
–	Edge angles[1]	lb	–	2.25	–
42	Reinforcing steel	lb	2,350,000	0.215	0.15
51	2 1/2 in. copper tube	lb	600	–	82.50
52	3 in. brass pipe	lb	250	–	20.00
53	3 in. carbon steel pipe	lb	6,500	–	4.60
.	.	.	.	.	.
Etc.	Etc.	Etc.	Etc.	Etc.	Etc.
.	.	.	.	.	.
73	Lube/govenor oil system	LS	–	–	50,000.00
74	Service water system	LS	–	–	31,250.00
.	.	.	.	.	.
Etc.	Etc.	Etc.	Etc.	Etc.	Etc.
.	.	.	.	.	.
92	1 in. Conduit	lf	–	–	10.00
93	1 1/2 in. Conduit	lf	–	–	12.00
94	2 in. Conduit	lf	–	–	18.00
.	.	.	.	.	.
Etc.	Etc.	Etc.	Etc.	Etc.	Etc.
.	.	.	.	.	.
96	115 kV circuit breakers	Each	–	–	11,500.00
97	2 V distribution panel	Each	–	–	9,500.00
.	.	.	.	.	.
.	.	.	.	.	.
.	.	.	.	.	.
Etc.	Etc.	Etc.	Etc.	Etc.	Etc.

[1]These items are required for each of a number of separate bid items.

FIGURE 5-6   Example format for permanent material and subcontract-assumed price libraries.

Description	Unit[1]	Unit price, $
Dimension lumber	MBF	300.00
¾ in. plywood	MSF	600.00
Miscellaneous form materials	sfca	0.15
Gelatin dynamite	lb	1.10
Presplit powder	lb	1.30
EBC (electric blasting caps)	Each	1.10
Soldier pile—buy	cwt	30.00
Soldier pile—salvage	cwt	(15.00)
Steel panel forms—buy	sf	35.00
Steel panel forms—salvage	sf	(20.00)
Lagging timber	MBF	250.00
Heavy timber—buy	MBF	350.00
Heavy timber—salvage	MBF	(150.00)

[1]MBF = 1,000 board feet; MSF = 1,000 square feet; sfca = square feet of form contact area.

FIGURE 5-7   Example format for an expendable material-assumed price library.

EXPENDABLE MATERIAL ASSUMED PRICE LIBRARY

In a manner similar to that employed for the permanent material and subcontract-assumed price libraries, the estimate sponsor should survey the plans and specifications and compile a list of the various important expendable materials that will be required for the work along with their appropriate units of measurement. Then, assumed unit prices or lump sum prices should be determined for each separate expendable material based on the best available information from quotations received for recent estimates. Most companies' estimating departments maintain extensive files of this type of information as well as files of permanent material and subcontract quotation information. An example of the format of a typical **expendable material library** appears as Figure 5-7.

USE OF LIBRARIES FOR MANUAL AND COMPUTER-AIDED ESTIMATES

Manual Estimates

For estimates to be prepared by manual methods, the various libraries as described above are simply printed out into convenient packets and distributed to the various estimators expected to make the direct cost estimate.

Computer-Aided Estimates

If the estimator plans to generate the estimate with a computer-estimating program, the information shown in the above libraries would be loaded electronically into the computer. The estimator would make each entry under a code that suggests the particular type of cost rate entered. It is important to realize that this information has all been generated with respect to one particular cost estimate and that the computer estimate database files so created should not be applied to other estimates without thorough review and probable revision. Every construction project is different, and a set of libraries appropriate for estimating one project seldom will be appropriate for use on another. Therefore, the estimator should create separate libraries for each new project.

CONCLUSION

This chapter described the steps that must be taken to properly establish a common set of ground rules that all estimators working on the estimate will follow when performing their estimating work. This process was referred to as "setting up the estimate," and, at the end of the chapter, the point was emphasized that these procedural steps should be taken and the appropriate libraries created specifically for each individual project. The library information created for one project will seldom be appropriate for another.

The following chapter describes the mechanics of converting all of the information previously developed into actual dollar and cents costs for a typical construction work item, commonly referred to in schedule-of-bid-item estimates as a bid item.

QUESTIONS AND PROBLEMS

1. What four procedural steps must be completed to "set up" the estimate?

2. What three examples of components of labor cost that are not included in the loaded hourly labor rate were mentioned in this chapter?

3. What two sources of labor-related data in a large construction company organization were discussed in this chapter?

4. Who makes the decision as to which components of labor cost are to be included in the hourly labor rate and which components are to be included elsewhere in the estimate?

5. Under what expected project circumstances does it make sense to include base pay, union or company-provided fringe benefits, payroll taxes, worker's compensation insurance premiums, and any applicable travel and/or subsistence payments in a loaded hourly labor rate? What additional circumstance would dictate that a third-party liability insurance premium component also be included in the loaded hourly labor rate?

6. Are payroll taxes, worker's compensation insurance premiums, and, when included in the loaded hourly labor rate, third-party liability insurance premiums always calculated on the basis of the labor base pay rate?

7. In the estimating system used in this book, by what method are components of labor cost resulting from shift differential, overtime premium, portal-to-portal pay, and working-through-lunch pay put into the estimate (don't attempt to explain the method, just identify it).

8. Why does the applicable worker's compensation rate vary for many projects? Name two examples discussed in this chapter where the rate would vary.

9. What three general tasks are required to be performed to establish the equipment rental library?

10. As part of the first of the three tasks (see question 9) is it necessary for the estimate sponsor to determine the required number of individual units of equipment? Is it even possible to do this at this stage of the estimate?

11. What four methods of equipment acquisition were discussed in this chapter?

12. What is meant by the following terms?
    (a) In house monthly rental rate
    (b) The write-off
    (c) O & M rental

13. What two sources for determining equipment operating expense and repair labor hourly rates are discussed in this chapter? Why must adjustments usually be made to hourly rates obtained from these sources? What additional factor should be taken into consideration when making these adjustments.

14. By what general procedure does the estimate sponsor compile lists of the various classifications of permanent materials, lists of the required work operations that are expected to be partially or totally performed by subcontractors, and lists of various important expendable materials likely to be required? What is the usual source for the assumed unit or lump sum prices for the above? How and when are these assumed prices later adjusted?

15. How are the various database libraries communicated to members of the estimating team for manual estimates and for computer-based estimates?

16. Why is a set of database libraries set up for one project often unsuitable for use on another project in the heavy construction segment of the industry?

17. Given the following information for craft labor, develop (manually or by computer spreadsheet) the fully loaded labor rates for use in the estimate for a new bid project. Use a format similar to Figure 5-1.

Craft wages classification	Base pay, $/hr	Union fringes		
		Heath and welfare, $/hr	Pension, $/hr	Other, $/hr
Carpenter foreman	18.34	0.60	1.13	0.02
Carpenter	17.84	0.60	1.13	0.02
Pile driver foreman	18.34	0.60	1.13	0.02
Pile driver	17.84	0.60	1.13	0.02
Cement mason	16.92	0.90	0.75	0.03
Iron worker foreman	18.38	0.83	1.13	0.03
Iron worker	17.25	0.83	1.13	0.03
Labor foreman	16.20	0.45	0.38	0.09
Laborer	15.83	0.45	0.38	0.09
Crane operator (>150 ft)	17.21	0.63	2.27	0.11
Crane operator (<150 ft)	16.46	0.63	2.27	0.11
Pump operator	16.46	0.63	2.27	0.11
Air compressor operator	16.46	0.63	2.27	0.11
Concrete conveyor operator	16.46	0.63	2.27	0.11
Oiler	14.09	0.63	2.27	0.11
Teamster	15.75	0.45	0.45	0.03

Worker's compensation insurance	Normal workers compensation, %	LSHW[2] additional, %	Compensation insurance including LSHW, %
Clerical[1]	2.80		2.80
Concrete work, land	13.95		13.95
Concrete work, water	13.95	5.77	19.72
Pile driving and cofferdam, land	29.63		29.63
Pile driving and cofferdam, water	29.63	5.77	35.40

[1]Applies to office personnel only.

[2]Long Shoreman's and Harbor Workers' Act (a federal law).

Third-party liability and property damage insurance premiums are paid separately and are *not* a component of the labor rates.

Payroll Taxes

FICA	6.95%
FUI	0.85%
SUI	5.75%
Total	13.55%

The following specific crafts are required to carry out work operations:

Concrete work on land

Carpenters and carpenter foreman

Laborers and labor foreman

Cement masons

Iron workers and iron worker foreman

All operating engineers

Teamsters

Concrete work on water

Carpenters and carpenter foreman

Laborers and labor foreman

Cement masons

Iron workers and iron worker foreman

All operating engineers

Pile driving and cofferdam work on land

Pile drivers and pile driver foreman

Crane operators and oilers

Compressor operator

Pile driving and cofferdam work on water

Pile drivers and pile driver foreman

Crane operators and oilers

Compressor operator

18. Given the following salaried personnel wage data, develop the monthly rates for use in the estimate. Use the same payroll tax rate as for problem 17 above and use a worker's compensation rate of 12.00%

Classification	Base, $/month	Health and welfare, pension, and other company benefits, %
Project manager	5,100	23
General superintendent	4,800	23
Carpenter/Concrete superintendent	4,200	23
Pile driver superintendent	4,200	23
Project engineer	3,900	23
Office engineer	3,300	23
Field engineer	3,000	23
Office manager	3,000	23
Secretary/clerk	2,100	23

19. A preliminary study for a new bid prospect indicates that the required equipment will likely be obtained as follows and that the anticipated usage is as indicated. On this basis, develop the equipment rental rates for use in the estimate for a new bid prospect.

Item	Equipment description	Months on job	Average hours per month	Rental, $/month	O & M rental, $/month	Buy, $	Salvage, $
1.	165 ton crawler crane	30	150			575,000	320,000
2.	Crane barge 90 ft × 50 ft	30	150			425,000	275,000
3.	Material barge 100 ft × 50 ft	30	150			195,000	50,000
4.	600 hp tugboat	30	150			395,000	100,000
5.	Ice 812 vibratory hammer	30	75			625,000	355,000
6.	45,000 ft-lb pile hammer	10	75	3,600			
7.	Pile leads and follower	10	75	2,100			
8.	1600 cfm diesel air compressor	10	75	4,200			
9.	185 cfm diesel air compressor	30	50	975			
10.	400 amp diesel welder	30	50	325			
11.	150 hp jet pump	9	75	9,600			
12.	10 in. electric submersible pump	6	75	6,150			
13.	3 in. electric submersible pump	30	744	675			
14.	Concrete conveyor system	24	75			495,000	150,000
15.	Truck-mounted concrete pump	24	75	6,750			
16.	High-cycle generator with concrete vibrators	24	75	825			
17.	Three-axle boom truck	30	173	3,750			
18.	Sedan	30	173	325			
19.	Pick-up	30	173	395			
20.	Clam shell bucket	6	75	950			
21.	30 ton high-bed truck-trailer	30	75		45.00		

20. On the basis of the following company records from a previous job, develop the equipment library rates for EOE and RL to be used for the estimate for a new bid prospect.

Equipment description	R L labor-hours per hour	FOG, $/hr	Power, $/hr	Parts, $/hr
1. 165 ton crawler crane	0.25	4.50		8.50
2. Crane barge 90 ft × 50 ft	0.10			1.50
3. Material barge 100 ft × 50 ft	0.05			0.50
4. 600 hp tugboat	0.15	2.00		3.50
5. ICE 812 vibratory hammer	0.60	3.25		3.25
6. 45,000 ft-lb pile hammer	0.45	0.50		5.75
7. Pile leads and follower	0.30			2.50
8. 1600 cfm diesel air compressor	0.20	6.00		2.25
9. 185 cfm diesel air compressor	0.15	2.10		0.85
10. 400 amp diesel welder	0.05	1.75		0.75
11. 150 hp jet pump	0.35	3.50		3.25
12. 10 in. electric submersible pump	0.10		3.00	0.60
13. 3 in. electric submersible pump	0.10		1.50	0.50
14. Concrete conveyor system	1.50		9.50	4.50
15. Truck-mounted concrete pump	1.50	4.50		4.75
16. High-cycle generator with concrete vibrators	0.25	0.95		3.00
17. Three-axleboom truck	0.25	1.75		1.50
18. Sedan	0.15	0.50		0.35
19. Pick-up	0.15	0.75		0.75
20. Clam shell bucket	0.15			2.50
21. 30 ton high-bed truck-trailer	0.25	2.50		3.50

At the time of the previous job:

Diesel	$0.60/gal
Power	$0.045/kwh
Parts index	125

For the present job:

Mechanic's wage rate	$24.90/hr (includes insurance and taxes)
Diesel	$1.25/gal
Power	$0.075/kwh
Parts index	275

6

Pricing Out the Direct Cost Estimate

Key Words and Concepts

General estimate direct cost summary sheet
Total project direct cost
Decision-making phase
Estimator's subjective judgment
Entry of information phase
Manual estimate
Computer-aided estimate
Purely mechanical phase
Subcontract costs
Permanent material costs
Costs incurred for work operations by own forces
Operations analysis to develop basis for the cost estimate

Typical crew
Crew productivity rate
Hours of productive work per shift
Number of shifts per work day
Number of work days per work week
Labor premium
Small tools and supplies
Crew method
Individual entry method
Lump sum bid item
Prorate factor

When the work of setting up the estimate, as described in the previous chapter, and the quantity takeoff described in Chapter 4 has been completed, the remaining task is to "price out" the direct cost estimate for the project. Each line item on the **general estimate direct cost summary sheet** (previously established according to the vertical format organization of the estimate) must be "priced"—that is, all elements of cost comprising that line item must be computed and entered on the general estimate summary sheet. When totaled, these line items yield a line total for the **total project direct cost.** This chapter deals with this pricing-out process.

The line item could be either a discrete part of the work for an estimate being prepared on a lump sum basis (see Chapter 3), or it could be a **bid item** under a schedule-of-bid-items form of a direct cost estimate. What follows in this chapter is presented in the context of the pricing-out process for such a bid item because several aspects of the process that apply to bid items do not apply when pricing out a line item for a lump sum bid.

THE PRICING-OUT PROCESS

For ease of explanation, the pricing-out process can best be described by breaking it down into the following three distinct phases:

- **Decision-making phase,** requiring exercise of the **estimator's subjective judgment**
- **Entry of information phase,** consisting of listing on a bid item extension sheet for a **manual estimate,** or, in the case of a **computer-aided estimate,** entering into a computer the results of the estimator's subjective judgments determined in the first part of the process as well as all of the takeoff information previously generated by the quantity takeoff work.
- **Purely mechanical phase,** consisting of multiplication, division, addition, and subtraction performed either by the estimator on the bid-item extension sheet for a manual estimate or performed electronically for computer-aided estimates.

THE DECISION-MAKING PHASE REQUIRING EXERCISE OF THE ESTIMATOR'S SUBJECTIVE JUDGMENT

In the general case, the costs incurred by the contractor in performing the work of the bid item will consist of **subcontract costs** for work expected to be performed by subcontractors, **permanent material costs** for all permanent materials the contractor must furnish, **expendable material costs** for all expendable materials the contractor must furnish, and finally, all **costs that will be incurred by the contractor in performing work with its own forces.** The quantities of subcontract work, permanent materials, and expendable materials would all have previously been determined as part of the takeoff process and should be ready to be entered into the estimate. No further **subjective judgment** on the part of the estimator is required at this point with respect to these cost items. However, to determine the cost of the work to be performed by the contractor's own forces, the estimator must make an **operations analysis to develop a basis for the cost**

estimate. In some cases, this analysis may have been required earlier to determine the quantities of some of the expendable materials as well. A number of decisions based on subjective judgment are necessarily required as an essential part of the operation's analysis process. That is, the estimator must provide answers to the following questions that will define the construction costs for each separate work operation:

- What is the composition of the **typical crew** that will perform the work of that operation?
- What will be the **production rate** in units of work per crew-hour for the typical crew?
- Under what project conditions will the work of the operation be performed? Three separate aspects of this question are

> What are the **hours of productive work per shift** that can be expected?
> What are the **number of shifts per work day** expected to be worked?
> How many **work days per work week** are expected to be worked?

> Answers to the above three questions are necessary to compute the **labor premium** to be added to the straight-time labor costs for

> Required shift premium pay
> Labor costs for lost time in portal-to-portal travel situations
> Premium pay for overtime work

- What is the percentage of labor costs to be included in the estimate to provide money for the cost of required **small tools and supplies (ST&S)?**

Composition of Typical Crew

For illustration, assume that the following work operation was required as part of a rock excavation bid item on a heavy construction project:

Name of work operation	Drill blast holes
Unit of work measurement	Linear feet (lf)
Work quantity	85,000

The estimator must establish the personnel and particular units of construction equipment believed to be required for this work operation. For instance, the estimator might establish the following combination of personnel and construction equipment:

Pneumatic air-track drills (ATD)	4
1200 cfm compressors	2
$2\frac{1}{2}$ ton flatbed truck	1
Drillers	4
Chucktenders	2
Compressor operator	1
Flatbed driver	1

A different combination of personnel and equipment usually will not result in a large variation in estimated costs if the relationship of the numbers of the various craft personnel to the type and quantity of equipment is reasonable.

Production Rate per Crew Hour

Closely related to the composition of the typical crew is this question: How many work units of the operation will the crew produce, on the average, in one crew-hour of work? Unlike the influence of a particular composition of the crew (as long as that composition is reasonable), construction costs vary greatly with the level of the crew's productive capacity. For instance, the estimator might decide that each ATD drill in the example crew would average 50 lf/hr, in which case the crew productivity would be 200 lf per crew-hour. The resulting cost of any work operation varies inversely with the crew productivity. Therefore, if the estimator had decided each drill would average 75 lf/hr for a crew productivity of 300 lf per crew-hour, the resulting costs for drilling blast holes would be two-thirds the costs for a drill productivity of 50 lf/hr.

Project Conditions under Which the Work Will Be Performed

Assume in the above example that the estimator decided that the work of drilling blast holes would be performed on a two-shift-per-day basis (one day shift consisting of 8 hr of work and one swing shift consisting of 7.5 hr of work), 6 days per work week. This decision means that a labor premium above the straight-time labor rate must be paid for work performed on the swing shift amounting to 8.0 hr pay for 7.5 hr of work, or (0.5 hr/7.5 hr) $\times$ 100 = 6.67%, as explained in Chapter 2. Further, if the estimator had decided that $\frac{1}{2}$ hr would be lost from the time the shift started until the work crew had been transported to the work location and that an additional $\frac{1}{2}$ hr would be lost at the end of the shift transporting the crew back to the point at which it had reported for work, an additional cost would be generated for lost time (portal-to-portal pay) equal to (2.0 hr/15.5 hr) $\times$ 100, or 12.90%. Finally, because the estimator determined that work would be performed 6 days per week, work on the sixth day (Saturday) would be performed at the overtime rate. Commonly, overtime after 5 days of work per week would be paid at time and one-half the straight-time rate on Saturday and double the straight-time rate on Sunday (see Chapter 2). Therefore, the 16.0 hr paid for work on Saturday would attract a 50% premium equivalent to an additional 8.0 hr of pay at the straight-time rate. This amounts to an additional labor premium of [8.0 hr/(6 $\times$ 15.5 hr)] $\times$ 100, or 8.60%. How these labor premiums are actually entered into the estimate will be explained later in this chapter.

Small Tools and Supplies Percentage

To perform work, work crews need small tools, many of which by industry practice are furnished by the contractor. Examples include powersaws, impact wrenches, cutting torches, and so on. Work crews also consume miscellaneous supplies, such as rigging slings and shackles, oxygen and aceytelene, and air and water hose and fittings—all of which are so small and varied that they can not reasonably be individually costed. Consequently, it is common to include a percentage of direct labor in the expendable materials column to generate money to pay for this kind of expense. The estimator must therefore decide what percentage of labor to include, depending on the type of work being performed. Work operations, such as loading and hauling common excavation and performing embankment construction, require a comparatively smaller allowance for ST&S than do operations such as pile driving and cofferdam work. A work operation such as "drill blast holes" requires considerable use of air hose and fittings, heavy tools for drill steels and bits, and similar items. An ST&S allowance of 10% of fully loaded labor costs (including union fringes and payroll insurance and taxes) would be appropriate. Note that when determining a particular ST&S percentage, the appropriate percentage is related to whether or not the labor costs in the direct cost estimate consist of bare labor with union fringes only or are fully loaded rates including payroll insurances and taxes. Also, the estimator should take the general level of the bare labor rate applicable in the project work location into consideration. Many construction companies keep accurate records of ST&S percentages from previously completed projects, which are helpful in pricing new estimates.

THE ENTRY OF INFORMATION PHASE

Once the estimator has made the above subjective judgment decisions and has assembled all takeoff quantities, this mass of information must be entered into the estimate in one of two different ways, depending on how the estimate is being made. For a manual estimate, the estimator must enter the information by hand onto one or more **bid-item extension sheets.** In the case of a computer-aided estimate, the information would simply be entered into the computer according to the format of the particular computer program used. All information to be entered for either a manual estimate or a computer-aided estimate falls into one of the following three general categories:

- Details pertaining to the bid-item description and bid-item quantity
- Details pertaining to the permanent material, expendable material, and sub-contract work quantities required under that particular bid item
- Details pertaining to each individual work operation to be performed by the contractor's own forces required by that particular bid item

The way to enter each of these categories of information into the estimate can best be explained in the context of a manual estimate.

Details Pertaining to the Bid-Item Description and Bid-Item Quantity

Assume that the previously mentioned "drill blast holes" work operation was part of the work of a bid item for which the following information had been assembled:

Bit-item number	12[1]
Bid-item description	Rock excavation[1]
Unit of measurement	Bank cubic yards (Bcy)[1]
Bid quantity	1,750,000[1]
Takeoff quantity for pay	1,691,300[2]

[1]Obtained from bid schedule of quantities in the bid documents.
[2]Obtained from the quantity takeoff.

All of this information pertains to the bid-item description and quantity, as distinct from information pertaining to what is contained within that bid item. The italicized entries at the top of Figure 6-1 illustrate the proper manner of entering this information onto a manual estimate bid-item extension sheet.

If the bid item is a **lump sum bid item,** or if the estimate is being made for a **lump sum bid** for the entire project, the unit of measurement would be written in as "lump sum," and a dash or n/a would be entered in the spaces for bid quantity and takeoff quantity.

Details of Permanent Material, Expendable Material, and Subcontract Work Quantities and Costs Required for the Work of the Bid Item

The rock excavation bid-item example would not require the purchase of any permanent materials. However, this type of work does require expendable materials and possibly some work to be performed by subcontractors. Assume, for explanation purposes, that in this case the estimator determined that the following expendable materials and subcontract work would be required:

Dynamite	75,000 lb at $1.20/lb.
Electric blasting caps	3,750 caps at $1.35 each.
Drill steels, 1.5 in. diameter × 16 ft. long	15 at $312 each.
Shanks[1]	15 at $150 each.
Bits, 3 in. diameter	150 at $155 each.
Deliver dynamite	A subcontractor to deliver the 75,000 lb of dynamite to a jobsite storage magazine at a subcontract price of $0.25/lb.

The italicized entries on Figure 6-1 illustrate the above expendable material and subcontract entries. If this example had been a bid item where permanent materials were required, the estimator would determine the various required quantities and the unit prices at which they could be obtained and enter them in the same manner.

Details of Each Work Operation to Be Performed by Contractor's Own Forces

Most bid items, such as the above example chosen for rock excavation, would require that a number of separate construction operations be performed by the contractor's own forces. For each work operation, the following details would need to be entered on the bid-item extension sheet. These details are illustrated below in terms of the pertinent information given for the "drill blast holes" operation discussed earlier:

- *Sequential number of the work operation:* Because the expendable materials and subcontracts were entered on Figure 6-1 under sequence number 1, the sequence number for "drill blast holes" would be 2.
- *Name of the operation:* Drill blast holes.
- *Work conditions in terms of hours per shift, shifts per day, and days per week:* This would be indicated as 7.75/2/6, meaning an average of 7.75 productive hours per shift, two shifts per day, 6 days per week.
- *Unit of work measurement:* The drill holes would be measured in linear feet.

---

[1] A heavy adapter required to connect the drill steel to the ATD drill hammer.

BI description _Rock excavation_   Unit _Bcy_   T.O. quantity _1,691,300_

BI# _12_

Bid quantity _1,750,000_

| Code | Description | Unit | Qty. | Crew | | | | REQ. hr | EOE | RL | Rental | Labor | EM | PM | Subs | Totals |
				EOE	RL	Rental	Labor(1)									
1	_Materials & subs_															
	Dynamite	lb	75,000				1.20						90,000			90,000
	EBC	ea.	3,750				1.35						5,063			5,063
	Drill steels 1.5 in. dia. × 16 ft	ea.	15				312.00						4,680			4,680
	Shanks	ea.	15				150.00						2,250			2,250
	Bits, 3 in.	ea.	150				155.00						23,250			23,250
	Deliver dynamite (sub)	lb	75,000				0.25								18,750	18,750

(1)Labor unit price column is utilized even though entries being made are not labor (we are simply "borrowing" the use of the column).

FIGURE 6-1 Example of entering details for the bid-item description and quantity and for PM, EM, and subcontract work required for a bid item onto the bid-item extension sheet for a manual estimate.

120

- *Quantity of work for the operation:* The estimator had determined a work quantity of 85,000 lf.
- *Work productivity per crew hour:* The estimator had determined the ATD productivity at 50 lf/hr, or a crew productivity of (4)(50) = 200 lf per crew-hour.
- *Personnel in crew by labor classification and the corresponding labor rate in dollars per hour for each:* For computer-aided estimates, the only information necessary to enter into the computer is the required number of workers for each classification. The balance of the information would have previously been stored in the computer labor library. For manual estimates, the estimator must enter all of the pertinent information.

    For our example, the estimator had established the following typical crew:

Drillers	4 each at $20.16/hr
Chucktenders	2 each at $20.01/hr
Compressor operator	1 each at $22.61/hr
Flatbed driver	1 each at $22.53/hr

- *Number of units of construction equipment by equipment classification and rates in dollars per operating hour for equipment operating expense (EOE), repair and service labor (RL), and rental (R) for each:* As was the case for details of the crew personnel, the only information necessary to enter into the computer for computer-aided estimates is the number of units for the kinds of equipment required. For manual estimates, all pertinent information must be entered.

    For our example, the estimator had determined the following units of equipment at the stated equipment hourly cost would be required:

    Air-track drills (ATD): Four each at $15.52/hr for EOE, $8.32/hr for RL, and $6.00/hr for R

    1200 cfm compressors: Two each at $11.36/hr for EOE, $1.78/hr for RL, and $16.67/hr for R

    $2\frac{1}{2}$ *ton flatbed truck:* One each at $2.88/hr for EOE, $1.90/hr for RL, and $5.53/hr for R

- *Labor premium percentage:* The estimator had determined that the work would be performed on a 7.75/2/6 basis (see above). In addition, the estimator had determined that 1/2 hr would be lost at the beginning and at the end of each shift while the crew was being transported to and from the point where the work was performed. Based on these facts, a combined labor premium percentage reflecting the shift differential, the portal-to-portal pay, and the overtime premium for working on Saturday can be computed as follows:

Work shift	Hours paid	Hours worked
Day shift, week days	(5)(8.0) = 40.00	(5)(7.0) = 35.00
Swing shift, week days	(5)(8.0) = 40.00	(5)(6.5) = 32.50
Day shift, Saturday	(1)(12.0) = 12.00	(1)(7.0) = 7.00
Swing shift, Saturday	(1)(12.0) = 12.00	(1)(6.5) = 6.50
Totals for week	104.00	81.00

$$\text{Labor premium } \% = \frac{(\text{hours paid} - \text{hours worked})}{\text{hours worked}} \times 100$$

$$= \frac{(104.00 - 81.00)}{81.00} \times 100 = 28.40\%$$

This relatively high labor premium illustrates the potential severe consequences of the conditions under which work operations are sometimes performed. Because the consequences can be so severe, and because this matter is frequently not clearly understood or overlooked, this chapter will conclude with a major section devoted to computing the appropriate labor premium under a number of different work conditions.

- *ST&S percentage:* For this example, the estimator had determined a figure of 10% of the fully loaded labor costs.

There are two basic methods by which the above information can be entered on the bid-item extension sheet: the **crew method** and the **individual entry method.** In the crew method, the estimator makes entries in such a way that the cost of the entire crew per crew-hour is determined, broken down into EOE/hr, RL/hr, R/hr, and L/hr. Then, the costs for the work operation are determined by multiplying the individual hourly crew costs by the number of required crew-hours. In the individual entry method, the entries are made in a manner such that the cost of each individual work classification and piece of equipment is extended individually, rather than being included in the combined cost for the entire crew.

Figure 6-2 illustrates the method of making all the above entries on the bid-item extension sheet by each of the two methods and shows that the extended estimated costs are the same. Note that regardless of which method is used, each piece of required information discussed above is entered on the bid item extension sheet in the particular position on the sheet shown in Figure 6-2. When this practice is followed systematically and consistently, a person reading the estimate knows exactly where to look to ascertain the value of any particular item of information without losing time searching through the entire sheet to find where that particular piece of information has been entered. For instance, note the position of the operation number on the far left and the operation description written just to the right in the description column and underlined. Just below the underlined description of the operation, note four vital pieces of information. First (on the next line in the description column) the work quantity of 85,000 lf, and the statement of work conditions (7.75/2/6), then (on the following line in the description column) the crew productivity of 200 lf/hr, and the computed number of crew-hours required to perform all of the work quantity at the stated crew productivity rate, in this case 85,000 lf/200 lf/hr = 425 hr.

Note also that, up to this point, only the entries to the left of the vertical double line near the middle of the bid-item extension sheet have been discussed. The extensions to the right of the vertical double line will be discussed in the following section of this chapter.

THE PURELY MECHANICAL PHASE OF THE PRICING-OUT PROCESS

With all essential information now entered into the estimate proper, the remaining part of the pricing-out process consists of the arithmetical operations required to convert this mass of information into dollars and cents. The estimator, or a properly instructed

BI description *Rock excavation*   Unit *Bcy*   T.O. quantity *1,691,300*   Bid quantity *1,750,000*

Code	Description	Unit	Qty.	Crew EOE	Crew RL	Crew Rental	Crew Labor	Req. hr	EOE	RL	Rental	Labor	EM	PM	Subs	Totals
2	*Drill blast holes*															
	85,000 linear-feet 7.75/2/6															
	200 linear-feet/hr 425 hr															
	ATD	ea.	4	15.52/62.08	8.32/33.28	6.00/24.00										
	1200 comp.	ea.	2	11.36/22.72	1.78/3.56	16.67/33.34										
	Flat bed	ea.	1	2.88/2.88	1.90/1.90	5.53/5.53										
	Drillers	ea.	4				20.16/80.64									
	Chucktenders	ea.	2				20.01/40.02									
	Compressor operator	ea.	1				22.61/22.61									
	Flat bed driver	ea.	1				22.53/22.53									
	Total crew			/87.68	/38.74	/62.87	/165.80	425	37.264	16.464	26.720	70.465				150.913
	ST&S - 10% Lab.	$	70.465										7.046			7.046
	Prem. pay - 28.40%	$	70.465									20.012				20.012
2	*Drill blast holes*															
	85,000/linear-feet 7.75/2/6															
	200 linear-feet/hr 425 hr															
	ATD	ea.	4	15.52	8.32	6.00		1,700	26.384	14.144	10.200					50.728
	1,200 compressor	ea.	2	11.36	1.78	16.67		850	9.656	1.513	14.170					25.339
	Flat bed	ea.	1	2.88	1.90	5.53		425	1.224	.808	2.350					4.382
	Drillers	ea.	4				20.16	1,700				34.272				34.272
	Chucktenders	ea.	2				20.01	850				17.008				17.008
	Compressor operator	ea.	1				22.61	425				9.609				9.609
	Flat bed driver	ea.	1				22.53	425				9.575				9.575
	ST&S 10% Labor	$	70.464											7.046		7.046
	Prem. pay 28.40%	$	70.464									20.012				20.012

FIGURE 6-2 Example of crew method and individual entry method for entering details of work operations performed by contractor's own forces.

assistant, must do this work for manual estimates. In the case of computer-aided estimates, everything in the following discussion occurs electronically, virtually instantaneously, to produce the estimated direct cost figures for the bid item. The following discussion, in the context of a manual estimate, delineates this process, which consists of making all arithmetical extensions, adding up the column totals, and prorating the column foot-totals to the bid quantity stated in the bid schedule.

The estimator must carry out all extensions, enter the results in the proper columns, total each line item, and enter that figure in the "Total" column. It is important at this point to ensure that the figure entered in the "Total" column is the exact sum of all the extensions in the various columns on that line. Otherwise, the estimate will not "box out" (that is, when additions made horizontally are compared to additions made vertically, the results will fail to agree). This extension process also involves computing the dollar amounts for ST&S and for the labor premium and entering them in the proper spaces on the bid-item extension sheet.

Once all the extensions and line totals have been completed, a heavy horizontal line should be struck across the entire sheet just below the last line extended. Should a number of bid-item extension sheets be required for lengthy bid items, this horizontal line should be struck just below the last line extended on the last bid-item extension sheet. Just below this line, two captions should be entered on the next two lines in the description column. The first caption should read "Total Takeoff." The second should read "Total Bid Item." Then for each, in the "Quantity" and "Unit" columns, the previously determined information for the bid item at the takeoff level and at the bid level, respectively, should be entered.

Once the previous step has been completed, all vertical columns, including the "Total" column, should be added and the foot-totals entered in the appropriate column on the line that has been captioned "Total Takeoff." Once again, the estimator should make certain at this point that the column foot-totals add to the exact foot-total shown in the "Total" column. If the foot-totals do not check, the estimator must find the cause of the discrepancy and correct it before proceeding. In carrying out this process for manual estimates, no page totals or subtotals of any other kind should be used. They contribute nothing but trouble and confusion in the event that later changes are required in the estimate. The use of subtotals and page totals does not present this problem in the case of computer-aided estimates because when changes are made all such subtotals and page totals automatically readjust.

The next step for a bid item in a schedule-of-bid-items estimate is the **prorate** process whereby the costs, which have been estimated at the takeoff level, are adjusted upward or downward to correspond to the bid-quantity level. Note that this step does not need to be carried out if the bid item is a **lump sum bid item,** or if the estimate is being made for a **lump sum bid** for the entire project.

First, the estimator must develop a **prorate factor** by dividing the bid quantity for the bid item by the takeoff quantity and expressing the result to five or six decimal places. Then, each column foot-total in the estimate is prorated from the takeoff level to the bid-item level by multiplying the takeoff level foot-total by the prorate factor. Following this process, the prorated foot-totals should again be added to make certain they exactly equal the prorated figure in the "Total" column. Often, minor discrepancies due to rounding occur and a minor adjustment is necessary in one of the figures so that the prorated figures check.

Finally, all unit costs for each vertical column should be computed and entered in the appropriate column on the line below the "Total Bid Item" line. These unit costs should be added to make certain they agree with the unit cost in the "Total" column. If the total does not match exactly because of rounding, a minor adjustment should be made in one of the unit costs so that this line will agree as well. This step completes the pricing-out procedure for the bid item.

This final part of the pricing-out process is illustrated by Figure 6-3. A large rock excavation bid item such as this illustration depicts would no doubt include a number of individual work operations. To shorten the length of the illustration, the drill-and-shoot portion of this hypothetical bid-item estimate is not shown but instead has been indicated at the top of the page of Figure 6-3 as a consolidated line total carried forward from separate preceding bid-item extension sheets. Also, the drill-and-shoot work in this hypothetical example is much more extensive than that described earlier in this chapter and depicted in Figures 6-1 and 6-2. The drill-and-shoot work shown in Figure 6-3, in a total amount of $2,329,500, as operation 1, is followed by three additional work operations described as follows:

Operation 2—Load haul and compact, haul 1

Operation 3—Load haul and compact, haul 2

Operation 4—Supervision and haul road maintenance

Note that the three additional work operations have been entered onto Figure 6-3 and extended by the crew method previously explained. Further, the total estimated direct cost at the takeoff level for 1,691,300 Bcy of rock excavation is $6,904,700, prorated to the bid-item level at the bid quantity of 1,750,000 Bcy for a total cost of $7,144,200. The total estimated unit cost (which applies at either the takeoff level or the bid-item level) is $4.082/Bcy. The prorate factor in this case is 1,750,000/1,691,300, or 1.034707.

It should be noted that the estimate shown in figure 6-3 "boxes out"; that is, the foot totals, when added horizontally, equal the vertical total in the Total column. This is also true for the unit costs.

California State University, Chico, Computer-Aided Estimating Program

A flexible database computer-aided estimating program developed for teaching purposes at California State University, Chico, is used for illustration a number of times in this book; for these illustrations the libraries included in the Appendix were preloaded into the program.

The California State University, Chico, program is typical of the operative part of many computer-aided estimating programs for pricing out construction estimates that are commercially available today. A PC computer disk containing this program has been supplied with this book and instructions for its use have been included in the Appendix.

LABOR PREMIUM COMPUTATIONS FOR VARIOUS OTHER PROJECT WORK CONDITIONS

The labor premium calculation for the "drill blast holes" work operation reflected how the particularly disadvantageous project work conditions assumed for that example resulted in a very high premium of 28.40%. Had that work been expected to be performed under normal day shift work conditions; (that is, 8 hr per day, 5 days per week), the labor premium would have been zero. Obviously, the particular work conditions will greatly affect the labor premium percentage and, thus, the magnitude of the labor cost for the work operation. The following labor premium calculations for frequently encountered work situations illustrate this point.

BI description _Rock excavation_   Unit _Bcy_   T.O. quantity _1,691,300_   BI# _12_

Bid quantity _1,750,000_

Code	Description	Unit	Qty.	Crew EOE	RL	Rental	Labor	Req. hr	EOE	RL	Rental	Labor	EM	PM	Subs	Total
1	Drill & shoot	Bcy	1,691,300	(C/F from a separate estimate sheet)					321,000	141,800	230,300	777,200	859,200			2,329,500
2	Load haul & compact—Haul #1 10/2/6															
	1,257,800 Bcy															
	500 Bcy/hr 2,516 hr															
	988 loader		2	12.50/25.00	13.50/27.00	35.00/70.00	23.34/46.68									
	D8 dozer		3	27.67/83.01	7.13/21.39	20.67/62.01	22.74/68.82									
	769C truck		6	27.45/164.70	5.95/35.70	33.33/199.98	22.96/137.76									
	Swamper		1				19.64									
	Total crew			272.71	84.09	331.99	272.90	2,516	686,158	211,570	835,287	686,616				2,419,611
	ST&S 10%		686,616									686,616	68,662			68,662
	Labor premium—20.4%		686,616									140,070				140,070
3	Load haul & compact—Haul #2 10/2/6															
	433,500 Bcy															
	268 Bcy/hr 1,618 hr															
	988 loader		1	12.50	13.50	35.00	23.34									
	D8 dozer		2	27.67/55.34	7.13/14.26	20.67/41.34	22.74/45.48									
	769C truck		5	27.45/137.25	5.95/29.75	33.33/166.65	22.96/114.80									
	Swamper		1				19.64									
	Total crew	$	328,875	205.09	57.51	242.99	203.26	1,618	331,836	93,051	393,158	328,875				1,469,920
	ST&S 10%	$	328,875										32,888			32,888
	Labor premium 20.4%											67,090				67,090
4	Supervision & haul road maint.															
	4,134 hr 10/2/6															
	1 hr/hr 4,134 hr															
	Crew from Bid item 20		286,817	24.96	8.33	45.33	69.38	4,134	103,185	34,436	187,394	286,817				611,832
	ST&S 10%		286,817										28,682			28,682
	Labor premium—20.4%		286,817									58,511				58,511
	Total Takeoff	Bcy	1,691,300						1,442,159	480,857	1,646,139	2,345,179	989,432			6,903,766
	Total Bid item	Bcy	1,750,000						1,492,212	497,546	1,703,272	2,426,573	1,023,772			7,143,375
	Unit prices								0.853	0.284	0.973	1.387	0.585			4.082
	Prorate factor		1,750,000 / 1,691,300 = 1.034707													

FIGURE 6-3 Example of the final bid-item extension sheet of a completed manual estimate for a rock excavation bid item.

10.0/1/5: One Extended Shift per Day, 5 Days per Week

The shortest time period that repeats itself in this situation is 1 work shift. For this condition, the labor premium calculation would be

Description	Hours paid	Hours worked
One typical shift	$8.0 + (2.0 \times 1.5) = 11.00$	10.0

$$\text{Labor premium} = \frac{(11.0 - 10.0)}{10.0}(100) = 10.0\%$$

10.0/2/5: Two Extended Shifts per Day, 5 Days per Week

The shortest time period that repeats itself is 1 work day.

Description	Hours paid	Hours worked
Typical day shift	$8.0 + (2.0 \times 1.5) = 11.00$	10.00
Typical swing shift	$8.0 + (2.5 \times 1.5) = \underline{11.75}$	<u>10.00</u>
Total work day	22.75	20.00

$$\text{Labor premium} = \frac{(22.75 - 20.0)}{20.00} \times 100 = 13.75\%$$

10.0/2/6: Two Extended Shifts per Day, 6 Days per Week

The shortest time period that repeats itself is 1 work week.

Description	Hours paid	Hours worked
Day shift, week days	$5[8.0 + (2.0 \times 1.5)] = 55.00$	50.00
Swing shift, week days	$5[8.0 + (2.5 \times 1.5)] = 58.75$	50.00
Day shift, Saturday	$(8.0 + 2.0) \times 1.5 = 15.00$	10.00
Swing shift, Saturday	$(8.0 + 2.5) \times 1.5 = \underline{15.75}$	<u>10.00</u>
Total work week	144.50	120.00

$$\text{Labor premium} = \frac{(144.50 - 120.0)}{120.00} \times 100 = 20.42\%$$

7.75/2/5: Two Straight-Time Shifts per Day, 5 Days per Week

The shortest time period that repeats itself in this case is 1 work day.

Description	Hours paid	Hours worked
Typical day shift	8.00	8.00
Typical swing shift	<u>8.00</u>	<u>7.50</u>
Total work day	16.00	15.50

$$\text{Labor premium} = \frac{(16.00 - 15.50)}{15.50}(100) = 3.23\%$$

7.50/3/5: Three Straight-Time Shifts per Day, 5 Days per Week

The shortest time period that repeats itself is 1 work day.

Description	Hours paid	Hours worked
Typical day shift	8.00	8.00
Typical swing shift	8.00	7.50
Typical graveyard shift	8.00	7.00
Total work day	24.00	22.50

$$\text{Labor premium} = \frac{(24.00 - 22.50)}{22.50}(100) = 6.67\%$$

7.50/3/5: Tunnel Project, Three Straight-Time Shifts per Day, 5 Days per Week (Change at Heading, 1/2 Hr Travel Time to Heading Paid at Overtime Rate, Crew Returns to Portal on Own Time; Work through Lunch Break 1/2 Hr Paid at Overtime Rate)

The shortest time period that repeats itself is 1 work day.

Description	Hours paid		Hours worked
Typical day shift	Straight-time pay	= 8.00	8.50
	Travel = 0.5 × 1.5	= 0.75	
	Lunch = 0.5 × 1.5	= 0.75	
Typical swing shift	Straight-time pay	= 8.00	8.00
	Travel	= 0.75	
	Lunch	= 0.75	
Typical graveyard shift	Straight-time pay	= 8.00	7.50
	Travel	= 0.75	
	Lunch	= 0.75	
Total work day		28.50	24.00

$$\text{Labor premium} = \frac{(28.50 - 24.00)}{24.00}(100) = 18.75\%$$

7.50/3/7: Emergency Utility Work, Three Straight-Time Shifts per Day, 7 Days per Week

For this situation, the shortest time period that repeats itself is 1 work week.

Description	Hours paid		Hours worked	
Day shift, week days	(5)(8.0)	= 40.00	(5)(8.0) =	40.00
Swing shift, week days	(5)(8.00)	= 40.00	(5)(7.50) =	37.50
Graveyard shift, week days	(5)(8.00)	= 40.00	(5)(7.00) =	35.00
Day shift, Saturday	(1)(8.00) × 1.5	= 12.00	(1)(8.00) =	8.00
Swing shift, Saturday	(1)(8.00) × 1.5	= 12.00	(1)(7.50) =	7.50
Graveyard shift, Saturday	(1)(8.00) × 1.5	= 12.00	(1)(7.00) =	7.00
Day shift, Sunday	(1)(8.00) × 2.0	= 16.00	(1)(8.00) =	8.00

Swing shift, Sunday	$(1)(8.00) \times 2.0 = 16.00$	$(1)(7.50) =\ 7.50$
Graveyard shift, Sunday	$(1)(8.00) \times 2.0 = \underline{16.00}$	$(1)(7.00) = \underline{\ 7.00}$
Total work week	204.00	157.50

$$\text{Labor premium} = \frac{(204.00 - 157.50)}{157.50}(100) = 29.52\%$$

CONCLUSION

This chapter has explained the entire procedure of making a direct cost estimate for a single major line item. The discussion, presented in the context of the line item being a single bid item for a schedule-of-bid-items bid, explained the principles to be followed when making a manual estimate with applicable commentary on how the principles would be applied in a computer-aided estimate. If the estimating team follows the principles presented in this chapter, the resulting estimate will be clear and easy to read by anyone who reviews it.

This was the last of a group of introductory chapters to this book explaining the underlining philosophy and mechanics of preparing heavy construction cost estimates. The next group of chapters will switch emphasis to **operations analysis**—that is, the application of engineering and construction technology required to develop **the basis of the cost estimate.** This kind of analysis is usually necessary to develop the takeoff quantities for expendable materials that must be furnished and work operations that the contractor's own forces must perform.

The field of heavy construction is so broad that only limited examples illustrating the central idea of operation analysis and the accompanying cost estimates can be included in this book. The topics presented in the following chapters have been selected on the basis of their overall importance to heavy construction and for their value in illustrating the general approach taken in heavy construction cost estimating. The following chapter on drill-and-blast operations is the first in this series of chapters.

QUESTIONS AND PROBLEMS

1. What actually occurs when a line item in an estimate is "priced out"? Is the pricing out process fundamentally different, depending on whether the line item is the discrete part of the work for a lump sum estimate or a bid item in a schedule-of-bid-items estimate?

2. What are the three distinct phases of the pricing-out process?

3. In general, what four categories of costs could be incurred in performing the work of a bid item?

4. What three basic questions about a work item to be performed by the contractor's own forces must be answered on the basis of the estimator's subjective judgment? What three additional questions about the project conditions applying to the work operation must be answered on the basis of the estimator's subjective judgment? What are three causations requiring a labor premium to be added to straight-time labor costs? What final question requiring the estimator's subjective judgment must be answered with respect to the cost of small tools and supplies required for performing the work of a bid item?

5. Once the estimator has made all of the necessary subjective judgments and has assembled all of the necessary takeoff information, what are the two separate ways the resulting information can be entered into an estimate, depending on how the estimate is being made?

6. What are the two distinct methods explained in this chapter for entering the necessary information pertaining to a work operation to be performed with the contractor's own forces into a manual estimate?

7. What is meant by "boxing out" a manual estimate?

8. What is the prorate process and what is the prorate factor, as those terms are used in this chapter?

9. What is a unit cost? How are unit costs obtained? Are the unit costs determined for the "Total Takeoff" column foot totals different from those determined for the "Total Bid Item" column foot totals?

10. What are the seven frequently encountered work situations for which labor premium calculations are explained in this chapter? Describe each situation notationally and in words.

11. Bid Item 21 for structure excavation work is described on the schedule of bid items as follows:

Bid Item 21   Structural Excavation   210,000 Bcy at $_____ per Bcy $_____

Assume that the correct takeoff quantities for this bid item were
(a)   Quantity of material that will actually be excavated = 327,444 Bcy
(b)   Quantity that engineer will measure for payment = 203,194 Bcy

The estimator's subjective judgments resulted in an estimated hourly production of 300 Bcy per crew-hour for the following work crew working a 10 hr day, 5 days per week:

6 cy hydraulic backhoe	1
35 ton end dump trucks	3
D8 dozer	1
Backhoe operator	1
Oiler	1
Dozer operator	1
Off-highway truck driver	3
Operator foreman	1
Small tools and supplies	5%
Overtime	1.5/1.5/2.0

Price out this bid item manually by the individual entry method shown in Figure 6-2 using the format shown in Figure 6-3 (except you are to use the individual entry method rather than the crew method). Your estimate will consist of only one operation entitled, "Excavate, Haul, Stockpile, and Dispose." Prorate to the bid quantity shown on the schedule of bid items. Determine all unit prices. Use the library rates included in the Appendix for the labor and equipment hourly rates.

Repeat the estimate using the California State University, Chico, computer-estimating program included with this book. The printout should agree with your manual estimate.

12. Bid Item 22 for structural backfill work is described on the schedule of bid items as follows:

Bid Item 22   Structural Backfill   36,900 Ecy at $ _____ per Ecy = $_____

Assume that the correct takeoff quantities for this bid item were
(a) Quantity of backfill that actually will be placed = 157,587 Ecy
(b) Quantity of backfill the engineer will measure for payment = 36,241 Ecy

The estimator's subjective judgments resulted in an estimated hourly production of 155 Ecy per crew-hour for the following work crew working a 9 hr day, 5 days per week:

7 cy wheel loader	1
D6 dozer	1
Cat 825 compactor	1
Loader operator	1
Dozer operator	1
Compactor operator	1
Operator foreman	1
Small tools and supplies	5%
Overtime	1.5/1.5/2.0

Price out this bid item using the computer-estimating program. There will be only one work operation entitled, "Rehandle, Spread, and Compact Backfill".

13. A reinforcing steel bid item is described on the schedule of bid items as follows:

Bid Item 24   Reinforcing Steel   5,675,000 lb at $ _____ per lb = $_____

Assume that the correct takeoff quantities for this bid item were:
(a) Quantity of reinforcing steel that must be furnished as permanent material = 5,580,000 lb
(b) Quantity of reinforcing steel that must be furnished as expendable material = 462,000 lb
(c) Quantity of reinforcing steel for which the installation subcontractor will be paid = 6,042,000 lb
(d) Quantity of reinforcing steel that the engineer will measure for payment = 5,580,000 lb

Your company as the general contractor will purchase fabricated reinforcing steel and have it delivered fob jobsite. All installation will be performed by the subcontractor. You as general contractor must furnish crane service for unloading the reinforcing steel delivery trucks and setting the reinforcing steel near the point of installation. A total of 250 crew-hours for the following crew working on an 8 hr per day, 5 days per week basis has been estimated to be required for unloading and handling the total quantity of reinforcing steel:

75 ton truck crane	1
Crane operator	1
Oiler	1
Small tools and supplies	5%

Price out this bid item using the computer-estimating program. Two operations should appear on the estimate:

Operation 1: "Materials and Subs"
Operation 2: "Unload and Handle"

7

Drill-and-Blast Operations

Key Words and Concepts

Drillers
Chuck tenders
Powdermen
Operating engineers
Laborers
Drill foreman
Powder foreman
Track drills
Down-the-hole drills
Rotary equipment
Drill jumbos
Sinker drill
Jack-leg drill
Stoper drill
Shanks
Sectional drill steels
Couplings
Drill bits
Nominal depth
Subdrill
Blast hole
Line hole
Drill pattern
Drill factor
Dynamite

Gelatin dynamite
Ammonium nitrate fuel oil mixture
Blasting slurries
Presplit explosives
Electric blasting caps
Non-el blasting caps
Primacord
Bottom loading
Deck loading
Stick count
Loading density
Loaded depth
Collar depth
Percent hole loaded
Burden
Stemming
Powder factor
Powder factor modifiers
Blast hole analysis for a single hole
Drill string component life
Collar time
Drill penetration time
Steel add and remove time
Drill productivity
Powder crew productivity

Before rock can be excavated, it must be loosened in some manner so that it can be loaded into haul vehicles. This is frequently accomplished by ripping the rock with powerful, hydraulically actuated ripper attachments on large crawler tractors. The stronger rock formations, however, cannot be loosened in this manner and must be broken by drilling and blasting prior to loading. This chapter will review the technology of this process in an elementary way and explain the analytical procedure employed by heavy construction cost estimators to determine the costs that will be incurred for drill-and-blast operations conducted above the ground surface. Similar operations for underground work are beyond the scope of this book.

CONSTRUCTION CRAFTS, EQUIPMENT, AND EXPENDABLE MATERIALS INVOLVED IN DRILL-AND-BLAST OPERATIONS

Construction Crafts

In terms of the trade union classifications commonly used in the United States, the construction crafts involved in drill-and-blast operations include **drillers, chuck tenders, powdermen,** and occasionally **operating engineers** as compressor operators when large air compressors are required. All of these personnel are **laborers,** with the exception of compressor operators. A **drill foreman,** who is a laborer, supervises drilling operations, where crews loading explosives into blast holes are supervised by a **powder foreman,** who is also a laborer.

Equipment Required

For most drilling operations conducted on the surface, drill holes are drilled by pneumatically or hydraulically powered drill hammers mounted on masts or booms controlled by hydraulically actuated cylinders. The entire unit is generally mounted on crawler tracks and commonly referred to as a **track drill.** Most drills of this type are capable of drilling holes ranging from about a 2 to a 6 in. diameter to depths in excess of 60 ft. A typical instantaneous penetration rate for a pneumatically powered track drill for a $2\frac{1}{2}$- to 3-in.-diameter hole ranges from $1\frac{1}{2}$ to $2\frac{1}{2}$ ft/min, depending on the hardness of the rock. Modern hydraulically powered drills advance holes in this size range at instantaneous penetration rates of $4\frac{1}{2}$ to 8 ft/min. Blast holes larger than 6 in. in diameter are usually drilled with larger equipment employing **down-the-hole** pneumatically powered drill hammers or are drilled by **rotary equipment.**

The equipment is carried on a track-mounted carriage. Large blast holes ranging from 24 to 36 in. in diameter and drilled to 50 to 60 ft depths are frequently employed in quarry operations or open-pit mining. Large-hole-diameter drilling-and-blasting operations are beyond the scope of this book.

Drills for advancing tunnels and other underground openings are similar to those utilized for surface work except that they are mounted on booms controlled by hydraulic cylinders. The entire arrangement is carried on either track-mounted, rail-mounted, or rubber-tire carriers called **drill jumbos.** For large tunnels, as many as six such booms may be carried on a single jumbo, although most underground drill-and-blast work is performed by jumbos with two or three drill booms.

PHOTO 7-1   Typical present day track drills. The upper photo shows a self-contained hydraulically powered drill, and the lower photo shows a pneumatically powered drill with trailing air compressor.

PHOTO 7-2 This two-boom hydraulic drill jumbo, shown outside a tunnel portal and while drilling horizontal blast holes at a tunnel face, is typical of the jumbo-mounted drilling equipment used today in underground drill-and-blast work.

Hand-held drilling equipment is widely used for small holes (1 to $2\frac{1}{2}$ in. in diameter) for both surface and underground work. One such drill type, called a **sinker,** or **jackhammer,** is intended to be operated by one person drilling vertical holes up to depths of 10 to 12 ft. Other one-person drills are **jack-leg drills** and **stopers,** both used in underground work. These drills have a pneumatically operated feed leg attached to them to assist the operator in holding and forcing the drill forward as the hole is advanced.

For a more complete description of the equipment used in drill-and-blast operations, see the several reference books listed in the bibliography to this text.

Expendable Materials

Because drill bits and drill steels encounter relatively rapid wear when drilling rock (which varies greatly by type, hardness, and abrasiveness), these items are generally considered expendable materials rather than spare parts for the drilling equipment. Principal drilling expendables consist of

- **Shanks** (sometimes called adapters) connecting the drill steel to the drill hammer
- **Sectional drill steels,** typically $1\frac{1}{2}$ to 2 in. in diameter, 12 to 20 ft long with threaded ends so they can be connected to form a drill string for deep holes
- **Couplings** for connecting sections of drill steel
- **Drill bits** that screw on to the end of the lowest drill steel and are typically fitted with tungsten-carbide inserts to better penetrate the rock

The explosives required to break rock and the blasting caps used to detonate them are also costed in the expendable material column because they are totally consumed in the operation. Detonating cord, connecting wire, and blasting batteries are also costed and carried as expendable materials.

DRILL-AND-BLAST TERMINOLOGY

Drill Hole Terminology

The following terms apply to the holes drilled in the rock to receive explosives:

- Nominal depth (lift depth)
- Subdrill
- Blast hole
- Line hole
- Drill pattern
- Drill factor

These drill hole terms are explained in terms of a rectangular rock excavation measuring 24 ft long $\times$ 12 ft wide $\times$ 10 ft deep, shown on Figure 7-1.

Nominal depth (lift depth): Often referred to as the depth expected to be "pulled" by the blast, this is the depth of the rock that is expected to be broken or loosened by a single blast. For the example shown in Figure 7-1, the nominal depth is 10 ft.

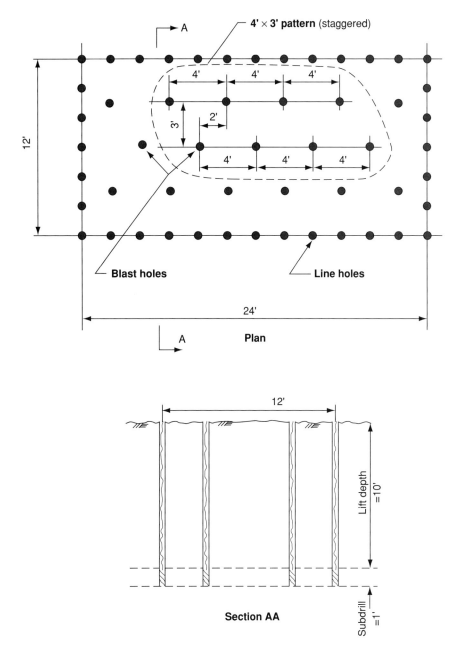

FIGURE 7-1 Small rock excavation illustrating drill hole terminology.

If this 10 ft depth was intended to be blasted in two 5 ft lifts, the nominal depth for each lift would be 5 ft. An excavation with the horizontal dimensions of Figure 7-1 that was deeper than 10 ft would have to be taken out in more than one lift, the exact number of lifts depending on the total depth of the excavation.

Subdrill: The distance below the bottom of the lift depth that a drill hole for loading explosives must be drilled to ensure that the rock will break at the high points between holes at an elevation no higher than the elevation of the nominal depth. For 10 ft nominal depths, such as those illustrated in Figure 7-1, the subdrill (shown as 1 ft) typically ranges from 10 to 20% of the nominal depth.

Blast hole: Holes drilled within the horizontal boundaries of the body of the excavation so that explosives may be loaded into them to break the rock.

Line hole: Holes drilled at a uniform spacing around the periphery of the excavation. Line holes are typically loaded with a special explosive (see below) to induce a splitting effect from hole to hole to form a reasonably neat surface at the boundaries of the excavation.

Drill pattern: The horizontal pattern in which the blast holes are drilled. The spacing between holes in a single line and the distance between parallel lines of holes is usually repetitive within the horizontal boundaries of the excavation.

Drill factor: The ratio of the total linear feet of blast holes drilled in a rock excavation (counting the subdrill) to the volume in bank cubic yards (Bcy) of rock intended to be broken, not counting the volume of rock below the nominal depth. The line holes are not included. The drill factor is expressed in feet per bank cubic yard.

Explosives Terminology

Terms pertaining to the explosives loaded in blast and line holes for a rock excavation include the following:

- Dynamite
- Gelatin dynamite
- Ammonium nitrate fuel oil mixture
- Blasting slurries
- Presplit explosives
- Electric blasting caps
- Non-el blasting caps
- Primacord
- Bottom loading
- Deck loading
- Stick count
- Loading density
- Loaded depth
- Collar depth
- Percent hole loaded
- Burden
- Stemming
- Powder factor

Dynamite: A cellulose material (often sawdust) impregnated with nitroglycerin and encased in a cylindrical paper cartridge, typically 1 to 3 in. in diameter. Dynamite (often referred to as "straight" dynamite) is typically furnished in various strengths depending on the nitroglycerin content by weight.

Gelatin dynamite: A dynamite gel obtained by dissolving cellulose material in nitroglycerin and packaging it in cylindrical paper cartridges, similar to those for straight dynamite described above. Gelatin dynamite has the property of remaining intact so that it can be reliably detonated under water. Straight dynamite is not usable under water or in wet environments. Both gelatin and straight dynamite are often referred to as "stick" powder.

Ammonium nitrate fuel oil mixture (ANFO): A mixture of commercial ammonium nitrate prills (typically marketed as an agricultural fertilizer) and fuel oil in the proportions of about 1 gal of fuel oil per 100 lb of ammonium nitrate. When fuel oil is mixed with ammonium nitrate in this manner, an extremely fast, powerful explosive is formed. ANFO has the advantage that it can be either poured or blown into blast holes. ANFO is one form or class of explosive commonly referred to as a "free-running" powder.

Blasting slurries: Water-based explosives that can be poured or pumped into blast holes. Slurries are normally used in quarry work in large-diameter blast holes. This form of explosive is frequently furnished in 2- to 4-in.-diameter plastic "baloney" bags approximately 24 in. long. Slurries packaged in this manner can be used under water or in wet conditions.

Presplit explosives: A special explosive intended for use in line holes. This form of stick powder is usually about 1 in. in diameter and approximately 24 to 30 in. long. The individual sticks are sometimes made so that they can be connected together to form a long vertical explosive train for use in deep vertical line holes.

Electric blasting caps (EBCs): Detonating devices approximately $\frac{1}{4}$ in. in diameter and approximately 2 in. long encased in a metal jacket and containing an explosive that will detonate when an electric current is passed through it. Each cap has two separate electric wires extending from the interior of the cap through one end in lengths of 6 to 15 or 20 ft, depending on the depth of the holes in which they are intended to be used. They serve to detonate the explosives in the particular hole in which they are placed. Typically, the caps are inserted in one end of a single stick of stick powder called a primer, which in turn is usually placed in the bottom of the hole before loading. Deep blast holes are sometimes loaded with a primer at the top as well as the bottom of the hole.

Detonator wires leading from the individual blast holes are connected together either in series or parallel to form closed circuits consisting of a number of blast holes. Each circuit is then connected in parallel to short bare wires known as "buzz" lines, which in turn are energized through a blasting line connected to an electric blasting battery when the blast is fired. Electric blasting caps are furnished in various millisecond[1] "delays" so that they can be arranged to successively detonate at various time intervals after an instantaneous electric current passes through them. This causes selected blast holes to fire before others to provide "relief" that aids in breaking the rock and "pulling" the blast.

Non-el blasting caps: A blasting cap similar to EBCs except that instead of two wires leading into the cap for the passage of electric current, a thin hollow tube coated on the inside with an explosive extends from the cap to propagate an explosion to the cap, which then further propagates the explosion into the surrounding explosives in the blast holes. An initial detonation must first occur somewhere in the hookup to propagate the explosion through the non-el blasting cap tube to the cap itself. This is accomplished by connecting the tube lead from the cap to a length of **shock tube** that, when detonated, will detonate the caps that have been connected to it. Non-el blasting caps have the advantage that, with the exception of the electrical means of inducing the initial detonating explosion in the hookup, no EBCs are involved that would be susceptible to stray electrical currents that could cause an unintended, premature explosion. Today, non-el caps are almost always employed for underground work, as well as for much surface work.

Primacord: An explosive in the form of a strong cord about 1/8 in. in diameter that is hollow and continuously filled with a very fast detonating explosive. Primacord has the appearance of other types of woven surface cord. It is furnished

---

[1] A millisecond is one-thousandth of a second.

PHOTO 7-3   This tunnel crew is loading stick powder into vertical blast holes for a bench cut in a large highway tunnel using primacord for detonation. Note the use of powder poles for tamping and the coil of primacord.

PHOTO 7-4   This shallow cut for a roadway widening project in a congested semiurban area, consisting of vertical blast holes, is being loaded with stick powder using non-el blasting caps for detonation. Note the powder poles for tamping and the close-up of the non-el caps.

on spools and can be spooled off and cut and safely tied together by common knots. When detonated, the explosion travels through primacord at the rate of about 22,000 ft/sec. Primacord is often used throughout the length of loaded holes to ensure complete detonation of the explosives in the hole, as well as a connector between holes to propagate the explosion throughout the blast.

Bottom loading: A method of loading blast holes where the explosives are placed continuously in the lower part of the hole from the bottom up to about two-thirds of the hole depth.

Deck loading: A method of loading blast holes where the explosives are placed intermittently from the bottom of the hole to about two-thirds of the hole depth, several feet of hole filled with explosive, alternating with several feet filled with stemming (see below), alternating with more explosive, and so on. Deck loading distributes a given weight of explosives over a longer length of hole.

Stick count: A term applying only to stick powder. The stick count is the number of individual explosive cartridges contained in a 50 lb box of stick power. This is another way of stating the weight of an individual stick of powder in pounds.

Loading density: The number of pounds of explosive that is loaded per foot of blast hole. The loading density will depend on the degree of tamping[2] employed.

Loaded depth: This refers to the total length of a blast hole continuously loaded with explosives.

Collar depth: The length of hole from the top of the explosive column to the collar of the hole. This length of hole is usually filled with stemming (see below).

Percent hole loaded: The ratio of the loaded depth of a blast hole to the total depth of the hole (including the subdrill) expressed as a percentage. Typically, blast holes are loaded to between 60 and 70% of their total depth.

Burden: A term that applies to a vertical (or near vertical) rock face that is being progressively blasted off. It is the distance from the exposed face back to the first line of blast holes drilled behind the face to blast the face outward.

Stemming: A material placed in blast holes, usually from the top of the explosive column to the collar of the hole. It usually is composed of drill cuttings resulting from the blast hole drilling operation. However, it can be almost any material that can be tamped tightly in the hole to contain the explosive.

Powder factor: The measure of the quantity of explosive required to break a given quantity of rock. It is expressed as the ratio of the pounds of explosive in a blast to the number of bank cubic yards intended to be broken by the blast.

Many of the explosives terminology terms explained above are illustrated in Figure 7-2.

DETERMINATION OF POWDER FACTOR, DRILL PATTERN, AND DRILL FACTOR

To prepare a cost estimate for drill-and-blast operations, the first thing the estimator must do is to determine the powder factor, drill pattern, and drill factor intended to be utilized. Today, blast design technology has reached a high degree of sophistication.

---

[2]Tamping is a process where the individual sticks of powder are crushed and expanded out against the inside of the blast holes by "mashing" them down by means of a wooden or plastic pole called a "powder pole."

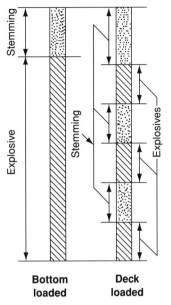
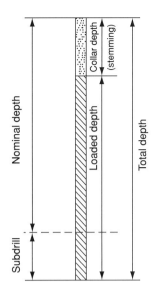

Bottom loaded **Deck loaded**

Bottom loading vs. deck loading   **Percentage hole loaded**

FIGURE 7-2 Explosives terminology.

Contractors usually secure the advice of qualified blasting specialists, including those employed as sales representatives by the companies that market explosives, to assist them in designing blasting patterns and selecting explosives to obtain optimum results. This book does not venture into that degree of sophistication. Rather, the object here is to acquaint readers with a general approach construction estimators can take in estimating the probable cost of drilling and blasting where rock excavation is required. Although the procedure described in the following paragraphs is simplistic and explained in the context of the less-sophisticated types of explosives, the principles discussed are applicable to more sophisticated approaches.

Determining the Required Powder Factor

Other things being equal, the powder factor (PF), expressed in pounds of explosive required to break 1 Bcy of rock is a function of the type and strength of the rock to be blasted. Figure 7-3 represents the approximate powder factor values required to break a number of the different kinds of rock commonly encountered in construction, and Figures 7-4 and 7-5 show modifiers to be applied to the powder factor to reflect particular conditions. The various rock types are displayed vertically, whereas two separate hardness scales, the Mohs and Shore Scleroscope scales, are displayed horizontally. As shown, the Mohs scale varies from a minimum hardness of 3 to a maximum hardness of 7.5, and the Shore Scleroscope scale varies from corresponding values of hardness ranging from a minimum of 30 to a maximum of 120. To put the meaning of these hardness scales into practical terms, the reader should note that a value of 1 on the Mohs scales indicates the softest rock found in nature (talc or chalk), and a value of 10 indicates the hardest rock found in nature (diamond). For each rock type, a horizontal bar indicates the range of hardness for that rock type that is commonly encountered. On these horizontal bars, vertical lines with numerical figures indicate approximate values of the required powder factor to break rock at the degree of hardness indicated by the horizontal position of the figure on the bar. For instance, Figure 7-3 indicates that a powder factor of 1.0 lb/Bcy

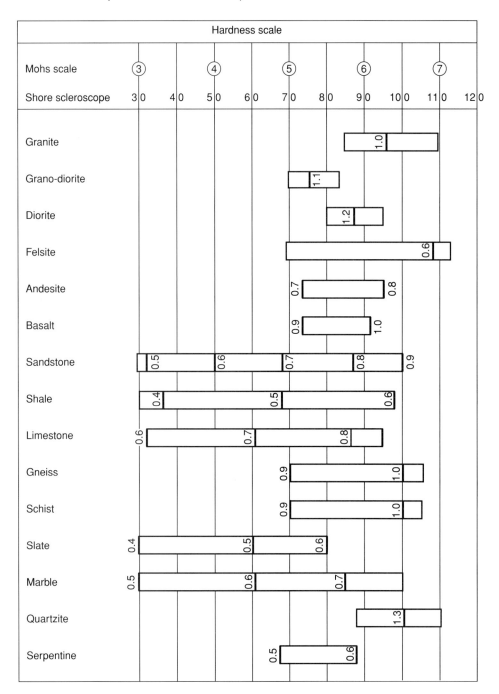

FIGURE 7-3   Unmodified powder factors for common rocks.

would be required to break granite with a Mohs hardness of 6.25 under average conditions, and a powder factor of 0.6 lb/Bcy would be required to break sandstone with a Mohs hardness of 4 under average conditions.

The special conditions shown in Figures 7-4 and 7-5 requiring **powder factor modifiers** involve the following three considerations:

- Degree of fragmentation desired
- Degree of confinement present
- Ratio of net to gross cubic yards (subdrill)

EXPLANATION OF TERMS

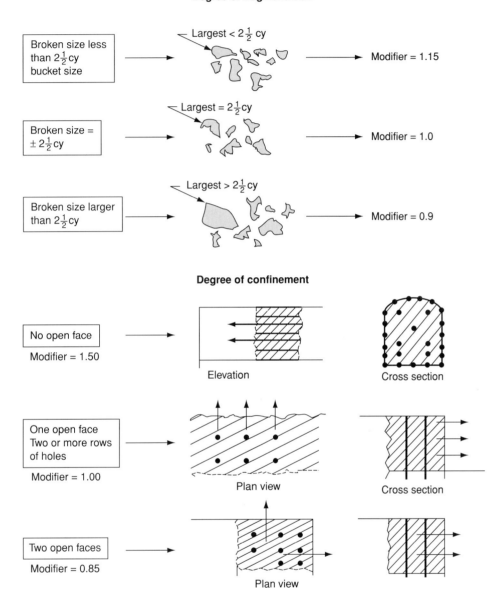

FIGURE 7-4   Powder factor modifiers.

To understand how these modifiers would be used, assume that a particular blasting application in sandstone with a Mohs hardness of slightly less than 5 requires an unmodified powder factor of 0.7 lb/Bcy. The degree of fragmentation desired is to break the rock to less than a 2.5 cy bucket size, there are no open faces in the configuration of the blast, and the intended subdrill is less than 25% of the nominal depth. Under these circumstances, the modified powder factor would be $(0.70)(1.15)(1.50)(1.0) = 1.21$ lb/Bcy.

There could be other circumstances regarding particular applications in which an estimator might use similar factors to adjust the unmodified powder factor up or down. For instance, the rock formation might be particularly massive with few joints (requiring a plus adjustment), or it might be horizontally bedded and highly jointed (where the adjustment would be minus).

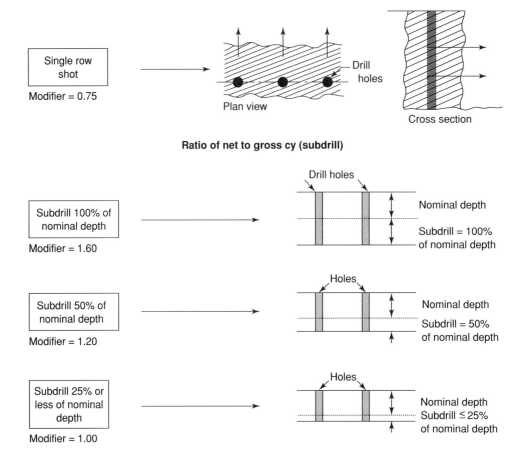

FIGURE 7-5 Powder factor modifiers, continued.

Required Steps to Determine the Required Drill Pattern and Drill Factor

Once the powder factor has been determined, the estimator must make the following four choices for the particular blasting application at hand:

- Lift depth in feet
- Hole size (diameter) in inches
- Subdrill depth in feet
- Type of explosive to be employed

The estimator presumably considered the relationship of subdrill depth to lift depth in determining the powder factor modifier, as explained above. In addition, the estimator must determine the lift depth, the blast hole diameter, and the type of explosive to be employed. For illustration, assume that the estimator chose a lift depth of 12 ft, a subdrill depth of 2 ft, and a 3-in.-diameter hole size and decided to use stick powder with reasonable tamping (somewhat less than maximum tamping possible) as the type of explosive.

Once the above choices have been made, the pounds of explosive required per hole on the basis of loading 70% of the hole can be determined. Figure 7-6 enables a

Bit size (in).	Pounds of explosive per foot loaded		
	Case I	Case II	Case III
1½	0.733	0.409	0.591
1⅝	0.92	0.52	0.72
1¾	1.08	0.53	0.78
1⅞	1.28	0.64	0.98
2	1.44	0.75	1.09
2½	2.33	1.42	1.88
3	3.38	2.33	2.85
3½	4.65	3.38	4.01
4	6.15	4.65	5.40
4½	7.80	6.15	6.98
5	9.75	7.80	8.78
5½	11.90	9.80	10.80
6	14.10	11.90	13.00
6½	16.70	14.10	15.40
7	19.40	16.70	18.00
7½	22.30	19.40	20.80
8	25.40	20.00	24.20
9	32.30	28.70	30.50

Case I: Free running powder or stick powder with tamping (hole completely filled)
Case II: Stick powder ½ in. less in diameter than hole (no tamping)
Case III: Stick powder with some tamping (average of I and II)

FIGURE 7-6   Pounds of explosive per linear foot of hole loaded.

reasonable estimate to be made based on the estimator's four choices stated above. Stick powder with some tamping (less than maximum) is represented on Figure 7-6 as Case III. Figure 7-6 indicates that 2.85 lb of explosive can be expected to be loaded for each foot of a 3-in.-diameter hole under the loading density condition represented by Case III. Therefore, the pounds of explosive per hole for 70% of the hole loaded would be $(12 + 2)(0.70)(2.85) = 27.93$ lb per hole.

The next step is to calculate the number of bank cubic yards that will be broken from each hole blasted. This can be easily done by dividing the number of pounds of explosive loaded per hole by the modified powder factor. Thus, for our example,

$$\text{Bcy broken per hole} = \frac{27.93 \text{ lb per hole}}{1.21 \text{ lb/Bcy}} = 23.08 \text{ Bcy per hole}$$

The estimator must now select the type of pattern to be used. Figure 7-7 indicates two pattern configurations: square and staggered. The radius R indicates the idealized zone of influence of the explosive in each hole, and the distance S indicates the hole spacing along each line of holes. It is clear from the size of the interstices between the circles of blast influence that the staggered pattern represents a more-efficient use of explosives. For the same size circles of blast influence, the distance between adjacent rows of holes decreases from $2R$ (or S) for the square pattern to $1.73R$ or $0.865S$ for the staggered pattern. Based on these facts, the pattern configuration for our example can be computed. Assuming that a staggered pattern is to be used, the horizontal area covered by a single blast hole can be thought of as a rectangle with dimensions of $S \times 0.865S$.

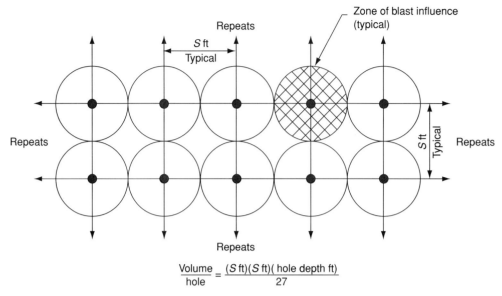

$$\frac{\text{Volume}}{\text{hole}} = \frac{(S\text{ ft})(S\text{ ft})(\text{ hole depth ft})}{27}$$

Square pattern

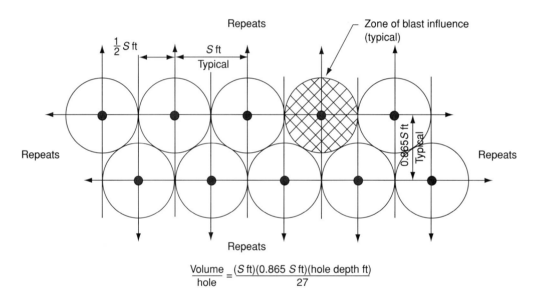

$$\frac{\text{Volume}}{\text{hole}} = \frac{(S\text{ ft})(0.865\ S\text{ ft})(\text{hole depth ft})}{27}$$

Staggered pattern

FIGURE 7-7   Square and staggered blast patterns.

Therefore, the volume of rock broken per blast hole down to the nominal depth of the blast in bank cubic yards is equal to

$$\text{Bcy broken} = \frac{(S\text{ ft})(0.865\ S\text{ ft})\ (12.0\text{ ft})}{27\text{ cf/Bcy}}$$

This expression can then be set equal to the 23.08 Bcy per hole figure previously determined and solved for the value of S. This computation yields the following values:

$$S = 7.75\text{ ft}\quad\text{and}\quad 0.865\ S = 6.70\text{ ft}$$

Thus, the required pattern would be 7.75 ft $\times$ 6.70 ft, staggered.

Finally, the estimator should calculate the drill factor (DF):

$$\text{DF (ft/Bcy)} = \frac{\text{feet drilled per hole}}{\text{Bcy broken per hole}} = \frac{(12 \text{ ft} + 2 \text{ ft})}{23.08 \text{ Bcy per hole}} = 0.61 \text{ ft/Bcy}$$

The above solution is only one of many that might be applied to break the rock while keeping the amount of powder consumed per bank cubic yard (the PF) constant. If the selected blast hole diameter (3 in. in the above example) is changed, the required pattern will change, resulting in the explosive being distributed within the rock mass differently. For instance, if the hole diameter is changed to 2 in., the above-described calculation yields the following results:

Explosives per hole = 10.68 lb

Bcy per hole   = 8.83

S = 4.79 ft, 0.865 S = 4.14 ft for a 4.79 ft $\times$ 4.14 ft staggered pattern

Drill factor   = 1.58 ft/Bcy

Similarly, if the hole size is changed to $1\frac{1}{2}$ in. in diameter, the results are

Explosives per hole = 5.79 lb

Bcy per hole   = 4.79

S = 3.53 ft, 0.865 S = 3.05 ft for a 3.53 ft $\times$ 3.05 ft staggered pattern

Drill factor   = 2.92 ft/Bcy

The above illustration clearly shows that, to get the same amount of explosives into the rock mass to be blasted for the smaller hole sizes, it is necessary to close up the drill pattern to get a sufficient number of linear feet of hole in which to load the powder. With decreasing hole diameter, the pattern dimensions decrease and the linear feet of drilling required per bank cubic yard broken increases dramatically. For instance, use of a $1\frac{1}{2}$-in.-diameter hole requires 2.92/0.61, or nearly 5 times as much drilling as required for a 3-in.-diameter hole. A 2 in. hole size requires 1.58/0.61, or 2.6, times as much drilling. Unless the horizontal dimensions of the rock mass to be broken are small compared to the depth of rock to be removed, using larger-diameter holes with larger hole spacing will result in the most economical operation due to the smaller number and lower total footage of blast holes that must be drilled. Small hole diameters and closely spaced patterns are usually used for narrow, deep excavations such as pipe trenches, small footings, and grade beam trenches.

Determination of Drilling and Explosives Quantity Required for Line Holes

The preceding **analysis for blast holes, performed for the rock mass of a single blast hole,** furnishes the basis for computing all required takeoff quantities to break the rock mass within a rock excavation by drilling and blasting. Controlled blasting work for rock excavations within which concrete structures are intended to be built usually requires that the periphery of the excavation be drilled with closely spaced line holes loaded with presplit explosives to produce a relatively smooth, unshattered surface. To price this part of the drill-and-blast operation, one must determine the size, number, and total linear feet of line holes that must be drilled, as well as the quantity of presplit explosives required. If the horizontal dimensions and depth of the rock excavation are known, these quantities are not difficult to obtain.

The previously described analysis, involving powder factors, drill factors, blasting patterns, and so on, does not apply to determining drilling and explosive quantities for line holes. Rather, all one needs to do is select a hole diameter and hole spacing and then divide the number of feet of periphery length by the hole spacing to obtain the number of line holes. The total number of feet to be drilled is the product of the number of holes required and the average depth. Ordinarily, no subdrill will be required. A few extra holes may need to be added to account for corners and such. The quantity of presplit explosives required is the product of the number of feet of line holes times the weight per foot of the particular explosive utilized (which can be obtained from the explosive supplier). A common line hole diameter is 2 in.; typical spacing varies from 1.5 to 2.0 ft.

For deep excavations that are being blasted in more than one lift, the line holes can be drilled, loaded, and fired in one of several ways. If the excavation is not too deep, the holes may be drilled from the surface to the bottom of the excavation and loaded and fired prior to the drilling and blasting of the rock in the body of the excavation. This procedure spawned the term "presplitting." Presplitting is intended to propagate a vertical crack around the periphery of the excavation so that when the rock from the body of the excavation is removed, it will break off smoothly to the presplit surface. It is difficult to maintain good vertical alignment of presplit holes drilled to depths greater than about 30 ft, so if the excavation is deeper, the presplit operation would have to be carried out in several steps. For instance, if the excavation were 60 ft deep, it might be possible to presplit the first 30 ft, drill and blast and remove the main body of the excavation to that depth, and then presplit the second 30 ft prior to drilling, blasting, and removing the balance of the rock. If this procedure is used, it is necessary to set back the top presplit line 6 to 9 in. from the theoretical boundary line of the excavation to permit enough clearance to properly collar the presplit holes for the second presplit lift. An alternative procedure for multilift excavation is to drill the line holes only to the depth of each lift and then load and fire them concurrently with the main blast in the body of the excavation, using delay detonators so that the line holes fire after the main blast holes had fired. This would provide some relief.

DETERMINATION OF TAKEOFF QUANTITIES FOR PRICING DRILL-AND-BLAST OPERATIONS

Drill-and-blast operations involve no permanent material quantities and ordinarily no subcontract work quantities. Also, drill-and-blast work is typically part of a rock excavation line item or bid item and is not paid separately by the owner, so a takeoff figure for pay quantity is usually not calculated as part of the estimate for drill-and-blast work. Out of the five general categories of takeoff quantities (see Chapter 4), the only two for which quantities must always be developed are the expendable material category and work operations performed by the contractor's own forces category.

Drill Expendable Material Quantities

Quantities must be developed for the number of sectional drill steels, shanks, couplings, and drill bits for each separate-sized drill hole required. In most instances, only two drill hole sizes will be required—one for blast holes and the second (usually smaller in diameter) for line holes. Development of these quantities for use in a typical air-track drill operation is explained below. Refer to the left-hand illustration of Figure 7-8 for typical air-track drill use for drilling deep presplit holes and to the right-hand illustration for drilling typical blast holes.

Sectional drill steels, shanks, and couplers wear out during drilling operations, become hung up and lost in difficult holes, or are otherwise lost. These components can be thought of as having a "life" before any of these events occur, requiring the component to be replaced. Although drill bits are resharpened during use, they too wear out and are lost. Experienced contractors usually maintain records that indicate the average **drill string component life,** expressed in total feet of hole drilled before the component is worn out or lost. The quantity of any one component required for a rock excavation job can then be estimated by the following expression:

Component quantity

$$= \left(\frac{\text{feet of hole required to be drilled}}{\text{Component life, ft}} \right) (\text{average number of components in drill string})$$

For blast hole drilling as illustrated on the right side of Figure 7-8, only one of each component is in use in the drill string at any one time. On the other hand, considering the left-hand illustration for deep presplit holes, the number of each component in use depends on the depth of the hole at any particular point during the operation. For instance, when the hole is collared, only one sectional steel and only one of each other

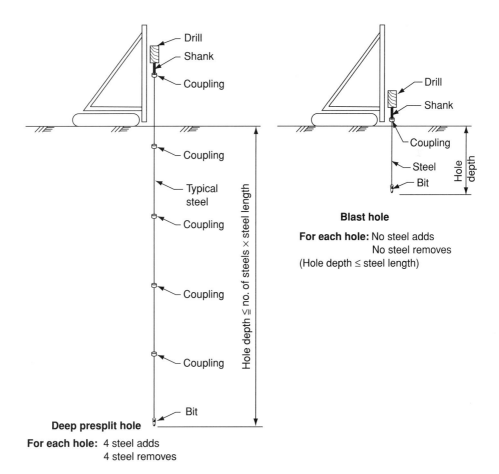

FIGURE 7-8   Deep- and shallow-hole drill string configurations for track drills.

component is in use. However, for the last few feet of hole drilled at depth, the following numbers of each component are in use:

Shanks	1
Sectional drill steels	5
Couplings	5
Drill bit	1

Therefore, the average number in use for each component is as follows:

Shanks   1

Drill steels $\dfrac{1 \text{ (first foot of hole)} + 5 \text{ (last foot of hole)}}{2} = 3$

Couplings $\dfrac{1 \text{ (first foot of hole)} + 5 \text{ (last foot of hole)}}{2} = 3$

Bits   1

Assume that a typical drill-and-blast situation required 69,000 ft of 2 in. line holes requiring five sectional steels to drill them to depth as indicated in Figure 7-8 and 58,000 ft of 3 in. blast holes that can be drilled to the required depth within the length of one sectional steel, as shown on the right-hand illustration of Figure 7-8. Assume further that the contractor's previous experience indicated the following lives for the drill string components under the expected job conditions:

Shank life	1,800 ft
Steel life	1,600 ft
Coupling life	1,300 ft
Bit life	850 ft

Under these circumstances, the required quantities of each component would be

$$\text{Required shanks} \quad = \quad \frac{69{,}000 + 58{,}000}{1800} = 71 \text{ each}$$

$$\text{Required steels} \quad = \quad \frac{(69{,}000)(3) + (58{,}000)(1)}{1{,}600} = 166 \text{ each}$$

$$\text{Required couplings} \quad = \quad \frac{(69{,}000)(3) + (58{,}000)(1)}{1{,}300} = 204 \text{ each}$$

$$\text{Required 2 in. bits} \quad = \quad \frac{69{,}000}{850} = 82 \text{ each}$$

$$\text{Required 3 in. bits} \quad = \quad \frac{58{,}000}{850} = 69 \text{ each}$$

Linear Feet of Drill Holes Required

Note that, to compute drilling expendable materials, the estimator must know how many linear feet of the various sized drill holes will be required. How does one determine this figure? The number of linear feet of line holes, as explained earlier, can be determined from the construction drawings by dividing the periphery length of the rock excavation by the line hole spacing and multiplying by the average depth of line hole. To obtain the linear feet of required blast holes, one must first perform the single-hole drill-and-blast analysis described earlier and determine the volume of rock excavation in bank cubic yards by a takeoff from the construction drawings.

In the case of excavations in the form of common geometrical shapes, such as square or rectangular openings or long narrow openings of constant width such as ditches, this determination is a simple matter of multiplication utilizing the governing dimensions. For more complicated excavations, one usually obtains the volume of excavation within the lines defined on the construction drawings by average end area volume computations utilizing either horizontal sections on the contour lines or vertical cross sections. Once the total volume of required excavation is known, the total linear feet of blast holes may be obtained by simply multiplying the rock excavation volume in bank cubic yards by the drill factor in feet per bank cubic yards obtained from the analysis performed for a single typical blast hole.

As explained previously, the single blast hole analysis involves subjective decisions by the estimator for the number of lifts, nominal lift depth, and subdrill depth to be utilized. Complicated excavations may have several typical situations in which the number of lifts, nominal depth, and subdrill depth vary. In such cases, the total linear feet of blast holes would equal the sum of the linear feet required for each of the several typical situations, which must be separately computed. Note that when computing quantities of drill expendable materials, situations can arise when the lift depth is such that more than one sectional steel is required to bottom out the blast holes. In this situation, the average number of steels and couplings in use at one time must be recognized in the same manner as when deep line holes are being drilled.

Explosives Expendable Material Quantities

Quantities that might have to be determined for explosives required in a particular situation could include

- Primary blasting agent measured in pounds, which may consist of either straight dynamite, gelatin dynamite, ANFO, slurries, or other special explosives, or combinations thereof
- Explosives to be used as primer cartridges measured in pounds, usually single sticks of straight dynamite
- Blasting caps, either EBCs or non-el caps
- Presplit explosives, measured in pounds or in linear feet, depending on how the explosive is quoted by the supplier
- Primacord, measured in linear feet

Determination of Primary Explosive and Primer Cartridge Quantities

Like the procedure for computing drill supplies quantities, the first step in determining explosive and primer cartridge quantities is to determine the total quantity in bank cubic yards of excavation to be blasted. Once this figure is known, the total quantity of explosives required can be calculated by multiplying the rock quantity figure by the previously determined modified powder factor:

Total explosives required (lb) = total rock quantity to be broken (Bcy) $\times$ PF (lbs/Bcy)

Blast holes are usually primed with one primer cartridge per hole placed in the bottom of the hole, which in turn is detonated by either a blasting cap or by a run of primacord wrapped around and securely tied to the primer cartridge. Deep blast holes may be primed with two primer cartridges, one at the bottom of the hole and the second at the top of the explosive column. Thus, the number of primer cartridges required equals the total number of blast holes times the number of primer cartridges per blast hole. The total number of blast holes is easily determined by dividing the total linear feet of blast holes required by the typical hole depth (including the subdrill). In other words,

$$\text{Number of blast holes} = \frac{\text{total blast hole quantity (ft)}}{\text{typical blast hole depth (ft)}}$$

If several different nominal lift situations are involved, the estimator must determine the number of blast holes for each, as explained above and then add them together to yield the total number required for the entire excavation. Then the estimator can calculate the total quantity of primer cartridges required in pounds by multiplying the number of required blast holes by the weight of a single cartridge of the type of stick powder to be used. This cartridge weight in pounds is determined by dividing 50 lb by the stick count:

$$\text{Primer cartridge quantity (lb)} = \text{no. of blast holes (each)} \times \frac{50 \text{ lb}}{\text{stick count}}$$

Once the quantity of primer cartridges is known, the required quantity of primary explosive is determined by subtracting the primer quantity from the total required explosive quantity:

$$\text{Primary explosive quantity (lb)} = \text{total explosive quantity (lb)} - \text{primer cartridge quantity (lb)}$$

Note that the primary explosive could be in the form of straight dynamite, gelatin dynamite, ANFO, slurries, or other special explosives.

Determination of Presplit Explosive Quantity

If the presplit explosive quantity is to be reckoned in linear feet, the required quantity is the same as the total length of presplit holes required for the excavation. If the explosive quantity is to be obtained in pounds, it is simply the total feet of required presplit holes times the unit weight per foot for the particular presplit explosives to be utilized. The unit weight can be obtained from the explosive supplier.

Determination of Required Blasting Cap Quantity

The quantity of blasting caps required for blast holes is the same as the number of primer cartridges, previously calculated. If primacord is to be used to detonate blast holes, blasting caps are not required.

Presplit explosives must also be detonated, either by a blasting cap at the top of the hole or by a primacord connection. If blasting caps will be used, the number required is simply the total number of presplit holes. If primacord is used, blasting caps will not be required.

Determination of Primacord Quantity

If primacord is to be used for priming blast holes, the required quantity can be taken as the total linear feet of required blast holes plus an allowance, say 10%, for horizontal runs and connections. The primacord quantity required for detonating presplit holes is approximately equal to the total periphery of the rock excavation to be presplit plus a small allowance for connections and other odds and ends.

Determination of Quantities for Work Operations Performed by the Contractor's Own Forces

The typical work operations required for drill-and-blast work include the following:

- Drill line holes for presplitting
- Drill blast holes
- Load and fire presplit holes
- Load and fire blast holes

Note that for large excavations the rock from each blast usually must be removed following the blast to clear the area for drilling out the following blast. Sometimes, for excavations of large horizontal extent, the drilling operations may progress ahead, followed some distance behind by loading and the blasting of the rock and intermittent removal of blasted rock between blast operations. These matters depend on the size and shape of the particular rock excavation and how the contractor decides to schedule the performance of the work.

Drill Line Holes for Presplitting

This work operation is required when the line holes are independently drilled and fired prior to drilling the blast holes in the main body of the excavation. In these situations, the line holes are usually a smaller diameter than the blast holes. In excavations where the line holes are drilled along with the blast holes to the same depth as the blast holes and fired in the same blast by the use of delay caps, the line holes usually will be the same diameter as the blast holes, and drilling them can be considered part of the work operation of drilling blast holes. In either case, both the number of individual holes and the total required drill footage would have previously been determined when the expendable material quantities were calculated, and no additional takeoff computation work is required.

Drill Blast Holes

As was the case for the presplit holes, the number of individual blast holes and the total required linear feet would have been previously determined when the expendable material quantities were calculated, and no additional takeoff computation work is required.

Load and Fire Presplit Holes

This work quantity is usually reckoned in the number of individual holes the powder crew must load, hook up, and fire. Alternately, the quantity may be reckoned as the number of linear feet of holes to be loaded and fired. In either case, the quantities are known from previously calculated expendable materials quantities.

Load and Fire Blast Holes

Like the presplit holes, the work quantity may be reckoned in either the number of individual holes to be loaded and fired or in the total number of linear feet. All quantities are known from previous calculations for expendable material quantities.

ESTIMATING DRILL CREW PRODUCTIVITY

Drill crews generally employ more than one drill, so crew productivity will depend on the average productivity achieved by each individual drill and the number of drills utilized. For open-cut rock excavation, either pneumatic or hydraulically actuated, track drills are typically employed. The rate at which such drills advance the drill hole depends on the type of drill used (pneumatic or hydraulic), the hole diameter and depth, the type and strength of rock being penetrated, and the general conditions of the work environment. Drill manufacturers furnish product literature to contractors that provides estimated penetration rates for their equipment for various sized holes in commonly encountered rocks. For large projects, technical representatives will also usually provide more specific advice. In general terms, instantaneous penetration rates of pneumatic drills range from $1\frac{1}{2}$ to $2\frac{1}{2}$ ft/min in the common rocks, depending on the hole diameter. These figures increase to $4\frac{1}{2}$ to 8 ft/min for hydraulically actuated drills. The general approach to estimating drill crew productivity, is best understood by reference to Figures 7-8 and 7-9.

Drill Cycle Time

The drill cycle time is the time in minutes required for the production of one complete drill hole. It is composed of the following four subcomponents.

Collar Time

The collar time is the time required to reposition the drill following completion of a drill hole to the location of the next hole to be drilled, engaging the ground with the drill string, and starting the hole. It ranges from less than 1 min for closely spaced holes to as much as 5 min (or more) for widely spaced holes in difficult terrain.

Drill Penetration Time

The drill penetration time is the time duration when the drill is advancing the hole through the rock. For a particular hole, this duration is equal to the hole depth in feet divided by the drill instantaneous penetration rate in feet per minute.

To Obtain:	Given By:
Drill cycle time, min	Collar time, min + (Hole depth, ft)/(Instantaneous penetration rate, ft/min) + (No. of steel adds)(Steel add time, min) + (No. of steel removes)(Steel remove time, min)
Drill productivity, ft/drill-hr	$\left(\dfrac{60 \text{ min/hr}}{\text{Drill cycle time, min}}\right)$(Hole depth, ft)(Efficiency, %)

FIGURE 7-9   Track drill productivity relationships.

Steel Add Time

Steel add time is the time required to add additional drill steels when the hole is too deep to be bottomed out within the length of a single drill steel. It includes the time required for the drill crew to disengage the hammer from the top of the steel near the ground surface, raise the hammer to the top of the mast, install a new drill steel in the string, set the hammer down and engage the top of the drill string, and commence drilling. This time ranges from about 1 min to as long as 2 or 3 min, depending on the terrain. The total duration for a single hole equals the number of steels that must be added times the time required to add a single steel. Note that in Figure 7-8, for the deep presplit hole illustrated, only four steel adds will be required even though five steels are required to bottom out the hole.

Steel Remove Time

Like steel add time, steel remove time is the total time required to remove steels following bottoming out the hole until only one steel remains prior to moving the drill to the next hole. It equals the number of steels that must be removed times the time required to remove one steel. The time required to remove a steel exceeds the time required to add one because the entire drill string must be supported once the drill hammer is disengaged from the drill string. Today, state-of-the-art drilling equipment is fitted out with attachments on the mast that greatly reduce the time required to add and remove drill steels.

Drill Productivity

Drill productivity, measured in feet per drill-hour, is calculated using the following formula:

Productivity (ft/drill-hr)

$$= \frac{60 \text{ min/hr}}{\text{drill cycle time (min per hole)}} \times \text{hole depth (ft)} \times \text{drill crew efficiency (\%)}$$

The drill crew efficiency term varies widely, depending upon the site conditions under which the holes are being drilled. A drill operation on level ground, without undue obstructions or interference, should range from 75 to 85%, whereas crew efficiency could be as low as 50 to 65% when working on a steep, uneven side hill or in other difficult circumstances.

   Drill-crew productivity is the individual drill productivity in feet per drill-hour times the number of drills employed in the drill crew.

ILLUSTRATIVE EXAMPLE FOR A DRILL-AND-BLAST OPERATION

A simple drill-and-blast operation illustrating the foregoing principles is depicted in Figure 7-10, showing a sizable deep excavation that might be required for a large concrete structure such as a subway station. The plan view and typical cross section shown reflect conditions from top of rock down after removal of any overlying common material.

   The presplit lines would have to be set back as shown on the cross-section sketch. The plan is to utilize 2-in.-diameter holes on 1 ft 6 in. centers, loaded with commercial presplit powder, and to carry out all presplitting prior to drilling and blasting the adjacent level of rock excavation. Note that the cross section indicates that the presplitting would be done in two stages all the way around the excavation. The first stage would

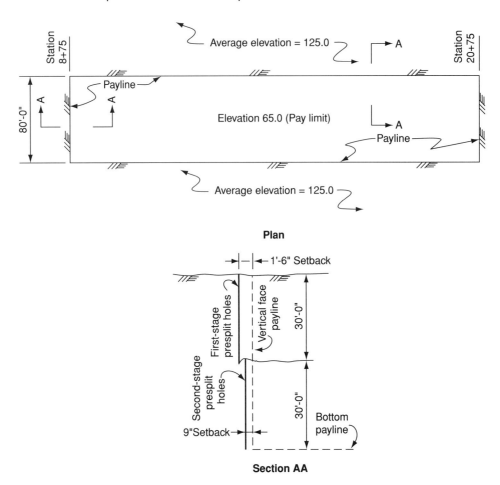

FIGURE 7-10   Example rock excavation for illustrative drill-and-blast estimate.

be 30 ft deep, and the second 30 ft section would then be presplit to the bottom of the excavation after removal of the main body of the first-stage excavation.

Assume further that the blast holes were to be 3 in. in diameter, loaded to 70% of their total depth. The total 60 ft of excavation is to be taken out in four 15 ft lifts with a 2 ft subdrill for each lift. The rock is limestone with a Shore Scleroscope hardness of 85. A subcontractor who insists that the rock be broken to a smaller than a $2\frac{1}{2}$ cy bucket size will perform the removal of blasted rock. The estimator has chosen to use ANFO as the primary blasting explosive and to prime each hole with one stick of straight dynamite at the bottom of the hole. The stick count for the straight dynamite is 105.

All holes are to be drilled using pneumatic track drills with $1\frac{1}{2}$-in.-diameter by 16-ft-long sectional steels. Prior experience indicates a steel life of 1,800 ft, a shank life of 2,000 ft, a coupler life of 750 ft, and a bit life of 750 ft.

Assume an instantaneous drill penetration rate of $2\frac{1}{2}$ ft/min for the 2-in.-diameter presplit holes and 2 ft/min for the 3-in.-diameter blast holes. Allow the following time durations for the indicated components of the drill cycle time:

- Time required to move, drill, and collar hole, 1 min for 2 in. presplit holes and 2 min for 3-in.-diameter blast holes
- Time required to add an additional steel, $1\frac{1}{2}$ min
- Time required to remove a steel, 3.0 min

Also consider a drill crew efficiency of 75% for presplit holes and 60% for blast holes.

PHOTO 7-5 The upper photo shows presplit powder being loaded into deep vertical line holes similar to those required for the example problem in the text. The second shows the quality of the resulting vertical wall (in a tunnel boring machine recovery shaft) to be expected after excavation. Note the one-half hole pattern in the resulting surface.

Quantity Takeoff Calculations

Before making the detailed quantity calculations, it is best to perform the analysis for a single blast hole required to break the rock in the excavation mass. The required steps are shown below:

> Unmodified PF = 0.80 lb/Bcy (from Figure 7-3)
>
> Fragmentation less than $2\frac{1}{2}$ cy, modifier $\qquad$ = 1.15 (from Figure 7-4)
>
> One free face,[3] modifier $\qquad$ = 1.00 (from Figure 7-4)
>
> Subdrill, modifier $\qquad$ = 1.00 (from Figure 7-5)
>
> Modified PF = (0.80 lb/Bcy)(1.15)(1.00)(1.00) = 0.92 lb/Bcy

The primary blasting explosive is ANFO, which is a free-running powder. This corresponds to Case I in Figure 7-6, indicating a loading density of 3.38 lb/ft.

$$\text{Explosives per hole} = (17 \text{ ft})(0.70)\,(3.38 \text{ lb/ft}) = 40.22 \text{ lb}$$

$$\text{Rock broken per hole} = \frac{40.22 \text{ lb per hole}}{0.92 \text{ lb/Bcy}} = 43.72 \text{ Bcy}$$

For a staggered pattern, S computes out to 9.54 ft, and 0.865 S = 8.25 ft. Therefore, a 9.54 ft by 8.25 ft staggered pattern will be used for the 17-ft-deep, 3-in.-diameter holes, to be 70% loaded:

$$\text{Drill factor} = \frac{17 \text{ ft}}{43.72 \text{ Bcy per hole}} = 0.39 \text{ ft/Bcy}$$

1. **Quantity of rock to be blasted:** This quantity should be determined out to the presplit lines and to an average of 12 in. below the bottom pay line to allow for reasonable overbreak beneath the bottom of the excavation:

$$\text{Top 30 ft} = \frac{(1203 \text{ ft})(83 \text{ ft})(30 \text{ ft})}{27} = 110{,}943 \text{ Bcy}$$

$$\text{Bottom 30 ft} = \frac{(1{,}201.5 \text{ ft})(81.5 \text{ ft})(31 \text{ ft})}{27} = 112{,}429 \text{ Bcy}$$

$$\text{Total} = 110{,}943 + 112{,}429 = 223{,}372 \text{ Bcy}$$

2. **Stick powder required for primer cartridges:** First, the total explosive quantity must be determined:

$$\text{Total explosives} = (223{,}372 \text{ Bcy})(0.92 \text{ lb/Bcy}) = 205{,}502 \text{ lb}$$

then,

$$\text{Total blast holes} = \frac{223{,}372 \text{ Bcy}}{43.72 \text{ Bcy per hole}} = 5{,}109 \text{ each}$$

---

[3]The first 80-ft-wide section of excavation removed from each lift will have no free faces. However, each additional 80 ft section removed in the 1,200 ft length of the excavation will have one free face. Typical sections will be about 50 ft long in the long direction of the excavation. Therefore, the predominate condition will be one free face.

and,

$$\text{Total stick powder quantity} = \frac{(5,109)(50 \text{ lb})}{105} = 2,433 \text{ lb}$$

3. **Total ANFO quantity:** Total ANFO required equals total required explosive quantity minus stick powder quantity, or,

$$\text{Total ANFO quantity} = 205,502 \text{ lb} - 2,433 \text{ lb} = 203,069 \text{ lb}$$

4. **Presplit explosive quantity:** Total holes $= (1,203 \times 2)/1.5 + (83 \times 2)/1.5 + (1,201.5 \times 2)/1.5 + (81.5 \times 2)/1.5 = 1,604 + 111 + 1,602 + 109 = 3,426$

$$\text{Presplit explosive quantity} = (3,426)(30 \text{ ft}) = 102,780 \text{ lf}$$

If the presplit explosive had a unit weight of 1.1 lb/lf, the quantity in pounds would be $(102,780 \text{ lf})(1.1 \text{ lb/lf}) = 113,058 \text{ lbs}$.

5. **Blasting cap quantity:** Assume EBCs are used, that each blast hole is single primed, and that each presplit hole is detonated by single cap:

$$\text{Total EBC quantity} = \text{total holes} = 5,109 + 3,426 = 8,535 \text{ each}$$

6. **2-in. bits:** Only one bit is required in the drill string at all times:

$$2 \text{ in. bit quantity} \quad = \frac{(3,426 \text{ holes})(30 \text{ ft per hole})}{750 \text{ ft per bit}} = 137 \text{ each}$$

7. **3-in. bits:** Only one bit is required in the drill string at all times:

$$3 \text{ in. bit quantity} \quad = \frac{(5,109 \text{ holes})(17 \text{ ft per hole})}{750 \text{ ft per bit}} = 116 \text{ each}$$

8. **Shanks:** Only one shank is required in the drill string:

$$\text{Shanks quantity} = \frac{(3,426 \text{ holes})(30 \text{ ft per hole}) + (5,109 \text{ holes})(17 \text{ ft per hole})}{2,000 \text{ ft per shank}}$$

$$= \frac{102,780 + 86,853}{2,000} = 95 \text{ each}$$

9. **Couplings:** In this case, the couplings must be figured separately for presplit holes and blast holes because more than one coupling is required for the presplit holes:

$$\text{Average number in use for presplit holes} \quad = \frac{(1 + 2)}{2} = 1.5$$

$$\text{Presplit hole quantity} \quad = \frac{(102,780 \text{ ft})(1.5)}{1,500 \text{ ft per coupler}} = 103 \text{ each}$$

$$\text{Blast hole quantity} \quad = \frac{86,853 \text{ ft}}{1,500 \text{ per coupler}} = 58 \text{ each}$$

$$\text{Total coupler quantity} = 103 + 58 \quad = 161 \text{ each}$$

10. **Drill steel quantity:** Again, separate calculations must be made for the presplit holes and for the blast holes because more than one steel is required in this string to bottom out the presplit hole:

$$\text{Average steels in use for presplit holes} = \frac{(1 + 2)}{2} = 1.5$$

$$\text{Steels required} = \frac{(102{,}780 \text{ ft})(1.5)}{1{,}800 \text{ ft per steel}} = 86$$

$$\text{Blast hole steels} = \frac{86{,}853 \text{ ft}}{1{,}800 \text{ ft per steel}} = 49$$

$$\text{Total steel quantity} = 86 + 49 = 135 \text{ each}$$

11. **Drill presplit hole work quantity:** This quantity is the same length of presplit holes utilized for calculating the quantity of presplit explosives, or 102,780 ft.

12. **Load and fire presplit holes work quantity:** This quantity is the total number of presplit holes, 3,426 holes $\times$ 30 ft deep, or 102,780 ft.

13. **Drill blast hole work quantity:** This quantity is the number of feet of blast holes utilized for calculating the drill bit and steel quantities, or 86,853 ft.

14. **Load and fire blast hole work quantity:** This quantity is the number of previously computed blast holes, or 5,109 holes.

Determining Crew Productivities

Typical crews for the drilling operations and the load and fire operation would consist of the following equipment and personnel:

Drill Crew

Pneumatic track drills	2 each
1200 cfm compressor	2 each
$2\frac{1}{2}$ ton flatrack truck	$\frac{1}{2}$ (truck shared with powder crew)
Drillers (one is the crew foreman)	2 each
Chuck tender	1 each

Powder Crew

Powder foreman	1 each
Powdermen	2 each
$2\frac{1}{2}$ ton flatrack truck	$\frac{1}{2}$ (truck shared with drill crew)

The drill crew productivity may be estimated as follows for the basic data furnished.

Presplit Hole Drilling

$$\text{Hole cycle time} = 1.0 \text{ min} + \frac{30 \text{ ft}}{2.5 \text{ ft/min}} + 1.5 \text{ min} + 3.0 \text{ min}$$

$$= 17.5 \text{ min per hole}$$

$$\text{Drill productivity} \ = \frac{(60 \text{ min})(30 \text{ ft per hole})(0.75)}{17.5 \text{ min per hole}} = 77.1 \text{ ft per drill-hour}$$

$$\text{Crew productivity} = (2 \text{ drills})(77.1 \text{ ft per drill-hour}) = 154.2 \text{ ft per crew-hour}$$

Blast Hole Drilling

$$\text{Hole cycle time} \ = 2.0 \text{ min} + \frac{17 \text{ ft}}{2 \text{ ft/min}} = 10.5 \text{ min per hole}$$

$$\text{Drill productivity} \ = \frac{(60.0 \text{ min})(17 \text{ ft per hole})(0.60)}{10.5 \text{ min per hole}} = 58.3 \text{ ft per drill-hour}$$

$$\text{Crew productivity} = (2 \text{ drills})(58.3 \text{ ft per drill-hour}) = 116.6 \text{ ft per crew-hour}$$

Presplit Load and Fire

The typical powder crew suggested above should be able to load and fire 300 ft of presplit hole per labor-hour. The powder foreman is a working foreman, so the crew consists of three people. Therefore, the crew productivity is $3 \times 300 = 900$ lf per crew-hour.

Load and Fire Blast Holes

The powder crew should be able to load and fire 10 blast holes of the size and depth in this example per labor-hour when using ANFO. On this basis, the crew productivity is $(3)(10) = 30$ blast holes per crew-hour.

Cost Estimate

The above takeoff quantities can be converted into dollars and cents costs based on labor, equipment, and expendable libraries that have been set up for the project by the pricing-out process discussed in Chapter 6. In addition to the information developed above, the estimator must also determine the working conditions under which the work will be performed and must determine a reasonable small tools and supplies (ST&S) percentage to be applied to the direct labor estimate. For instance, the estimator might choose to plan the performance of the work on two 9.0 hr shifts per work day, 5 work days per week (9.0/2/5). With overtime at 1.5/1.5/2, these working conditions result in a labor premium percentage of 9.72% (see Chapter 6). A 7.5% ST&S allowance would be appropriate.

The drill-and-shoot work in the foregoing example is usually required as part of the work of a rock excavation bid item, which also includes the loading, hauling, and disposal of the blasted rock. In this illustration, a subcontract price had been secured for this work.

To complete the illustration of the complete drill-and-blast estimating process, including the dollars and cents cost estimate, assume that the excavation illustrated was part of a rock excavation bid item described as follows:

Bid item 12   Rock excavation   223,500 Bcy at $ _____ per Bcy = $_____

Assume further, that the specification measurement and payment provisions state that only excavation within the pay lines shown will be measured for payment and that payment will be made at the bid price per bank cubic yards for the complete work of presplitting, blasting, and excavating the rock.

BI description _Rock excavation_   Unit _Bcy_   T.O. quantity _213,334_   BI# _12_

Bid quantity _223,500_

Code	Description	Unit	Quantity	Crew EOE	Crew RL	Crew Rental	Crew Labor	Required hr	EOE	RL	Rental	Labor	EM	PM	Subs	Totals
1	_Drill materials_															
	2 in. bits (EM)	ea.	137				75.00						10,275			10,275
	3 in. bits (EM)	ea.	116				175.00						20,300			20,300
	Shanks (EM)	ea.	95				150.00						14,250			14,250
	Couplings (EM)	ea.	161				60.00						9,660			9,660
	16 ft drill steels (EM)	ea.	135				325.00						43,875			43,875
2	_Blasting materials (EM)_															
	Gelatin dynamite (EM)	lb	2,433				1.35						3,285			3,285
	ANFO (EM)	lb	203,069				0.22						44,675			44,675
	Presplit powder (EM)	lf	102,780				1.75						179,865			179,865
	EBC (EM)	ea.	8,585				1.25						10,669			10,669
3	_Drill presplit holes_															
	102780 linear feet 9.0/2/5															
	159.2 linear ft/hr 667 hr															
	EQ 59 Airtrack drill	ea.	2	17.610	2.220	10.110		1,334	23,492	2,961	13,487					39,940
	EQ 63 1200 Compressor	ea.	2	26.090	1.610	13.670		1,334	34,804	2,148	18,246					55,198
	EQ 81 Flatbed	ea.	0.5	7.520	1.390	5.530		334	2,512	464	1,847					4,823
	LAB 8 Foreman	ea.	1				33.070	667				22,058				22,058
	LAB 5 Driler	ea.	1				29.840	667				19,903				19,903
	LAB 6 Chucktender	ea.	1				28.140	667				18,769				18,769
	ST&S @ 7.5%	$	60,730											4,555			4,555
	Labor Premium @ 9.72%	$	60,730									5,903				5,903

FIGURE 7-11 Cost estimate for rock excavation bid item 12, sheet 1 of 3.

BI description _Rock excavation_   Unit _Bcy_   T.O. quantity _213,334_   BI# _12_   Bid quantity _223,500_

Code	Description	Unit	Quantity	Crew EOE	Crew RL	Crew Rental	Crew Labor	Required hr	EOE	RL	Rental	Labor	EM	PM	Subs	Totals
4	_Load & Fire Presplit holes_															
	102,780 LF 9.0/2/5															
	900 lf/hr 114 hr															
	EQ 81 Flatbed	ea.	0.5	7.52	1.39	5.53		57	429	79	315					823
	LAB 9 Powder-foreman	ea.	1				31.07	114				3,542				3,542
	LAB 7 Powderman	ea.	2				29.84	228				6,804				6,804
	ST&S @ 7.5%	$	10,346										776			776
	Labor Premium @ 9.72%	$	10,346									1,006				1,006
5	_Drill blast holes_															
	86,853 lf 9.0/2/5															
	116.6 lf/hr 745 hr															
	EQ 59 Airtrack	ea.	2	17.610	2.320	10.110		1,490	26,239	3,457	15,064					44,760
	EQ 63 1200 Compressor	ea.	2	26.090	1.610	13.680		1,490	38,874	2,399	20,383					61,656
	EQ 81 Flatbed	ea.	0.5	7.520	1.390	5.530		373	2,805	518	2,063					5,386
	LAB 8 Foreman	ea.	1				33.070	745				24,637				24,637
	LAB 5 Drillers	ea.	1				22.840	745				22,231				22,231
	LAB Chucktender	ea.	1				28.140	745				20,964				20,964
	ST&S @ 7.5%	$	67,832										5,087			5,087
	Labor Premium @ 9.72%	$	67,832									6,593				6,593
6	_Load & Fire Blast Holes_															
	5109 ea. 9.0/2/5															
	30 ea./hr 171 hr															
	EQ 81 Flatbed	ea.	0.5	7.520	1.390	5.550		86	647	120	476					1,243
	LAB 9 Powder-foreman	ea.	1				31.070	171				5,313				5,313
	LAB 7 Powderman	ea.	2				29.840	342				10,205				10,205
	ST&S @ 7.5%	$	15,518										1,164			1,164
	Labor Premium @ 9.72%	$	15,518									1,508				1,508

FIGURE 7-12   Cost estimate for rock excavation bid item 12, sheet 2 of 3.

BI description *Rock excavation*   Unit *Bcy*   T.O. quantity 213,334

BI# 12

Bid quantity 223,500

Code	Description	Unit	Quantity	Crew EOE	Crew RL	Crew Rental	Crew Labor	Required hr	EOE	RL	Rental	Labor	EM	PM	Subs	Totals
7	*Load & haul blasted rock (sub)*	*Bcy*	*223,732*				*7.50*								*1,677,990*	*1,677,990*
	Total Takeoff	*Bcy*	*213,334*						*129,802*	*12,146*	*71,881*	*169,436*	*348,436*		*1,677,990*	*2,409,691*
	Total Bid item	*Bcy*	*223,500*						*135,994*	*12,725*	*75,310*	*177,519*	*365,056*		*1,758,080*	*2,524,634*
	Unit Costs		*—*						*0.608*	*0.057*	*0.337*	*0.794*	*1.633*		*7.866*	*11.295*
	Prorate Factor	=	*223,500*													
		=	*213,334*													
			1.0477													

FIGURE 7-13 Cost estimate for rock excavation bid item 12, sheet 3 of 3.

On the basis of the above, the estimator must calculate a takeoff figure for the bid-item pay quantity according to the stated measurement provisions. According to the pay lines shown on Figure 7-9, the pay quantity takeoff figure is

$$\text{Pay quantity} = \frac{(1200 \text{ ft})(80 \text{ ft})(60 \text{ ft})}{27 \text{ ft/Bcy}}$$

$$= 213{,}334 \text{ Bcy}$$

Using the labor, equipment, expendable material and subcontract libraries included in the Appendix to this book, the cost estimate for the rock excavation bid item 12, most of which consists of the drill-and-blast work, is shown on Figures 7-11, 7-12, and 7-13. The italicized figures are the handwritten entries that would be made, extended, and totaled by the estimator for a manual estimate (see Chapter 6). These italicized figures would also appear on the printout for a computer-aided estimate.

CONCLUSION

This chapter, the first in a series on the subject of operations analysis for various types of construction operations, illustrated the technology involved and the takeoff procedures necessary to price the drilling and blasting of rock excavation.

The following chapter will deal with a related subject, applicable to excavation load and haul operations in either common materials or rock for projects consisting of a linear succession of excavations (cuts) and embankments (fills). Such a succession of cuts and fills is common in railroad and freeway construction, canals, and similar civil works structures.

QUESTIONS AND PROBLEMS

1. What alternative method of breaking and loosening rock prior to excavation other than by drilling and blasting was mentioned in this chapter?

2. What are the generic names of drill-and-blast crew members and, for work performed under collective bargaining agreements, what two unions are involved?

3. What range of hole diameters are typical for track drills? What maximum hole depths? What are typical instantaneous penetration rate ranges for $2\frac{1}{2}$- to 3-in.-diameter drill holes for present day pneumatically and hydraulically powered track drills, respectively?

4. What three types of hand-held drills were mentioned in this chapter?

5. What are the four components of the drill string utilized by track drills discussed in this chapter?

6. Define the following terms as they pertain to drill holes:
   (a) Nominal depth   (b) Subdrill
   (c) Blast hole   (d) Line hole
   (e) Drill pattern   (f) Drill factor

7. Define the following terms as they pertain to explosives used in drill-and-blast work:

(a) Dynamite
(b) Gelatin dynamite
(c) Ammonium nitrate fuel oil mixture (ANFO)
(d) Blasting slurries
(e) Presplit explosives
(f) Electric blasting caps (EBC)
(g) Non-el blasting caps
(h) Primacord
(i) Bottom loading
(j) Deck loading
(k) Stick count
(l) Loading density
(m) Loaded depth
(n) Collar depth
(o) Percent hole loaded
(p) Burden
(q) Stemmig
(r) Powder factor

8. What three modifying considerations for the powder factor determination were discussed in this chapter?

9. What four choices must the estimator make to determine the required drill pattern and the drill factor? Name two types of drill patterns. Which is more commonly used?

10. A long rectangular excavation in rock is required for a project as shown below.

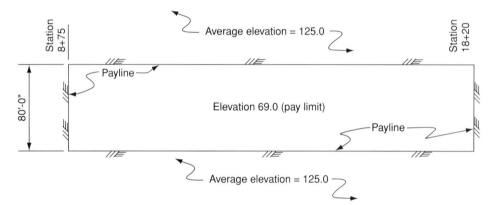

The schedule of bid items describes the work as follows:

Bid Item 12   Rock Excavation   162,500 Bcy at $ _____ per Bcy = $ _____

The measurement and payment provisions state that all excavation within the pay lines will be measured for payment and that payment will be made at the bid price per cubic yard for the complete work of presplitting, blasting, and excavating the rock.

Presplit Requirements

Presplit line to be set as shown on cross-section sketch
2-in.-diameter holes on 15 in. centers
Use presplit powder, full depth of holes
Complete all presplitting prior to starting drilling for rock excavation

Blast Hole Requirements

3-in.-diameter holes, 70% loaded depth
Four each, 14 ft lifts for the total 56 ft depth
2 ft subdrill (for each lift)

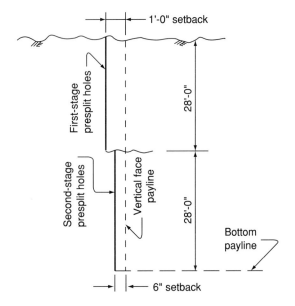

Rock is limestone with 60 hardness.

Fragmentation smaller than $2\frac{1}{2}$ cy size required

Use ANFO for the blasting agent.

Prime each hole with one stick gelatin dynamite.

Stick count for powder = 105 sticks per 50 lb box.

Drill Bit and Steel Information

Use 16 ft, $1\frac{1}{2}$-in.-diameter steels.

Steel life, 3,200 ft

Shank life, 2,500 ft

Bit life, 600 ft

Coupler life, 3,000 ft

Productivity Information

Instantaneous drill penetration rate is 3.0 ft/min (2-in.-diameter bit), 2.5 ft/min (3-in.-diameter bit).

Time required to move over a new hole and collar hole, 1 min (presplit holes) and 2.5 min (blast holes).

Time required to add an additional steel is 1.0 min.

Time required to remove a steel is 2.5 min.

Drill crew efficiency is 75% (presplit drilling) and 60% (blast hole drilling).

Work Crew Compositions and Productivities

Drill crew

2 Air track drills (ATDs)

2 1200 cfm compressors

2 Drillers (one is the foreman)

1 Chucktender

$\frac{1}{2}$ Flatrack truck

Productivity depends of whether drilling presplit or blast holes.

Powder Crew

1 Powder foreman	Can load and fire ten blast holes per labor-hour when using ANFO.
2 Powdermen	
$\frac{1}{2}$ Flatrack truck	Can load and fire 300 lf of presplit hole per labor-hour.
	All labor-hour figures include making up primers and take efficiency into account.

For the problem stated above, determine required takeoff quantities as follows:

(a) Quantity of blasted rock to be loaded and hauled by a subcontractor (actual quantity, out to the presplit lines and to an average of 12 in. below bottom pay line).
(b) Gelatin dynamite (lb)
(c) Ammonium nitrate (lb)
(d) Presplit powder (lf)
(e) EBC (each)
(f) 2-in.-diameter bits (each)
(g) 3-in.-diameter bits (each)
(h) Shanks (each)
(i) Couplings (each)
(j) Drill steels (each)
(k) Pay quantity within the neat excavation lines shown
(l) Drill presplit holes (lf)
(m) Load and fire presplit holes (lf)
(n) Drill blast holes (lf)
(o) Load and fire blast holes (each)

Prepare a direct cost estimate for bid item 12 using the CSU, Chico, estimating program. Use a total of six operations as follows:

1 Drill materials
2 Blast materials and subs
3 Drill presplit holes
4 Load and fire presplit holes
5 Drill blast holes
6 Load and fire blast holes

Cost the work on the following work hours and overtime premium basis:

Work hours, 10/2/5
Overtime premium, 1.5/1.5/2

Use $7\frac{1}{2}\%$ of direct labor for the ST&S allowance.

8

Mass Diagrams

Key Words and Concepts

In situ natural soil condition
Compacted soil condition in fill
Where is the dirt?
Profile drawing
Cross-sections
Slope lines
Profile grade
Mass diagram
Primary balance stations
Borrow excavation

Waste excavation
Haul
Take the short hauls first
Auxiliary balance lines
Consideration of waste/borrow excavation to
    minimize haul
Theoretical average haul distance
Haul distance approximation
Elevation differences and other considerations

Mass diagrams greatly aid visualization and planning of excavation and embankment construction operations requiring that material be removed from cut sections along a linear alignment and utilized for the construction of embankments along the same alignment. Excavation and embankment operations of this type typically occur in highway and railroad construction, canals and levees, and other similar linear projects. The construction estimator faces many problems in determining the cost of such work, including the following:

- What is the nature of the material in the cuts? Will it shrink or swell from its **in situ natural condition** to its **compacted condition in the fills?** Put another way, "How many bank cubic yards (Bcy) of excavation from the cut are required to provide enough material to make 1 embankment cubic yard (Ecy) in the fills?

- Where along the alignment can the material from cuts be obtained for use in fill construction, and how much material will be available from each of the various cut locations? In other words, **Where is the dirt?**

- Is enough material available in the cuts to enable construction of the fills without obtaining additional material (borrowing) from off-site? Or does the material in the cuts greatly exceed that required to construct the fills, making removal of excess cut material (wasting) from the site necessary?

- How far must the material be transported, or hauled, from cut to fill?

- What is the best operational plan for excavation and embankment to minimize the distance that material must be hauled from cut to fill?

All these questions, and more, can be answered through the intelligent use of mass diagrams.

EARTHWORK PROFILE DRAWINGS

In most instances, the designer of linear cut-and-fill projects will include in the construction drawings a **profile** along the centerline of the project and a series of **cross sections** at regular stations along the alignment, usually every 100 or 200 ft. Both the profile and the cross sections will show the outline of the original ground in its undisturbed condition, and the cross sections will show the **slope lines** of the cuts and fills to the **profile grade** (the finished grade line at the bottom of the cuts and the top of the fills). Today, both the profile and cross sections are usually drawn to a convenient horizontal and vertical scale so that digital software may be utilized to quickly compute the volume of cuts and fills along the alignment.

By simply examining the profile drawing, one may visually determine some "broad-brush" information about the general nature of the project. For instance, consider the greatly simplified profile drawing for a hypothetical highway project indicated in Figure 8-1. This profile, taken along the centerline of the project, indicates that the work starts at station 0+00 and ends at station 56+00, over a mile away. Two medium-sized cuts are shown, as well as one intervening large fill and a smaller fill near the end of the project. If the profile is to scale, as most are, it provides a general idea of where the material must come from to construct the fills and even a very rough indication of the distance that material must be moved from cut to fill.

However, a profile drawing such as Figure 8-1 does not give an accurate indication of the volume of cuts and fills or a very definite idea of the distances excavated materials must be hauled along the alignment to construct the fills. Profile drawings, by themselves, can be very misleading in this respect. Without reference to the cross sec-

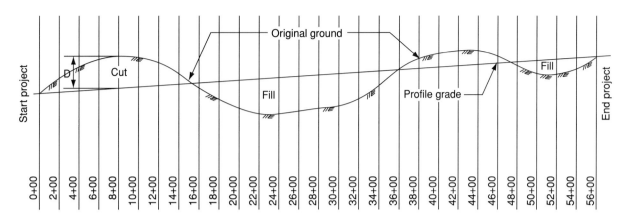

FIGURE 8-1 Typical profile drawing showing cuts and fills.

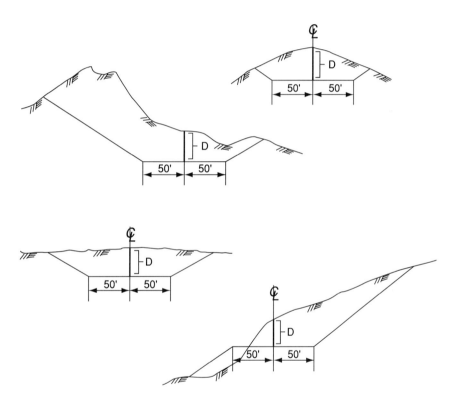

FIGURE 8-2 Variations in size and shape of possible cross sections at Figure 8-1 Station 8+00.

tions, one cannot obtain a clear idea of the project. For instance, consider the cut at station 8+00. The cross section at this station could be any of the four very different cross sections indicated on Figure 8-2 or one of any number of additional cross sections.

Obviously, the nature of the project work will vary considerably depending on what the cross section at station 8+00 (and at all the other stations) actually looks like. Moreover, even after careful study, a profile drawing and the cross sections, in and of themselves, are insufficient to properly visualize and plan the required load and haul operations. Something more is needed.

Profile drawings such as Figure 8-1 with accompanying cross sections at reasonably spaced stations (the spacing depending on the terrain) permit the total quantity of excavation (cuts) and embankment (fills) to be calculated by manual or computer-aided methods. However, without a mass diagram, one cannot get an accurate sense of the project or devise an intelligent plan for performance of the work.

MASS DIAGRAMS

A **mass diagram** is nothing more than a plot of the accumulative net incremental quantities of cut and fill along the alignment, starting with zero at the beginning station of the project. The units used in the plot are bank cubic yards of excavation, regardless of whether a cut or a fill is being referred to. In the case of a fill, the quantity referred to is the incremental quantity between two adjacent stations in bank cubic yards of excavation required to construct the fill between those two stations, taking shrinkage (or in the case of blasted rock, swell, into consideration). In the case of cuts, the incremental quantity being expressed is the actual volume of the cut in bank cubic yards between two adjacent stations.

Cut quantities are considered plus, or positive, and fill quantities are minus, or negative. Thus, a net positive accumulative quantity will be plotted above the horizontal axis of the mass diagram (which represents a zero net accumulative quantity). Net negative accumulative quantities are plotted below the horizontal axis. A mass diagram may be likened to a bank account in which cut quantities in bank cubic yards are considered deposits and fill quantities, also expressed in bank cubic yards, are considered withdrawals. If the accumulative plot lies above the horizontal axis of the mass diagram, the bank account has a positive balance, whereas if the plot falls below the horizontal axis, it is overdrawn.

Example of Mass Diagram Construction in Common Materials That Shrink from Cut to Fill

Refer to Figure 8-3. The first three columns represent hypothetical cut quantities in bank cubic yards and fill quantities in embankment cubic yards between the stations shown in Figure 8-1. These quantities are hypothetical. For instance, Figure 8-3 indicates that 22,534 Bcy of cut lie between stations 0+00 and 2+00 and 167,321 Bcy of cut are between stations 6+00 and 8+00. Similarly, 18,231 Ecy of fill are required between stations 15+00 and 16+00 and 136,489 Ecy of fill are required between stations 26+00 and 28+00, and so on.

The mass diagram calculations illustrated by Figure 8-3 are predicated on the basis that the material in the cuts is common earth believed to be subject to a shrinkage factor from cut to compacted embankment in the fills of 16.4%. This means that the number of bank cubic yards required to construct 1 Ecy equals $1 / (1.0 - 0.164) = 1.1962$ Bcy. The fourth column in Figure 8-3 expresses the conversion of each fill quantity in embankment cubic yards to the equivalent number of bank cubic yards required to construct that volume of fill. The last column, the mass diagram accumulative quantity in bank cubic yards, is simply a running accumulative total starting from zero at station 0+00 and ending with −202,451 Bcy at station 56+00.

Figure 8-4 is the mass diagram plot for the mass diagram quantities computed in Figure 8-3, shown below the profile drawing, Figure 8-1. The mass diagram plot rises to a maximum of 833,694 Bcy at station 16+00, crosses the horizontal axis at approximately station 27+00, then descends to a minus quantity of 503,094 Bcy at station 35+50, again crossing the horizontal axis at station 46+00, rising to a small peak of 35,739 Bcy at station 47+75, again crossing the horizontal axis at approximately station 49+00, and finally, descending to −202,451 Bcy at station 56+00. Note that

Station	Cut (Bcy)	Fill (Ecy)	Equivalent[1] (Bcy)	Mass diagram (Bcy)
0+00				0
2+00	22,534			22,534
4+00	76,598			99,132
6+00	142,258			241,390
8+00	167,321			408,711
10+00	174,298			583,009
12+00	162,556			745,565
14+00	62,734			808,299
15+00	25,395			833,694
16+00		18,231	21,808	811,886
18+00		49,792	59,561	752,325
20+00		79,924	95,605	656,720
22+00		138,251	165,376	491,344
24+00		171,293	204,901	286,443
26+00		160,568	192,072	94,371
28+00		136,490	163,269	−68,898
30+00		129,168	154,511	−223,409
32+00		124,968	149,487	−372,896
34+00		87,493	104,659	−477,555
35+50		21,357	25,547	−503,102
36+00	15,179			−487,923
38+00	75,137			−412,786
40+00	110,252			−302,534
42+00	125,573			−176,961
44+00	95,894			−81,067
46+00	85,231			4,164
47+75	31,567			35,731
48+00		8,296	9,924	25,807
50+00		37,057	44,328	−18,521
52+00		67,349	80,563	−99,084
54+00		59,282	70,913	−169,997
56+00		27,138	32,462	−202,459

[1]Common excavation to engineered embankment shrinkage factor is 16.4%. Bcy required for 1 Ecy = $1 / (1.0 − 0.164) = 1.1962$.

FIGURE 8-3   Mass diagram calculations for profile shown in Figure 8-1 if material in cuts is common earth excavation.

when the profile is in cut, the mass diagram will always be ascending, whereas in fill, it will always be descending. Also, whenever the mass diagram plot crosses the horizontal axis, the cut-and-fill quantities between that point and the previous point where the mass diagram crossed the horizontal axis, balance. That is, the volume of material generated by the cuts in bank cubic yards is the exact quantity required to construct the fills. Thus, Figure 8-4 depicts balances of cuts and fills between the following stations:

- Station 0+00 to station 27+00
- Station 27+00 to station 46+00
- Station 46+00 to station 49+00

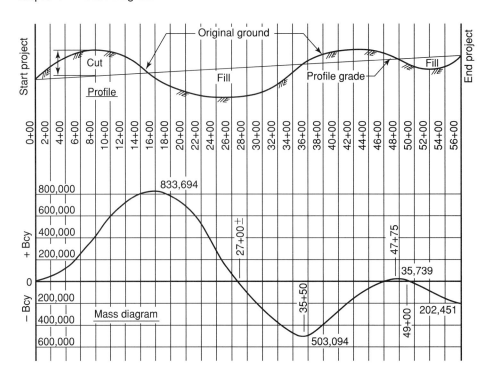

FIGURE 8-4 Mass diagram for profile shown in Figure 8-1 if material in cuts is common earth excavation with a 16.4% shrink factor.

Stations 27+00, 46+00, and 49+00 are called **primary balance stations.**

The diagram also clearly illustrates that no cut material is available from which to construct embankment beyond station 49+00. Therefore, to complete the embankment from that station to station 56+00, a total of 202,451 Bcy of similar common material must be obtained from off-site. In other words, 202,451 Bcy of **borrow excavation** must be obtained. Additional important items that can be developed from this mass diagram will be discussed later in this chapter.

Example of Mass Diagram Construction for Rock That Will Swell from Cut to Fill

Refer to Figure 8-5 for a tabulation of cut-and-fill quantities and mass diagram calculations similar to those shown in Figure 8-3. The cut-and-fill quantities between the various stations in Figures 8-3 and 8-5 are hypothetical values that could result from manual or computer-aided average-end area calculations for the profile shown in Figure 8-1. The difference between Figures 8-5 and 8-3 begins with the fourth column from the left (equivalent bank cubic yards) because the material (hypothetically speaking) is now rock, not soil. Whereas this column on Figure 8-3 reflected the fact that the material from the cuts would shrink from the in situ natural condition to a compacted condition in the fill, the corresponding column in Figure 8-5 reflects the opposite—that is, the rock material from the cuts will swell, and less than 1 Bcy of cut will be required to produce 1 Ecy of fill. As noted on Figure 8-5, the calculations were made on the basis that the rock was believed to swell by a factor of 36%, meaning that the bank cubic yards required for 1 Ecy = 1 / 1.36 = 0.735 Bcy.

The mass diagram calculations from Figure 8-5 are plotted on Figure 8-6 in the same manner as the mass diagram calculations from Figure 8-3 were plotted on

Station	Cut (Bcy)	Fill (Ecy)	Equivalent[1] (Bcy)	Mass diagram (Bcy)
0+00				0
2+00	22,534			22,534
4+00	76,598			99,132
6+00	142,258			241,390
8+00	167,321			408,711
10+00	174,298			583,009
12+00	162,556			745,565
14+00	62,734			808,299
15+00	25,395			833,694
16+00		18,231	13,400	820,294
18+00		49,792	36,597	783,697
20+00		79,924	58,744	724,953
22+00		138,251	101,614	623,339
24+00		171,293	125,900	497,439
26+00		160,569	118,018	379,421
28+00		136,489	100,319	279,102
30+00		129,169	94,939	184,163
32+00		124,961	91,846	92,317
34+00		87,493	64,307	28,010
35+50		21,357	15,697	12,313
36+00	15,179			27,492
38+00	75,137			102,629
40+00	110,252			212,881
42+00	125,573			338,454
44+00	95,894			434,348
46+00	85,231			519,579
47+75	31,567			551,146
48+00		8,296	6,098	545,048
50+00		37,057	27,237	517,811
52+00		67,349	49,502	468,309
54+00		59,282	43,572	424,737
56+00		27,138	19,946	404,791

[1]Rock excavation to engineered embankment results in a swell factor of 36%. Bcy required for 1 Ecy = 1 / (1.36) = 0.735.

FIGURE 8-5   Mass diagram calculations for profile shown in Figure 8-1 if material is rock excavation.

Figure 8-4. The first important fact to note from Figure 8-6 is that now there will be an excess of cut over that needed to construct fills of 404,789 Bcy that must be removed from the alignment, or **wasted.** The mass diagram rises to a peak of 833,694 Bcy at station 16+00, then descends to a valley of +12,310 Bcy at station 35+50, rising again to a peak of 551,143 Bcy at station 47+75, and descending to a residual balance of +404,789 Bcy at station 56+00. Note also that, although the cuts and fills nearly balance between stations 0+00 and 35+50, the mass diagram plot does not cross the zero line throughout the entire length of the alignment. Thus, Figure 8-6 depicts no primary balance stations. From station 35+50 onward, the amount of material generated in the

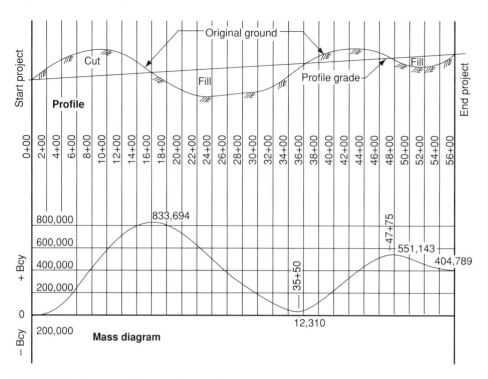

FIGURE 8-6   Mass diagram for profile shown in Figure 8-6
if material in cuts is rock with a 36% swell factor.

rock cut so greatly exceeds that required for fill that upon completing the last fill from
approximately station 47+75 to station 56+00, the balance of 404,789 Bcy remains.

In practice, highway and railroad designers try to establish a profile grade in relation to the original ground so that the cuts and fills will balance, or nearly balance, to
minimize the necessity for either borrowing excavation or wasting material. The situations reflected in Figures 8-4 and 8-6 seldom arise in actual roadway or railroad design; they are intended to illustrate the extremes that mass diagram plots of that particular profile could reflect.

USE OF A MASS DIAGRAM TO DETERMINE
THE MOST ECONOMICAL LOAD-AND-HAUL PLAN

Ordinarily, the most economical load-and-haul plan is the one that results in the material from the cuts being moved the shortest possible distance. A principal use of mass
diagrams is to help the estimator quickly devise this plan.

Take the Short Hauls First Principle

The movement of excavated material from a particular cut to a particular fill, or a delineated part of a particular cut to a delineated part of a particular fill, is called a **haul.**
If the load-and-haul operation is carried out starting with the shortest possible hauls,
and when those are completed, moving to the next longer hauls, then the next longer,
and so on, none of the material from the cuts will be hauled a greater distance than ab-

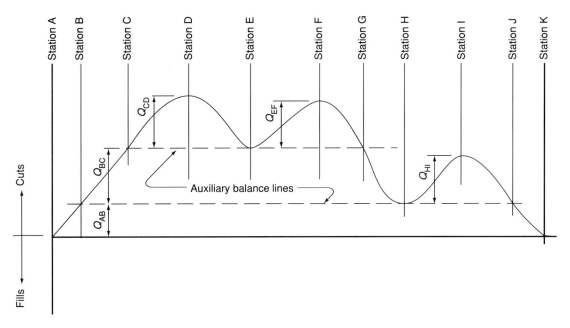

FIGURE 8-7   Illustrating use of mass diagram to determine the most economical load-and-haul plan based on taking the short hauls first.

solutely necessary to deposit it in a fill. This principle, which can be stated as **take the short hauls first,** can be easily grasped by examining Figure 8-7.

This mass diagram plot indicates a situation where the project designer selected a profile grade line so that the cuts and fills balance. Project work begins at station A with a long cut extending to station D, then a fill extending to station E, a cut extending to station F, a fill extending to station H, a cut extending to station I, and finally a fill extending to the end of the project at station K. The horizontal axis of the mass diagram can be thought of as the primary balance line. **Auxiliary balance lines** are drawn as shown in Figure 8-7 in the positions shown. The lower auxiliary balance line intersects the mass diagram at stations B and J, whereas the upper auxiliary balance line intersects the mass diagram at stations C and G.

Also indicated in Figure 8-7 are all of the cut quantities: Q_{AB}, Q_{BC}, Q_{CD}, Q_{EF}, and Q_{HI}. By examining the diagram, one can easily see that the quantity of cut between stations A and B (Q_{AB}) is just sufficient to make the fill between stations J and K. Similarly, the cut between stations B and C (Q_{BC}) is just sufficient to make the fill between stations G and H, the cut between stations C and D (Q_{CD}) is just sufficient to make the fill between stations D and E, the cut between stations E and F (Q_{EF}) is just sufficient to make the fill between stations F and G, and finally the cut between stations H and I (Q_{HI}) is just sufficient to make the fill between stations I and J.

Note that if the work is performed in the sequence of moving Q_{CD} from cut to fill first, followed by moving Q_{EF}, followed by Q_{BC}, followed by Q_{HI}, and finally taking the long haul quantity Q_{AB} last, not only will there be no unnecessary movement of cut to fill, but after Q_{CD} and Q_{EF} are moved, successive hauls of cut to fill can traverse the alignment on the profile grade line, which is the best haul profile from cut location to fill location.

Note also that material from the first long cut from station A to station D is used in three different fill locations—first the fill between stations D and E (Q_{CD}), then the fill between stations G and H (Q_{BC}), and finally the fill between stations J and K (Q_{AB}). Without using a mass diagram, one would find it impossible to determine this load-and-haul plan from the profile drawing and cross sections.

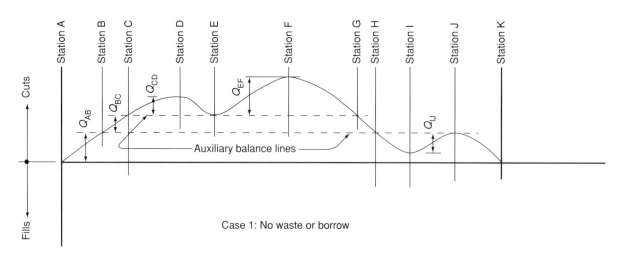

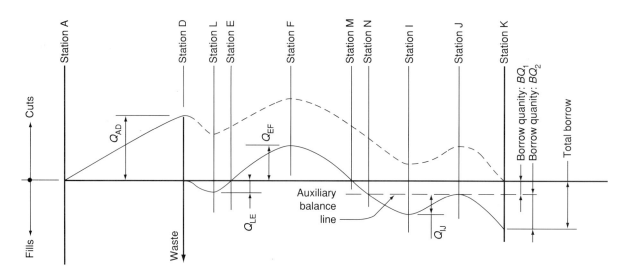

FIGURE 8-8   Use of mass diagram to consider use of waste and borrow
to reduce haul distance.

Consideration of Waste and Borrow Excavation to Minimize Haul Distance

Consider Figure 8-8. The upper mass diagram, case 1, is very similar to Figure 8-7. Two auxiliary balance lines are used to define the most economical load-and-haul plan involving five separate hauls with cut quantities Q_{AB}, Q_{BC}, Q_{CD}, Q_{EF}, and Q_{IJ}, respectively. In this plan, the haul distances for Q_{CD}, Q_{EF}, and Q_{IJ} are relatively short, and the haul distances for Q_{BC} and Q_{AB} are very long. As long as no cut material is wasted off the alignment and no borrow material from off the alignment is utilized, the load-and-haul plan illustrated by case 1 is the most economical way to accomplish the work. However, depending on the length of the alignment between stations A and K, the long hauls for Q_{BC} and Q_{AB} could be costly.

If it is permissable to waste cut material adjacent to the alignment at station D and obtain borrow material adjacent to the alignment near station K, one might effect cost savings. To investigate this possibility, the estimator could consider the scenario illustrated by case 2 in Figure 8-8 and make comparative cost estimates to determine which of the case 1 or case 2 scenarios is the least expensive. Such a cost estimate would determine whether the money saved by shortening the extremely long hauls was greater than the extra costs resulting from the waste dump at station D and from the extra loading of borrow material near station K.

The plan reflected by case 2 involves wasting all the material from the long cut between stations A and D (Q_{AD}) off the alignment near station D. In terms of the diagram, this has the effect of moving the portion of the mass diagram beyond station D from the dotted position shown to the solid line commencing at the mass diagram axis at station D. One auxiliary balance line is used to define the most economical load-and-haul plan beyond station D. In this plan, cut between stations L and E (Q_{LE}) is used to build the fill between stations D and L, the cut between stations E and F (Q_{EF}) is used to build a fill between stations F and M, the cut between stations I and J (Q_{IJ}) is used to build the fill between stations N and I; and borrow quantity BQ_1 obtained near station K is used to construct a fill between stations M and N, and borrow quantity BQ_2 is used to construct a fill between stations J and K. To make a valid cost comparison between cases 1 and 2, one would have to take into consideration the haul distance from the alignment to the waste dump at station D and the haul distance from the alignment to the borrow area at station K.

USE OF MASS DIAGRAMS TO DETERMINE AVERAGE LENGTH OF HAUL

Theoretical Average Haul Distance

Consider the haul of the cut quantity Q_{CD} from between stations C and D to the fill between stations D and E for case 1 of Figure 8-8. Clearly, the material near the end of the cut at station D does not have to move very far to reach the fill area, whereas the material at the beginning of the cut near station C has to move much farther to reach the fill area. The average distance that material must be moved lies between these two extremes. The movement of this material, expressed in bank cubic yard feet (Bcy-ft.) can be shown by the principles of integral calculus to be represented on the diagram by the cross-hatched area shown in Figure 8-9(a). Because the total bank cubic yard feet representing this material movement can also be expressed as the quantity in bank cubic yards (Q_{CD}) times the average distance moved in feet, it follows that

$$\text{Average haul distance (ft)} = \frac{\text{cross-hatched area (Bcy-ft)}}{Q_{CD} \text{ (Bcy)}}$$

Similarly, one would determine the average haul distance for the other hauls of the case 1 alternative of Figure 8-8 as indicated in Figure 8-9 by sketches (b), (c), (d), and (e).

The average distances for the hauls in the case 2 alternative shown in Figure 8-8 would be determined as indicated in Figure 8-10. Sketches (a) through (f) depict the appropriate areas to be used for the six hauls involved.

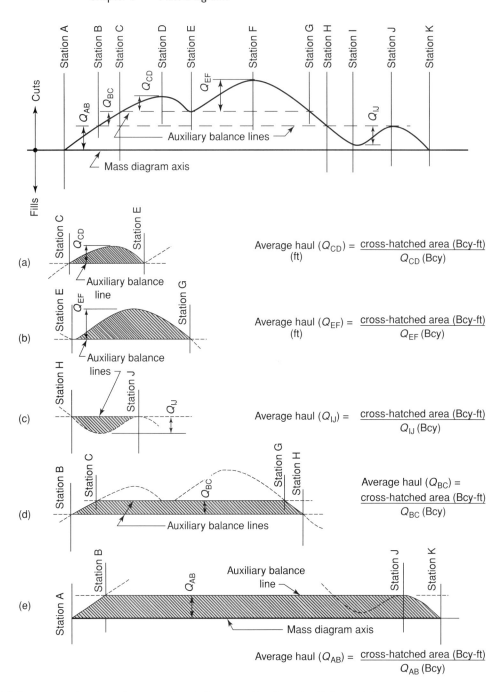

FIGURE 8-9   Determination of average haul distances for Case 1 of Figure 8-8.

Haul Distance Approximation

To determine the irregular areas required for the precise calculation of average haul distances, such as those indicated in Figures 8-9 and 8-10, an estimator must use a polar planimeter or a similar method. Most estimators do not bother to make the precise calculations and take the distance between the limiting stations instead as the horizontal

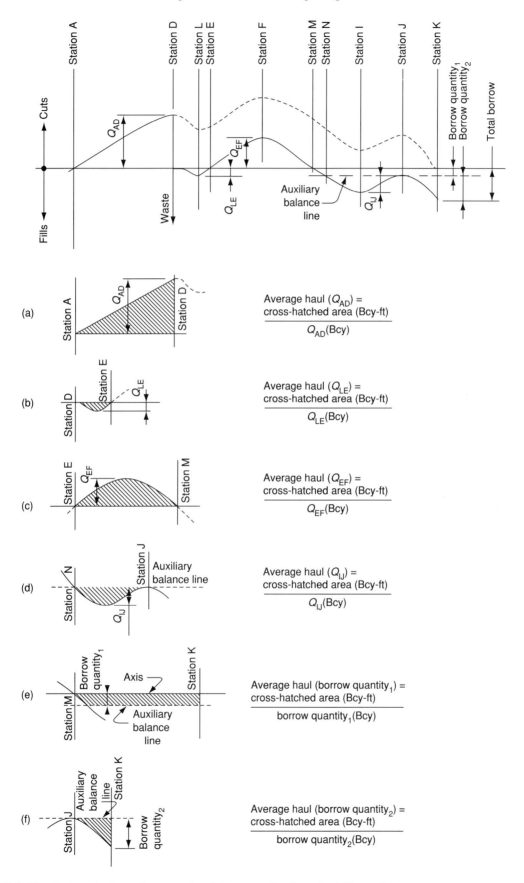

FIGURE 8-10   Determination of average haul distances for Case 2 of Figure 8-8.

haul distance. For instance, referring to the haul for quantity Q_{CD} for case 1 of Figure 8-8, the haul distance would be taken as the distance between stations C and E. This latter approach is conservative.

Consideration of Elevation Difference Affecting Grade of Haul Route and Other Considerations Affecting Haul Route Length

All the above discussion of haul length refers to the centerline horizontal distance between stations on the profile drawing. Depending on the terrain, there could be a considerable difference in elevation between the center of the cut mass and the center of the fill mass to which the cut material is to be moved. In this case, the horizontal distance along the centerline of the alignment between these two centers of mass can be determined from the mass diagram, but the actual haul route between these two points is on an incline. This fact will not seriously affect the haul length unless the terrain is very rugged and the difference in elevation between the two centers of mass is large. Also, haul road grades are limited by safety considerations and by the gradeability[1] of the haul units used. Therefore, when the elevation difference is large, switchbacks on and off the alignment might be necessary, which would add considerable length to the haul length determined from the mass diagram above. Site environmental protective requirements or other factors might also restrict haul route location and alignment, which could increase haul route length significantly.

   Although the estimator must take into consideration the above additional factors when they are present (see Chapter 9), the first step required in preparing a load-and-haul cost estimate is the mass diagram analysis determining the horizontal centerline haul lengths along the alignment.

REPRESENTATION OF LOAD-AND-HAUL PLAN ON THE PROFILE DRAWING

Once the mass diagram analysis is complete, summarizing the information developed on the profile drawing where it can be conveniently referred to for preparation of the cost estimate is helpful. Also, for projects that have already been obtained, such summaries on the profile drawing are very useful in explaining and discussing the planned work with the field personnel who are going to carry it out. Field personnel are usually conversant with profile drawings, but some may not be familiar with mass diagrams. Thus, to provide information in a form that field crews will readily understand, the profile drawing should indicate each material movement including the quantity involved, the beginning and ending station where the material is obtained, the beginning and ending station where it is to be used for constructing fill, and the average haul distance.

   Figure 8-11 is a hypothetical representation of the results of mass diagram analysis for a roadway project on the profile drawing and the corresponding mass diagram from which the information was developed.

---

[1]"Gradeability" refers to the ability of a haul vehicle to climb steep grades or traverse very soft haul road surfaces (or a combination of both) when fully loaded.

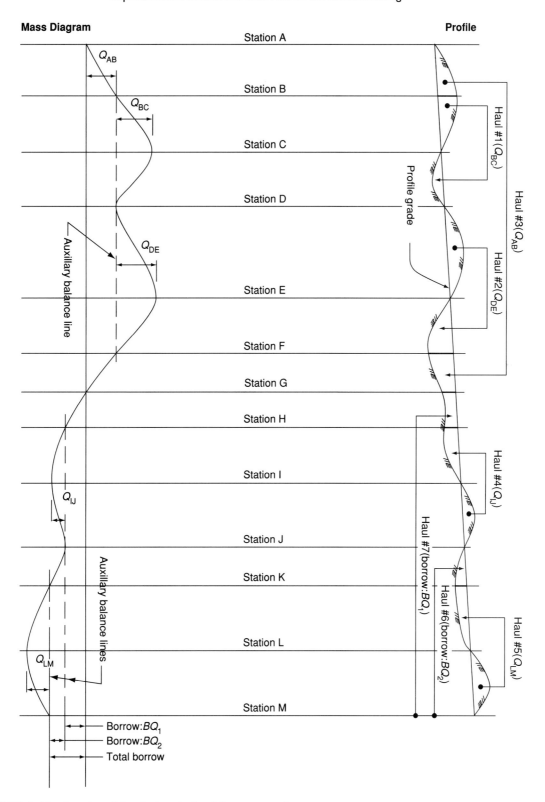

FIGURE 8-11 Load-and-haul plan derived from mass diagram.

CONCLUSION

This chapter discussed the relationship of mass diagrams to profile drawings, described how to construct mass diagrams, and explained how to use them in planning load-and-haul operations for highways, railroads, canals, levees, and other linear projects. Mass diagrams play an important part in the operation analysis phase of preparing cost estimates for such projects.

Once the load-and-haul plan is determined and the haul distances are known, the estimator can complete the remaining aspects of load-and-haul operations analysis and prepare the dollars and cents cost estimate. These topics are the focus of the following chapter.

QUESTIONS AND PROBLEMS

1. What are four problems or questions faced by an estimator in determining the cost of linear cut and fill projects that are discussed in this chapter?

2. Explain the following terms used in connection with mass diagram analysis.
   (a) Profile drawing   (g) Primary balance stations
   (b) Cross sections   (h) Borrow excavation
   (c) Slope lines   (i) Waste excavation
   (d) Profile grade   (j) Haul
   (e) Mass diagram   (k) Shrinkage factor
   (f) Balance lines   (l) Swell factor

3. A section of state freeway construction is out for bid that extends from station 148+00 to station 250+00, a distance of 10,200 ft or 1.93 miles. A bidding contractor has plotted up the cross sections from the plans and has computed the volumes of cut or fill between stations along the alignment. For each station along the alignment, the respective volume of cut or fill between that station and the preceding station is listed below.

   The bid documents state that if the cut and fill quanities do not balance, the state will provide either a borrow area containing suitable material or, alternatively, a waste disposal area to accommodate any excess, both 1500 ft off the alignment at station 250+00. Further, no borrow or waste will be permitted anywhere else along the alignment.

   Soils tests along the alignment indicate that the average dry density of the in situ material in the cuts is 98.3 lb/cf. Furthermore, the technical specifications require that the fills must be compacted to 95% of modified Proctor test maximum density. The modified Proctor test maximum density for the soils in the cut areas has been determined by soil tests to average 124.5 lb/cf.

   Plot a mass diagram reflecting the above given data on cross-section paper. Use a vertical scale of 1 in. = 250,000 Bcy and a horizontal scale of 1 in. = 500 ft. You must determine the average shrinkage factor from cut to fill to plot the mass diagram.

   (a) Do the quantities balance? If not, what is the excess excavation in bank cubic yard that must be wasted at station 250+00, or alternatively, what is the required borrow quantity to be obtained at station 250+00 in bank cubic yards?
   (b) There is one set of hauls that will result in the most economical load-and-haul plan for the data and conditions stated. For each of these hauls, determine the haul quantity in bank cubic yards and the haul distance in feet.

(c) Construct a load-and-haul plan similar to that shown in Figure 8-11. For each haul, show the haul number, the quantity in bank cubic yards, and haul distance in feet. Also, show on the diagram the stations where cuts change to fills (and visa versa) and the stations for the "balance" points.

| | Volume between stations | | | Volume between stations | |
Station	Cut, Bcy	Fill, Ecy	Station	Cut, Bcy	Fill, Ecy
148+00			226	32,390	
150+00	31,800		228		40,000
152	95,400		229+10		145,500
154	153,600		230		217,500
156	257,100		232		292,700
158	396,300		234		320,200
160+00	417,600		236		241,600
162	314,800		238		92,000
164	204,900		238+70		46,500
166	103,800		240+00	47,600	
166+61	7,800		242	109,100	
168		32,900	244	107,100	
170+00		136,600	246	123,900	
172		181,000	248	108,600	
173+33		62,300	250+00	56,000	
174	19,100				
176	142,600				
178	244,400				
179+23	58,800				
180+00		81,600			
182		245,300			
184		407,600			
186		668,900			
188		620,200			
190+00		518,600			
191+46		236,300			
192	39,500				
194	305,500				
196	270,100				
198	188,500				
199+00	38,900				
200+00		31,900			
202		163,300			
204		237,200			
206		227,600			
207+17		52,600			
208	107,100				
210+00	111,100				
212	127,600				
214	160,000				
216	193,300				
218	219,500				
220+00	312,000				
222	325,300				
224	150,000				

9

Load-and-Haul, Spread, and Compact Operations

Key Words and Concepts

Schedule of haul quantities
Average haul
Haul road length and profile
Switchbacks
Percent grade
Rolling resistance
Total haul road resistance
Sustained travel speed
Rimpull chart
Pushcat maneuver time
Loaded travel time
Return travel time
Scraper payload
Scraper struck capacity
Scraper load factor
Scraper productivity
Efficiency
Scraper spread production
Balance number
Computer-aided solution for scraper
   load-and-haul analysis
Truck load time

Truck payload
Truck rated capacity
Bank cubic yards per loader bucket
Rated bucket size
Material swell
Bucket fill factor
Truck load time
Loader cycle time
Truck spot and maneuver time
Truck/loader haul unit and spread productivity
Thickness of the compacted layer
Compaction unit productivity
Compactor travel speed
Width and number of compacter passes
Load- and haul-plan
Haul road characterization
Equipment selection
Amount of equipment required to meet
   construction schedule
Work crew composition
Estimated costs

Almost all surface heavy construction projects involve load, haul, and compaction operations. Whether the project simply requires large excavations to be made for the construction of structures where the material must be transported off the site and wasted or both excavation and embankment operations are involved on the site, large quantities of soil, rock, or both often must be loaded, transported, and either compacted into embankments or wasted. This chapter presents appropriate procedures for analyzing these types of construction operations and for converting the results of such analyses into dollar and cents estimates of the probable cost of accomplishing the actual work.

CONSTRUCTION CRAFTS AND EQUIPMENT

Construction Crafts

In terms of traditional union trade classifications, three separate construction trades are usually involved. The heavy equipment will be operated by either operating engineers or teamsters, depending on the type of equipment. All haul vehicles, with the exception of scrapers, will be driven by teamsters, whereas all other equipment, including scrapers, will be operated by operating engineers. Laborers are employed for flagging and similar duties in connection with loading and dumping activities. Crew foreman and surveying and grade-checking duties are functions performed by operating engineers.

Loading Equipment

Loading equipment falls into two basic categories, crawler tractor (pushcat) equipment for loading scrapers and top-loading equipment for loading trucks. Top-loading equipment consists of front-end loaders, front shovels, backhoes, cable-operated clamshells and draglines, large rotary wheel excavators, and conveyor belt loaders fed by bulldozers. Today, these loading units are very large. Front-end loaders, front shovels, and backhoes commonly carry buckets rated at 12 to 15 cy. Clamshells and draglines used in construction work commonly are fitted with 5 to 6 cy buckets. Front shovels, backhoes, and draglines used in mining work are much larger.

Haul Equipment

There are two basic categories of haul equipment commonly used in heavy construction—scraper equipment and trucks. On very large projects, rail haulage utilizing large bottom-dump rail cars has been occasionally used.

Two separate types of scrapers are common, the conventional type that requires pushcat assistance in loading and the so called paddle-wheel scraper, which is self-loading. Scrapers are used for off-highway haul situations only. Push-loaded scrapers carry bowls ranging from 14 to 32 cy struck capacity (meaning "water-level" capacity when the scraper is on level ground). Some are powered by one engine only, and others have two engines to shorten loading time and to improve gradeability. On level haul roads, scrapers commonly attain haul speeds in excess of 30 mph.

Two common types of trucks are used for construction work: end-dump trucks, which are usually mounted on a single chassis, and bottom-dump trucks, usually consisting of a wagon coupled to a separate tractor unit. Both are manufactured for either

PHOTO 9-1   Examples of two distinctly different load-and-haul systems. Upper photo, front-end loader with $12\frac{1}{2}$ cy bucket top-loading 50 ton end-dump truck. Lower photo, push cat loading 21 cy twin-engine scraper.

on- or off-highway use. Trucks are rated by their payload capacity expressed in 2,000 lb tons. End-dump trucks typically range from 35 to 85 ton payload capacity, whereas bottom-dump trucks with payloads as high as 130 tons are commonly used. End-dump trucks used in the mining industry are much larger. Bottom-dump trucks generally have poor gradeability but attain high haul speeds on long flat hauls and are generally more economical in that type of use. End-dump trucks are designed to combine good gradeability with high haul speeds when on the level. Both end- and bottom-dump trucks commonly attain loaded haul speeds in excess of 35 mph on unrestricted level hauls.

Compaction Equipment

Two broad classes of compaction equipment are commonly used: compactors that must be towed (generally by a crawler tractor) and self-powered compactors. Each of these broad classes can be furnished in one of several subtypes—sheepsfoot, or spiked tooth rollers, pneumatic rollers, or smooth drum vibratory rollers. The larger sizes of self-powered rollers are commonly articulated to enhance their steering and maneuverability characteristics.

Both towed and self-powered compactors have become quite large, particularly the towed compactors. Towed pneumatic rollers weighing as much as 50 tons are not uncommon. Both classes of compaction equipment are furnished in many different configurations.

Support Equipment

Load-and-haul operations are supported by a number of supplementary heavy equipment units. This equipment consists of track-mounted bulldozers (some fitted with powerful hydraulically actuated ripping equipment, slope-trimming blades, etc.), rubber-tired bulldozers, motor graders, water trucks, and lightplants. Track-mounted bulldozers and rippers have become very large and powerful. They are principally used to augment and assist loading operations in rock, although track-mounted bulldozers are also used for slope trimming operations, haul road construction, assisting haul units that have become stuck, and spreading material on fills prior to compaction. Rubber-tired bulldozers have the advantage of much faster speed and are commonly used for spreading material on fills. Motor graders are utilized for spreading material on fills, final grade operations, haul road maintenance, and similar tasks. Water trucks fitted with spray bars are used to control moisture content and dust. Although haul equipment commonly is fitted with headlights, large portable lightplants consisting of diesel-powered generators and light towers fitted with luminaires are required for load-and-haul operations conducted at night.

Sources of Equipment Information

Specifications for the various types of heavy construction equipment utilized in load, haul, and compact operations are available in the equipment handbooks published by the various manufacturers. Also, full description of the common usage of this equipment is available in the several excellent reference books listed in the bibliography to this text.

PHOTO 9-2   Compaction equipment placing impervious clay core material in a dam. Upper photo, rubber-tired tractor pulling spreader box manipulating material to reduce moisture content. Lower photo, crawler tractor pulling 50 ton pneumatic-tired roller, compacting material to required density.

PHOTO 9-3   Upper photo, self-powered, articulated, smooth drum vibratory roller compacting flow breccia (naturally fractured rock). Lower photo, crawler tractor pulling double-drum sheepsfoot roller compacting fine-grained common earth material.

PHOTO 9-4   Two examples of support equipment. Upper photo, track-mounted bulldozer shoving up alluvium into muck pile for top loading into trucks. Lower photo, large motor grader manipulating impervious clay material to reduce moisture content before compaction in core zone of a dam.

BASIC CONSIDERATIONS COMMON TO ALL LOAD-AND-HAUL OPERATIONS

The first thing that must be done to analyze load-and-haul operations is to organize the format of the analysis. In the case of lump sum estimates where the total project cost is not broken down by bid items, organization of the load-and-haul part of the project estimate is simply a matter of performing sufficient takeoff work to develop a **schedule of haul[1] quantities** by source of excavation and destination to which the material is to be moved, either for the construction of a fill or to be wasted. The quantity of each individual haul in the schedule should be the largest quantity for which the load-and-haul conditions are similar enough so that they can be reasonably modeled or characterized by an average haul road length and profile (i.e., the **average haul** for that particular quantity of excavation, even though it is recognized that in the actual movement of that haul quantity, many different haul situations will occur, some longer than the average haul, and some shorter).

Properly defining average hauls is a matter requiring considerable experience and judgment. The use of mass diagrams, described in the previous chapter, is a valuable first step for linear projects such as roads and railroads and, for those projects for which the haul is more or less level, is usually all that is required. However, both the preparation of material movement schedules and defining the hauls for other types of projects, such as an earth or rockfill dam are complicated matters. Such an analysis involves study of excavation and embankment drawings by means of horizontal and/or vertical cross sections and careful planning to determine what quantities of material from required excavations can be directly placed into required embankments, what excavation material is to be used for what portion of the embankment, how much, if any, of the excavated materials must first be stockpiled and then recovered later for embankment construction at the proper time, and so on. Such a schedule for large projects of this type could easily involve dozens of individual material movements, each of which requires an individual load-and-haul analysis for that particular average haul.

In the more common case where the excavation and embankment work of the project is organized by the owner into a series of individual bid items, the procedure described above must be carried out for each of the individual excavation bid items. The schedule of haul quantities for an individual bid item will be only a part of the schedule for the entire project, but all the elements of preparation of the schedule and the characterization of the hauls will be precisely the same.

Haul Road Length and Profile

After the schedule of all excavation hauls (whether for an individual bid item or for an entire lump sum project) has been organized and tabulated, the **length and profile** of each haul must be defined. For an accurate analysis, this means determining the length and grade (or inclination) of each separate segment of the haul road.

Remember that the schedule of hauls was organized on the basis that each haul quantity in the schedule could be modeled or characterized by a so-called average haul, that is, a hypothetical haul road that could be considered to represent the average hauling condition for that haul quantity. For linear projects, mass diagram analysis as described in the previous chapter is often sufficient to determine the average haul length. However, as Figure 9-1 illustrates this will not always be the case. In this

---

[1]See Chapter 8 for the definition of a "haul."

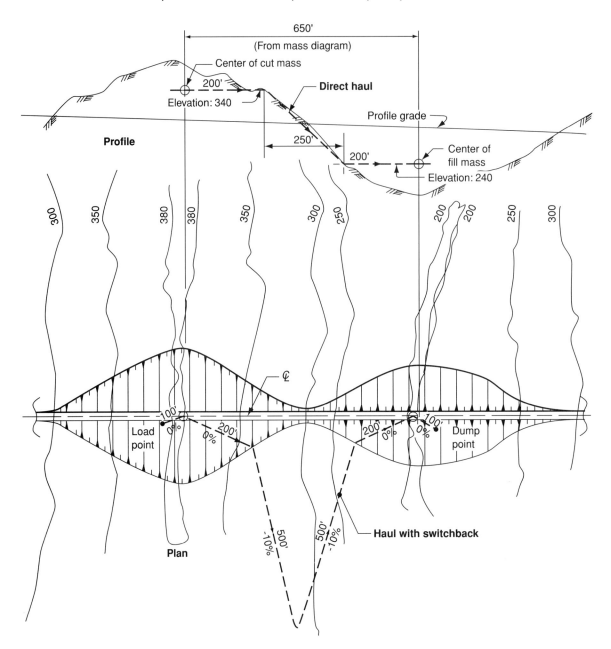

FIGURE 9-1   Illustrating requirement for additional haul length
due to centers-of-mass elevation difference.

situation the terrain is very rugged and the difference between the centers of mass of
cut (at elevation 340) and the fill (at elevation 250) is 100 ft. The average total one-
way haul from cut to fill determined by mass diagram analysis is 650 ft. This means
that the central portion of the direct haul road along the centerline of the alignment between these two centers of mass would have a loaded haul grade equal to (100 ft/250
ft) × 100, or −40% (the 200 ft level haul segments result because the cut-and-fill areas would normally be maintained approximately level as the cut is brought down and
the fill built up).

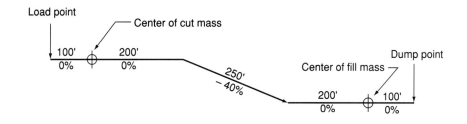

Direct haul on centerline

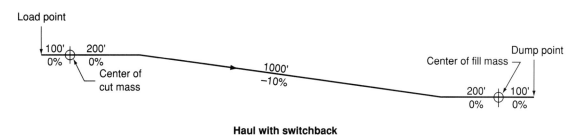

Haul with switchback

FIGURE 9-2   Haul road profiles for direct haul and switchback hauls shown in Figure 9-1.

Even if twin-engine scraper equipment with superior gradeability ($\pm 30\%$) were employed, a **switchback** on the return haul to reduce the adverse grade would be necessary. If the material was rock, requiring end-dump trucks for hauling, the haul road grade for both the loaded and return hauls should be limited to $\pm 10\%$. To achieve this, the haul road must be lengthened. Such a lengthened haul road with a 10% maximum grade can be achieved by use of a switchback as shown in the plan view on Figure 9-1.

Both the direct centerline haul and the switchback hauls shown in Figure 9-1 are represented in simplified form in Figure 9-2. In each case a 100 ft addition was added at each end to allow for some lateral travel in the cut and fill between the centers of mass and the load and dump points, respectively. Each haul consists of five individual segments. Each haul profile defines the haul by showing the length and grade of each segment. As noted earlier, the direct haul on centerline is too steep to be practical and would not be used. Figures 9-1 and 9-2 illustrate the effect that extreme differences in elevation of centers of cut-and-fill mass can have in requiring the lengthening of the haul road from that indicated by mass diagram analysis alone.

The haul road length and profile definition illustrated by the switchback haul of Figure 9-2 represents a greatly simplified case. Considerable thought and study must be given to idealized average haul road definition for more complicated situations such as that illustrated by Figure 9-3. Here an average haul road configuration is shown for the placement of a quantity of impervious embankment in the core trench on the right abutment of a large earthfill dam. This particular average haul is an example of the technique used to analyze the excavation and embankment operations for projects of this type where the total embankment to be placed in the dam from all material sources (borrow pits and/or required excavations) is broken down into discrete individual masses for which an average haul condition can be modeled.

A fully detailed discussion of the study and takeoff work required to carry out this process for an entire project such as a large dam is beyond the scope of this book. The

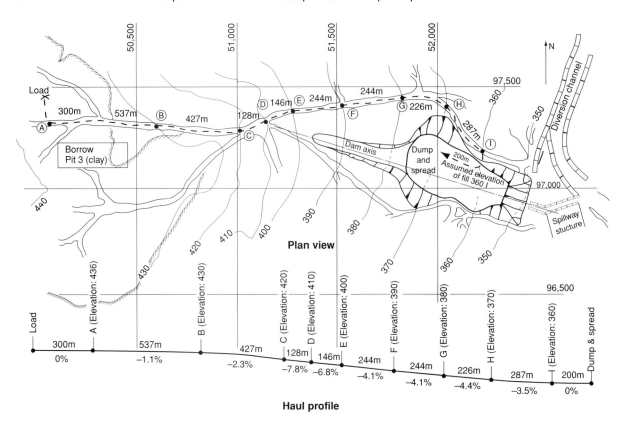

Plan view

Haul profile

FIGURE 9-3   Haul road profile for placing initial right abutment embankment in an earthfill dam from clay borrow pit (all elevations and horizontal distances in meters).

intent here is to explain the analytical and computational procedures that would be applied to each of the average hauls determined from the takeoff process so as to set up the appropriate crews to accomplish the work. When each of the discrete haul situations has been analyzed and "crewed up," the overall requirements for the personnel and equipment to meet the project schedule requirements can be determined and the dollar and cents costs can be estimated.

The 2,739 m (8,984 ft) average haul depicted in plan and profile in Figure 9-3 consists of 10 separate segments, varying in length from the shortest of 128 m (420 ft) to the longest of 537 m (1,761 ft). The loaded haul grades are either level (in the borrow pit or on the fill) or downhill (ranging from −1.1 to −7.8%). The horizontal configuration of the haul was chosen to utilize a general pattern of semipermanent haul routes established when the overall scheme of material movement for the project was planned. Because the quantities to be hauled in projects of this kind are so large (often totaling many millions of cubic yards), a great deal of thought would have been devoted to the material movement scheme and the establishment of the semipermanent network of haul roads.

Grade Resistance

The inclination of individual haul road segments in Figures 9-2 and 9-3 has been shown in **percent grade,** which is simply the rise (or fall) of the haul road per unit of horizontal projection, expressed as a percentage. For example, a +4.5% grade would rise 4.5 ft in a 100 ft horizontal run, whereas a −6.5% grade would fall 6.5 ft in a 100 ft horizontal run. Naturally, a positive grade resistance is an impediment to vehicle forward motion, whereas a negative grade resistance is an assist.

Rolling Resistance

Any haul vehicle propelled by an onboard engine transmitting rotational power though the transmission and final drive to the wheels in contact with the haul road surface encounters a resistance to forward motion, depending on the flexing of the tires and the general condition of the haul road surface. This resistance to motion, expressed in units of pounds per ton of vehicle weight, is called **rolling resistance.** It always acts opposing the direction of motion, and it always acts in a direction parallel to the haul road surface regardless of whether the direction of motion is horizontal, uphill, or downhill. The Caterpillar, Inc., equipment handbook suggests the values of rolling resistances shown in Table 9-1 for haul analysis purposes.

Total Haul Road Resistance

The two previously discussed haul road characteristics that have an influence on vehicle forward motion, that is, grade resistance and rolling resistance, can be combined into a single haul road characteristic called **total haul road resistance (THRR),** expressed as a percentage. It can easily be shown that 20 lb/ton of rolling resistance is equivalent to an adverse grade resistance of 1%, insofar as its effect on haul vehicle forward motion is concerned. Therefore,

$$\text{THRR } (\%) = \text{grade resistance } (\%) + \frac{\text{rolling resistance (lb/ton)}}{20}$$

Thus, a +3.5% grade resistance combined with a 100 lb/ton rolling resistance would equal a THRR of 3.5 + 100/20 = +8.5%, whereas a −6.0% grade resistance combined with a 100 lb/ton rolling resistance = −6.0% + 100/20 = −1.00% THRR.

TABLE 9-1   Rolling Resistance Values

A hard, smooth, stabilized surface roadway with no tire penetration; watered, maintained	40 lb/ton
A dirt roadway, rutted, flexing under load, little maintenance, no water, 1 to 2 in. tire penetration	100 lb/ton
Rutted dirt roadway, soft under travel, no maintenance, no stabilization, 4 to 6 in. tire penetration	150 lb/ton
Loose sand and gravel	200 lb/ton
Soft, muddy, rutted roadway, no maintenance, tire penetration over 6 in.	200 to 400 lb/ton

PHOTO 9-5 Examples of tire penetration affecting rolling resistance. Upper photo, slight to moderate tire penetration (100 to 150 lb/ton) in core zone in a dam. Lower photo, excessive tire penetration on top of a waste fill (+400 lb/ton).

Travel Time Determination

A haul vehicle traversing a hypothetical haul road of infinite length will be able to attain a **sustained travel speed,** usually reckoned in miles per hour depending on

- Whether the haul vehicle is loaded or empty
- Whether the haul vehicle is going uphill or downhill
- The rolling resistance of the haul road surface
- The horsepower of the engine and the gear range the transmission is in

However, because construction haul roads are usually characterized in segments of finite length and because the speed of the vehicle entering a segment may be less or more than the speed at which the vehicle would travel on a sustained basis if the segment was of infinite length, the average speed the vehicle attains in each segment may be less or more than the theoretical sustained speed.

For instance, a haul vehicle starting from rest and then traveling a distance say of 400 ft in a borrow pit might not be able to reach the speed it would theoretically reach on a sustained basis on a level surface with the rolling resistance value of the borrow pit. Even if it could reach the sustained speed, it would not do so until the latter part of the segment, so the average speed would be far less than the sustained speed.

Another example would be a haul vehicle coming down a hill at a relatively high speed entering onto the following segment, which is flat. Once the haul unit hits the flat segment, it will start to slow down, but if the segment is short, its momentum from the downhill segment will cause it to traverse the flat segment at an average speed that could exceed the speed at which the vehicle could traverse that section on a sustained basis.

Thus, when the sustained speeds are known and the acceleration and deceleration effects are considered, the average speed at which a haul unit can be expected to traverse a haul road segment will depend on

- The speed at which the haul vehicle exits the previous segment
- The length of the segment being traversed
- The sustained speed that could be attained on the traversed segment assuming it was of infinite length

This last factor, the sustained speed for the traversed segment, can be readily determined from the **rimpull chart** that all equipment manufacturers publish for the scraper and truck haul units they build and market.

Rimpull charts usually are in the form of Figure 9-4. Each of the scalloped intersecting curves represents a different vehicle transmission range. Each of the inclined rays represents a different THRR. By projecting horizontally from the intersection of a ray representing a particular THRR with either of the vertical dotted lines (one representing a loaded vehicle and the other an empty one), an intercept on one of the scalloped curves is obtained. Projecting horizontally to the vertical axis yields the developed sustained rimpull (usually expressed in pounds) that produces forward motion, whereas projecting vertically downward to the horizontal axis yields the sustained travel speed, usually expressed in miles per hour.

Today, both scraper and truck haul units are furnished with built-in retarding gearing that can be engaged on downhill hauls to limit the vehicle speed to a safe limit. The performance of such retarders is published in charts similar to the rimpull chart shown in Figure 9-4, from which the sustained retarded speed on any downhill segment can be similarly obtained.

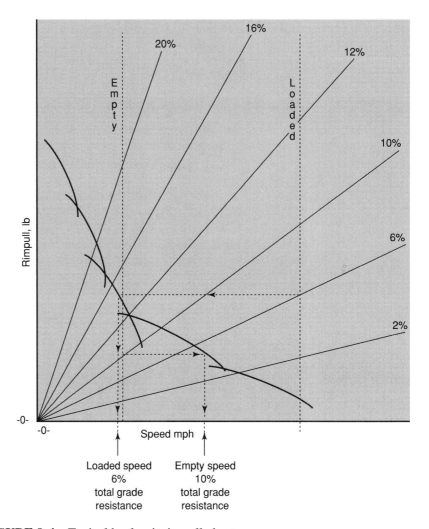

FIGURE 9-4   Typical haul unit rimpull chart.

Computer-Aided Solutions for Load-and-Haul Analysis

Formerly, hauls were analyzed, segment by segment, as described above, to obtain sustained speeds. Then, by applying an **acceleration/deceleration adjustment** depending on the entrance speed from the previous segment and the segment length, the average speed for the segment was obtained. The travel time in minutes for the segment was then calculated on the basis that

$$\text{Segment travel time (min)} = \frac{\text{Segment length (ft)}}{\text{Average travel speed (mph)(88)}}$$

The travel time (in minutes) for the total haul equals the sum of the loaded and return travel times of the individual segments.

Today, computer programs based on the rimpull and retarder characteristics of individual commercially available haul units are used to greatly simplify and speed up this tedious process. An example of such a computer-generated solution for a Caterpillar 631E scraper obtained by use of a computer program developed at California State University, Chico, is illustrated in Figure 9-5 for a 4,800 ft one-way haul with the haul road characteristics shown.

Loaded/Return Description	Seg Dist Feet	R.R. Lb/Ton	G.R. %	THRR %	Sus Spd mph	Dif Sus Spd	Spd Factor	Av Spd mph	Travel min
Travel in Cut	300	200	0	10.0	7.5	-7.5	0.91	6.8	0.50
Haul Segment	1000	100	5	10.0	7.5	0.0	1.00	7.5	1.52
Haul Segment	2500	40	3	5.0	15.3	-7.8	0.98	15.0	1.89
Haul Segment	800	100	-6	-1.0	31.0	-15.7	0.80	24.8	0.37
Travel on Fill	200	200	0	10.0	7.5	23.5	1.25	9.4	0.24
Travel on Fill	200	200	0	10.0	12.5	-12.5	0.81	10.1	0.22
Return Segment	800	100	6	11.0	11.7	0.8	1.00	11.7	0.78
Return Segment	2500	40	-3	-1.0	31.0	-19.3	0.91	28.2	1.01
Return Segment	1000	100	-5	0.0	30.0	1.0	1.01	30.3	0.38
Travel in Cut	300	200	0	10.0	12.5	17.5	1.18	14.8	0.23
Travel Data	9600							15.29	7.13

FIGURE 9-5 Travel time analysis for a caterpillar 631E scraper on a 4,800 ft one-way haul.

SCRAPER LOAD-AND-HAUL ANALYSIS

Scraper and Pushcat Cycle Times

Scraper operation can be visualized by the aid of Figure 9-6, which diagramatically portrays the movement of a single pushcat and the scrapers that are loaded by that single pushcat unit. The diagram graphically represents the length of time required for the entire scraper cycle (the large oval loop) and the time required for the much shorter pushcat cycle (the smaller rectangular loop). All the elements of the scraper and pushcat cycles are graphically identified.

Note that the pushcat and scraper are physically in the same place at the same time for only a small part of the scraper cycle, that is, the length of time the pushcat is maneuvering in the borrow pit to engage the push block on the rear of the scraper preparatory to push-loading and then the time the pushcat is push-loading the scraper and boosting it onto its haul route out of the loading area. After the loaded scraper is on its way, the pushcat usually backs up and prepares to engage the next scraper in the spread that enters the loading area.

The key time elements involved in the scraper cycle are the following:

- Pushcat maneuver time
- Load and boost time
- Loaded travel time
- Turn and dump time
- Return travel time

The key elements of the pushcat cycle time are:

- Pushcat maneuver time
- Load and boost time
- Pushcat return time

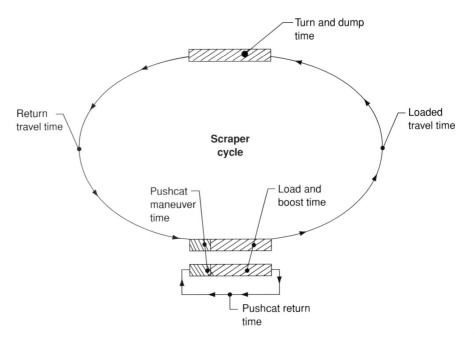

$$
\begin{aligned}
\text{Scraper cycle time} \atop \text{(Min)} \quad &= \quad \begin{aligned}&\text{Load and boost + loaded travel + turn and dump +}\\&\text{return travel + pushcat maneuver time}\end{aligned}\\[2ex]
\text{Pushcat cycle time} \atop \text{(Min)} \quad &= \quad \begin{aligned}&\text{Load and boost + pushcat return +}\\&\text{pushcat maneuver time}\end{aligned}\\[2ex]
{\text{No. of scrapers to} \atop \text{balance pushcats}} \quad &= \quad \frac{\text{Scraper cycle time}}{\text{pushcat cycle time}}
\end{aligned}
$$

FIGURE 9-6   Graphical depiction of scraper load-and-haul cycle.

Note that the first two time elements (pushcat maneuver time and load and boost time) are common to both the scraper and pushcat cycle times.

The loaded and return travel times are obtained by means of the travel time analysis previously described, yielding results such as those shown in Figure 9-5. The balance of the time elements depend on the size of the scraper and the size and number of pushcats loading it. Equipment manufacturers' handbooks contain this type of recommended information for three scraper/pushcat combinations. Typical values (suitable for estimating purposes) are listed in Table 9-2.

TABLE 9-2   Scraper and Pushcat Cycle Time Elements

Time element	Self-loading scraper, min	21 cy and less capacity, min	32 cy capacity, min
Pushcat maneuver		0.15	0.15
Load and boost	1.00	0.60–0.80	0.70–0.90
Pushcat return[1]		0.24–0.32	0.28–0.36
Turn and dump	0.70–0.90	0.60–0.70	0.60–0.70

[1]Taken as 0.4 × load and boost time.

Scraper Haul Unit and Spread Productivity

The remaining relationships necessary to determine the production capacity of a single scraper or of a pushcat/scraper "spread" are shown in Figure 9-7. The definition of the elements involved in these relationships and where or how the required numerical constants for estimating purposes can be obtained are as follows:

- **Scraper pay load:** The average load in bank cubic yards that will be carried by the scraper per cycle
- **Scraper struck capacity:** The water level capacity of the scraper bowl in cubic yards
- **Scraper load factor:** A dimensionless factor representing the decimal equivalent of the percentage of the scraper struck capacity represented by the number of bank cubic yards in the scraper payload. It is a measure of the degree to which the scraper bowl is filled during loading.
- **Scraper productivity:** The number of bank cubic yards that is being moved in one scraper-hour
- **Efficiency:** A decimal equivalent of the portion of the total work-hour that full theoretical production can be considered to be attained
- **Scraper spread production:** The number of bank cubic yards that is being moved in one pushcat/scraper spread-hour

As indicated on Figure 9-6, the **balance number,** or the number of scrapers in the pushcat/scraper spread that are required to just balance the productive capacity of the pushcat, will be given by dividing the scraper cycle time by the pushcat cycle time. If more scrapers than this balance number are used, they will just queue up behind the pushcat and contribute nothing to additional spread productivity. If fewer scrapers than the balance number are used, the pushcat will be underutilized and spread productivity will be decreased.

To obtain:	Given by:
Scraper payload, Bcy	(Scraper struck capacity)[1] (Scraper load factor)[2]
Scraper productivity, Bcy/hr/scraper	$\left(\dfrac{60 \text{ min}}{\text{cycle time, min}}\right)$(Payload, Bcy)(Efficiency)[3]
Scraper spread[4] production, Bcy/hr	$\left(\begin{array}{c}\text{Scraper productivity,} \\ \text{Bcy/hr/scraper}\end{array}\right)$(No. of scrapers in spread)[5]

[1]Obtain from scraper manufacture.

[2]Varies from 0.8 to 1.2 depending on nature of material and number of pushcats.

[3]Varies from 45 min. hr = 0.75 to 50 min. hr. = 0.83.

[4]A "spread" is one pushcat or pair of two pushcats working together and a given number of scrapers being loaded by that single pushcat or pair of two pushcats working together.

[5]The number of scrapers should never exceed the "balance number" of scrapers for one pushcat.

FIGURE 9-7 Pushcat/scraper spread productivity relationships.

PHOTO 9-6   Examples of factors affecting the elements of scraper load-and-haul analysis. Upper photo, 21 cy scraper bowl nearly full of blocky clay at end of load cycle (load factor approximately 0.90). Bottom photo, 21 cy scraper at start of turn-and-dump cycle carefully discharging material for compaction (turn-and-dump time = 0.50 to 0.75 min).

Usually, the theoretical balance number will not be a whole number, and the spread must operate at either the whole number below or the whole number above the balance number. It generally will be found to be more economical to operate the spread with the lower number of scrapers, or fewer, as long as the number used is not so few that the pushcat is badly underutilized. On very long hauls or when enough scrapers are not available, there is no choice, and the spread must operate with the pushcat being underutilized.

Computer-Aided Solution for Scraper Load-and-Haul Analysis

The printout for a complete computer-generated load-and-haul analysis using the California State University computer program for the pushcat/scraper spread shown in Figure 9-5 is shown in Figure 9-8. The assumed figures for load factor, load-and-boost time, pushcat return time, pushcat maneuver time, and turn-and-dump time are all as shown on the printout.

Note that in this case, the balance number is given by the haul unit cycle time of 8.83 min divided by the pushcat cycle time of 1.60 min, or 5.5 scrapers. The maximum

Project Information

Project Name EXAMPLE
Estimator ....SWB
Load Haul # 1
Ld. Material ....COMMON

Haul Vehicle ....CAT 631E
Efficiency (%) 83
Load Equipment 1 D9
Load Factor 0.85

Load & Boost (min) 1.00
P.C. Return (min) 0.40
P.C. Maneuver (min) 0.20
Turn & Dump (min) 0.50

Loaded/Return Description	Seg Dist Feet	R.R. Lb/Ton	G.R. %	THRR %	Sus Spd mph	Diff Sus Spd	Spd Factor	Av Spd mph	Travel min
Travel in Cut	300	200	0	10.0	7.5	-7.5	0.91	6.8	0.50
Haul Segment	1000	100	5	10.0	7.5	0.0	1.00	7.5	1.52
Haul Segment	2500	40	3	5.0	15.3	-7.8	0.98	15.0	1.89
Haul Segment	800	100	-6	-1.0	31.0	-15.7	0.80	24.8	0.37
Travel on Fill	200	200	0	10.0	7.5	23.5	1.25	9.4	0.24
Travel on Fill	200	200	0	10.0	12.5	-12.5	0.81	10.1	0.22
Return Segment	800	100	6	11.0	11.7	0.8	1.00	11.7	0.78
Return Segment	2500	40	-3	-1.0	31.0	-19.3	0.91	28.2	1.01
Return Segment	1000	100	-5	0.0	30.0	1.0	1.01	30.3	0.38
Travel in Cut	300	200	0	10.0	12.5	17.5	1.18	14.8	0.23
Travel Data	9600							15.29	7.13

Worksheet Calculations

Haul Unit Struck Capacity (cy)			21.0
Haul Unit Payload (bcy)			17.8
P.C. Cycle Time (min)			1.60
Haul Unit Cycle Time (min)			8.83
Haul Unit Prod bcy/hr			101
Balanced # Haul Units			5.52
Spread Prod bcy/hr using	5	Haul Units	503
Spread Prod bcy/hr using	6	Or > Haul Units	556

FIGURE 9-8 Computer-generated load-and-haul analysis for a pushcat/scraper spread operating on the 4,800 ft one-way haul depicted by Figure 9-5.

spread productivity that could be obtained is either 503 Bcy/hr or 556 Bcy/hr depending on whether five or six scrapers are used. Also, if three scrapers were used, the spread productivity would be 3×101 Bcy/hr = 303 Bcy/hr. Also noteworthy is the level of haul speeds achieved (maximum 24.8 mph loaded and 30.3 mph return). With well laid out haul roads and proper haul road maintenance, these speeds are reasonable.

TRUCK LOAD-AND-HAUL ANALYSIS

Truck load-and-haul analysis is very similar to that for pushcat/scraper spreads. The procedures for characterizing the haul road and determining loaded and return travel times are exactly the same. The principal difference is in determining the loading cycle. This is a considerably more complicated matter when trucks being loaded by some type of load unit are involved.

Whereas scrapers are either self-loading or, more commonly, push-loaded by pushcats, truck spreads are always top-loaded, usually by a loading unit equipped with some sort of a bucket. For instance, trucks are commonly loaded by the following bucket-equipped loading units:

- Front-end loaders
- Front shovels
- Backhoes
- Cable-operated draglines
- Cable-operated clamshells

In all these cases, the load cycle will require a number of passes of the loader bucket to fill the truck. Also, time must often be included for the truck to pull in under the loading unit following the loading of the previous truck.

Another variable is that truck spreads are often employed to move blasted rock or large natural cobble or talus deposits. These materials are very difficult, if not impossible, to move with scrapers.

Finally, trucks are sometimes loaded by conveyor belts or overhead muck bins. These relatively rare loading situations (which require separate analysis) are not encountered with scrapers.

Truck Load Time Determination

The relationships necessary to determine the time duration for loading a truck are shown in Figure 9-9.

The elements of **truck loading time** and the definition of the terms involved in its calculation include the following:

- **Truck payload (Bcy):** The quantity of material in bank cubic yards moved in one complete truck haul cycle.
- **Truck rated capacity (tons):** Unlike scrapers, where the capacity is rated by the manufacturer on a volumetric basis, truck manufacturers in the United States rate truck capacity in U.S. tons. Unless the material being handled is extremely light, the truck payload calculated on the basis of the truck-rated capacity in tons will be accurate. Truck manufacturers also rate their trucks on the basis of the struck (water level) capacity and heaped (2:1 slopes) capacity in cubic yards for determination of payload on a volumetric basis. If the truck pay-

To obtain:	Given by:
Truck payload, Bcy	$\dfrac{\text{(Rated capacity, tons)}^{(1)}\text{(2000 lb/ton)}}{\text{(In-situ material weight, lb/Bcy)}^{(2)}}$
Bcy per loader bucket	$\dfrac{\text{(Rated bucket-size, cy)}^{(3)}}{\left(1 + \dfrac{\text{swell \%}}{100}\right)^{(4)}}\text{(Bucket fill factor)}^{(5)}$
Truck load time, min	$\dfrac{\text{(Truck payload, Bcy)}}{\text{(Bcy per loader bucket)}}\text{(Loader cycle time, min)}^{(7)} + \left[\begin{array}{cc}\text{Truck spot \&} & \text{Loader} \\ \text{maneuver} \;-\; & \text{cycle} \\ \text{time, min}^{(6)} & \text{time, min}\end{array}\right]^{(8)}$

(1) From manufacturer.

(2) Computed from site tests.

(3) From manufacturer.

(4) Soil: 10–50%; rock: 50–100%.

(5) Soil: 0.80–1.20; rock: 0.60–0.90.

(6) Varies from 0.25–1.50 min.

(7) Front-end loaders: 0.30–0.60 min; Hydraulic backhoes/shovels: 0.25–50 min; Draglines & clamshells: 0.75 min plus.

(8) Bracketed duration cannot be less than zero.

FIGURE 9-9   Relationships for determination of truck load time duration.

load in bank cubic yards calculated on the basis of the relationships shown in Figure 9-9 [when converted to loose cubic yards (Lcy) by taking the material swell into consideration] exceeds the heaped capacity listed by the manufacturer, the payload must be reduced accordingly.

For instance, Caterpillar Inc., rates their 773B end-dump truck at 50 tons payload capacity and also lists the volumetric capacity (heaped at 2:1 slopes) as 44.6 cy. If a material with an in situ unit weight of 3,200 lb/Bcy and a swell factor of 30% were being hauled, the payload based on weight would be (50 tons)(2000 lb/ton) / 3200 lb/Bcy = 31.3 Bcy. On a volumetric basis, the volume of 1 Bcy after it had been loosened during excavation would be 1.30 Lcy, and the truck payload would be 44.6 cy / 1.3 = 34.3 Bcy. The determination based on weight (31.3 Bcy) should be used. However, if the material was extremely light, with an in situ weight of 2,200 lb/Bcy and with a swell factor of 15%, the payload based on weight would be (50 tons) (2,000 lb/ton) / 2,200 lb/Bcy = 45.5 Bcy. However, based on the volumetric capacity of the truck, the payload would be 44.6 / 1.15 = 38.8 Bcy, which would govern.

- **In situ material weight (lb/Bcy):** The unit weight of the material being moved, including the weight of any natural moisture content, must be known to calculate truck payload. This unit weight is often obtained during the geotechnical investigation made for important projects by the owner's engineer and reported in the project bidding documents. Otherwise, it must be estimated from other geotechnical information contained in the bidding documents or from contractor-performed field tests.

    For instance, if the geotechnical report had indicated that the material to be moved had an average in situ dry density of 92.5 lb/cf and an average in situ moisture content of 23.5%, the weight of 1 Bcy to be hauled would be (27 cf/cy)(92.5 lb/cf)(1.235) = 3,084 lb/Bcy. If the in situ dry unit weight and moisture content are not stated, they can be determined by contractor-performed prebid field tests or can be estimated from past experience.

Alternately, if the material to be hauled was solid rock that was to be loosened by blasting prior to being loaded, the in situ weight would be $(27 \text{ cf/cy})(62.4 \text{ lb/cf}^2)(2.65^3) = 4,465$ lb/Bcy.

- **Bank cubic yards per loader bucket:** The content in bank cubic yards of the bucket of the loading equipment being used that, on the average, will be loaded into the truck by one loader bucket cycle.
- **Rated bucket size (cy):** The manufacturer's rated heaped capacity in cubic yards of the bucket being used.
- **Material swell (%):** The percentage volumetric increase when the material being moved is loosened from the in situ density to the much lower density of broken up, loose material during the loading operation. Swell factors vary considerably, depending on the nature of the material being excavated.
- **Bucket fill factor:** A dimensionless factor representing the decimal equivalent of the percentage of the full-rated bucket capacity that, on the average, will be retained in the bucket as the bucket moves through the material during each bucket pass, or cycle. It depends on the type of bucket and the nature of the material being excavated.
- **Truck load time (min):** The time duration, in minutes, that, on average, is required to fill the truck to its payload quantity. It is measured from the point in time at which the loader bucket has dumped the first bucket load and started its return swing to the material being loaded to the similar point in time for the last bucket loaded.
- **Loader cycle time (min):** The time duration, in minutes, that, on average, is required for the loader to complete one loader bucket cycle, measured from the point in time at which the loader dumps its contents into the truck, moves back to the material being loaded, obtains another bucket of material, moves back to the truck, and completes dumping the material into the truck. It depends greatly on the type of loading equipment being utilized and on the configuration of the loading situation.

    For instance, contrast the difference between a front-end loader filling trucks from a material bank where the loader

    - Backs up a short distance after dumping its bucket over the side of the truck bed
    - Turns 90°, moves a short distance into the material bank, and obtains another bucket of material
    - Backs up, turns 90°, moves forward to the truck, and dumps its bucket over the side of the truck bed as before

    to the situation where a large cable-operated dragline operating with a 150 ft boom

    - Lifts its bucket free of the truck after dumping over the side of the truck bed, swings the crane boom and bucket over a 180° arc, and casts it outward to the area that is being excavated
    - Retrieves the bucket with the drag cable toward the crane, loading the bucket
    - Lifts the bucket, swings the boom and bucket 180°, lowers the bucket over the side of the truck bed, and dumps it

    The loader cycle could be as fast as 0.30 min in the first situation, whereas the second would require 3 to 4 times as much time. Or, in case of a cable-operated

---

[2]The unit weight of water.

[3]The average specific gravity of soil or rock solids.

PHOTO 9-7   Examples of factors affecting the elements of loader-truck haul analysis. Upper photo, $12\frac{1}{2}$ cy loader bucket dumping large rock (swell = 50% plus; bucket factor = ±0.90). Lower photo, 50 ton end-dump tuck dumping common earth material under direction of a dumpman in random fill zone on a dam (turn-and-dump timc = 0.50 min plus).

To obtain:	Given by:
Truck cycle time, min	$\left(\begin{array}{c}\text{Truck load}\\\text{time, min}\end{array}\right) + \left(\begin{array}{c}\text{Loaded travel}\\\text{time, min}\end{array}\right) + \left(\begin{array}{c}\text{Turn \& dump}\\\text{time, min}\end{array}\right)^{(1)} + \left(\begin{array}{c}\text{Return travel}\\\text{time, min}\end{array}\right)$
No. of trucks to balance loader	$\dfrac{\text{(Truck cycle time, min)}}{\text{(Loader cycle time, min)}}$
Truck unit productivity, Bcy/hr	$\dfrac{60 \text{ min}}{\text{(Truck cycle time, min)}} \text{(Truck payload, Bcy)(Efficiency)}^{(2)}$
Truck spread$^{(3)}$ productivity, Bcy/hr	(No. of trucks in spread)$^{(4)}$ (Truck unit productivity, Bcy/hr)

$^{(1)}$Varies 0.25–1.5 min.

$^{(2)}$45-min hr = 0.75%; 50-min hr = 0.83%.

$^{(3)}$A "spread" consists of 1 load unit + a given number of trucks being loaded by that load unit.

$^{(4)}$Should never exceed the balanace number. May be less.

FIGURE 9-10   Relationships for determination of truck/loader haul unit and spread productivity.

clamshell loading material from the bottom of a deep shaft or from within a congested cofferdam, the required loader cycle time could be even longer.

- **Truck spot and maneuver time (min):** The time duration in minutes that, on the average, is required for a truck when loaded (after the final loader bucket has been dumped) to pull out of the loading location to the start of the haul route and for the next truck (assumed to have just arrived) to pull into the loading position. Its duration will depend on the configuration of the loading situation.

Truck and Loader Haul Unit and Spread Productivity

Once the truck load time duration has been determined (as explained above) and the loaded and return travel time has been determined by the procedure explained earlier in this chapter (which procedure is exactly the same for a truck/loader operation as for a pushcat/scraper operation), **truck/loader haul unit** and **spread productivity** are easily determined. The necessary relationships are shown in Figure 9-10. The key elements all have parallel definitions and significance to the similar elements previously discussed for pushcat/scraper operations and need not be explained further.

Computer-Aided Solution for Truck and Loader Load-and-Haul Analysis

Just as for the manual computations necessary to analyze pushcat/loader operations, the analysis of truck/loader operations by manual methods is tedious and time consuming. Computer programs in which the necessary rimpull and other equipment manufacturers' information for specific trucks are incorporated are easy to use and generate complete solutions very quickly. The printout for the analysis of the operation of a 12 cy front-end loader and Caterpillar 773B 50 ton end-dump trucks moving rock excavation on a 11,900 ft one-way haul obtained by use of a truck/loader computer program developed at California State University, Chico, is shown as Figure 9-11.

Project Information

```
==========================================================================================
```

Project Name EXAMPLE	Haul Vehicle ....CAT 773B	Bucket Factor 0.90
Estimator ....SWB	Mtrl Wt. (#/Bcy) 4400	Loader Cycle (min) 0.40
Load Haul # ROCK.1	Mtrl Swell (%) 60	Spot & Mnvr (min) 0.50
Efficiency (%).............83	Load Bckt (cy) 12.0	Turn & Dump (min) 0.50

Loaded/Return Description	Seg Dist Feet	R.R. Lb/Ton	G.R. %	THRR %	Sus Spd mph	Diff Sus Spd	Spd Factor	Av Spd mph	Travel min
Travel in Cut	300	150	0	7.5	14.1	-14.1	0.77	10.9	0.31
Haul Segment	1000	100	5	10.0	10.0	4.1	1.03	10.3	1.10
Haul Segment	8500	40	3	5.0	21.3	-11.3	1.00	21.3	4.55
Haul Segment	1800	100	-6	-1.0	40.0	-18.8	0.84	33.6	0.61
Travel on Fill	300	150	0	7.5	14.1	25.9	1.27	17.9	0.19
Travel on Fill	300	150	0	7.5	27.8	-27.8	0.40	11.1	0.31
Return Segment	1800	100	6	11.0	20.0	7.8	1.04	20.8	0.98
Return Segment	8500	40	-3	-1.0	40.0	-20.0	1.00	40.0	2.41
Return Segment	1000	100	-5	0.0	38.0	2.0	1.02	38.8	0.29
Travel in Cut	300	150	0	7.5	27.8	10.3	1.11	30.8	0.11
Travel Data	23800							24.88	10.87

Worksheet Calculations

```
==========================================================================================
```

Haul Unit Payload (tons)			50.0
Haul Unit Payload (bcy)			22.7
Loader Bucket Payload (bcy)			6.75
Number of Loader Buckets			3.37
Haul Unit Load Time (min)			1.45
Haul Unit Cycle Time (min)			12.82
Haul Unit Prod. bcy/hr			88
Balanced # Haul Units			8.86
Spread Prod bcy/hr using	8	Haul Units	706
Spread Prod bcy/hr using	9	Or > Haul Units	782

FIGURE 9-11 Computer-generated load-and-haul analysis for a truck/loader spread moving blasted rock on a 11,900 ft one-way haul.

In this case, the rock in situ weight was estimated to be 4,400 lb/Bcy (nearly solid rock) and assumed to swell 60% when loosened by blasting. Other assumed figures for bucket factor, loader cycle time, spot and maneuver time, and turn-and-dump time were as shown in Figure 9-11, resulting in a haul unit productivity of 88 Bcy/hr and a spread productivity of 706 Bcy/hr, based on utilizing eight trucks.

A computer-aided solution generated by the California State University, computer program for the operation of Dart 5130 130 ton bottom-dump haul units loaded by a 15 cy front-end loader moving common excavation on a 20,467 ft one-way haul is shown in Figure 9-12.

For this operation, the in situ weight of the material (including the natural moisture content) was estimated to be 2,850 lb/Bcy and the assumed bucket factor and time duration constants were as shown in Figure 9-12. The haul unit productivity was 177 Bcy/hr, and the spread productivity using six haul units was 1,062 Bcy/hr. These production figures illustrate the suitability of these very large haul units for long hauls with little or no adverse grade throughout most of the haul length.

Project Information

```
=================================================================================================
Project Name  COMPARE              Haul Vehicle .......DART 5130      Bucket Factor ..............1.25
Estimator ...............KEY        Mtrl Wt. (#/Bcy) ............2850   Loader Cycle (min) ........0.50
Load Haul # ..................2     Mtrl Swell (%) .................30  Spot & Mnvr (min) .........0.75
Efficiency (%).............75       Load Bckt (cy) ..............15.0   Turn & Dump (min) .......0.75
```

Loaded/Return Description	Seg Dist Feet	R.R. Lb/Ton	G.R. %	THRR %	Sus Spd mph	Diff Sus Spd	Spd Factor	Av Spd mph	Travel min
Travel in Cut	800	200	0	10.0	5.5	-5.5	0.96	5.3	1.72
Haul Segment	18000	40	-2	0.0	33.0	-27.5	1.00	33.0	6.20
Haul Segment	667	100	6	11.0	4.8	28.3	1.27	6.0	1.26
Travel on Fill	1000	150	0	7.5	6.8	-2.0	0.99	6.7	1.70
Travel on Fill	1000	150	0	7.5	18.5	-18.5	0.77	14.2	0.80
Return Segment	667	100	-6	-1.0	30.5	-12.0	0.85	25.9	0.29
Return Segment	18000	40	2	4.0	31.5	-1.0	1.00	31.5	6.49
Travel in Cut	800	200	0	10.0	13.5	18.0	1.16	15.7	0.58
Travel Data	23800							24.43	19.04

Worksheet Calculations

```
=================================================================================================
```

Haul Unit Payload (tons)			130.0
Haul Unit Payload (bcy)			91.2
Loader Bucket Payload (bcy)			14.42
Number of Loader Buckets			6.33
Haul Unit Load Time (min)			3.41
Haul Unit Cycle Time (min)			23.20
Haul Unit Prod. bcy/hr			177
Balanced # Haul Units			6.80
Spread Prod bcy/hr using	6	Haul Units	1062
Spread Prod bcy/hr using	7	Or > Haul Units	1203

FIGURE 9-12   Computer-generated load-and-haul analysis for a truck/loader spread moving common material on a 20,467 ft one-way haul.

SPREAD AND COMPACT OPERATIONS

Closely related to load-and-haul operations is the work of handling the material after it is dumped at the end of the loaded haul. If the material is to be wasted or stockpiled (as distinct from utilizing it in constructing some kind of engineered embankment or fill), all that is necessary is to shove the material over the edge of the dumpfill (at the edge of a steeply sloped ravine, for instance) or to push it over the edge of a rising fan-shaped stockpile on level ground. In this instance, the only equipment on the dump would be a track or rubber-tire-mounted bulldozer, possibly assisted by a laborer to flag trucks to keep them clear of the soft dump shoulder when they dump.

On the other hand, when engineered embankments are being constructed from common earth materials, the dumped material will have to be spread in uniform horizontal layers of a controlled thickness, disked or otherwise manipulated to reduce the moisture content (or watered and manipulated to increase the moisture content), and then compacted in place to increase its density to some specified minimum figure. In this case, in addition to a flag person to control traffic, the required crew will include some sort of spreading equipment (either a bulldozer or motor grader), compaction units, and if moisture control is required, a water tanker or towed disking units (or some other type of soil manipulating equipment). Also, bulldozers (with or without special slope board attachments) are usually necessary to finish exposed slopes to required tolerances.

Embankments constructed from naturally occurring broken rock or sand gravel mixtures or from ripped or blasted rock are constructed similarly except that (1) moisture control is seldom required, (2) the thickness of the layers prior to compaction is thicker than for common earth materials, and (3) either vibratory compaction or very heavy towed pneumatic compactors will be used.

The various possible combinations are so numerous that they cannot be discussed within the scope of this text. However, the basic relationships applying to compaction operations will be briefly explained below.

Shrink and Swell

Practically all materials **swell** when excavated from the in situ natural condition (bank condition) to a loose condition in which they are hauled, dumped, and spread on an embankment prior to compaction. The material in this loose condition will be less dense than the in situ condition in the bank. If there were no swell, a layer removed from the bank of thickness t in. when spread on an embankment would still be t in. thick. Because swell does occur, the layer will be t (1 + % swell/100) in. thick when spread on the embankment prior to any compaction. If the layer were then compacted by rolling over it with some sort of compactor unit until its density was increased to the original density of the in situ bank condition, its thickness would be reduced back to t in.

Most engineered embankments constructed from common earth materials are required to be compacted to a density that exceeds the natural in situ density of the materials prior to being excavated. Whereas the swell of a material reflects its volumetric increase when passing from the in situ bank condition to the loose condition, the **shrink** reflects the volumetric decrease between the in situ bank condition and the final compacted condition in the engineered embankment.

The extent to which common earth materials will shrink is highly dependent on the density of the material in its natural in situ condition and on the degree of compaction required. For instance, if project bidding documents stated that the average in situ dry density of the material specified for use in constructing an engineered embankment was 94.5 lb/cf and the required minimum dry density of the compacted embankment was to be 110.0 lb/cf, the cubic feet of in situ material to construct 1 cf of embankment would be 110.0/94.5 = 1.16 cf and the shrink would be

$$\frac{(1.16 \text{ cf} - 1.00 \text{ cf})}{1.16 \text{ cf}} \times 100 = 13.8\%$$

It follows that a layer removed from the bank of thickness t in., after being hauled, spread on the embankment, and then compacted, would be reduced in thickness to t (1 − shrink %/100) in.

Whereas sand and gravel and fine-grained soil materials swell between less than 10 to 30% (occasionally as much as 50%) and shrink from less than 10% to as much as 20 to 25%, respectively, blasted rock will swell from 50 to 100% and will exhibit a negative shrink. In other words, regardless of how much compactive effort is expended, blasted rock can never be compacted to a density equal to that of the original solid rock. This negative shrink (which amounts to a swell) could be as high as 25 or 30%.

Thickness of the Compacted Layer

Compaction specifications generally state the thickness to which the material going into an engineered fill must be spread prior to compaction. The condition of the material, spread in this uncompacted layer, will be the loose condition. During compaction to the specified density, the uncompacted layer thickness will gradually decrease to the final thickness of the fully compacted layer, which by application of the swell and shrink factors discussed above, will be

$$\frac{(\text{Thickness of layer before compaction, in.})(1 - \text{shrink }\%/100)}{(1 + \text{swell }\%/100)}$$

For instance, an 8-in.-thick uncompacted layer, composed of a material with a 25% swell and a 12% shrink, would reduce to

$$\frac{(8 \text{ in.})(1 - 12/100)}{(1 + 25/100)} = 5.6 \text{ in.}$$

and a 2-ft-thick uncompacted layer of blasted rock with a swell of 60% and a shrink of −20% would reduce to

$$\frac{(2 \text{ ft})[1 - (-12)/100]}{(1 + 60/100)} = 1.4 \text{ ft}$$

Compaction unit productivity estimates always involve an estimate of the thickness of the compacted layer for the particular material being spread and compacted to a specified final density.

Compaction Unit Productivity

Based on the preceding discussion, the relationships necessary to estimate compaction productivity are shown on Figure 9-13. The elements involved include

- Thickness of the compacted layer (ft)
- The compactor travel speed (mph)
- The width of one pass (ft)
- The number of passes required (each)
- General efficiency of the compaction operation, %

Methods for estimating the thickness of the compacted layer in a particular case were previously discussed. The other elements and how to determine them are explained below.

Compactor Travel Speed

The range of normal travel speeds in miles per hour for self-powered compactors is listed in the manufacturers' handbooks or other literature. Depending on the configuration of the fill area, it may not be possible for the compactor to ever attain the top-rated speed and the average speed will be much less. Using an average speed of one-half the top-rated speed is reasonable.

Towed compactors are usually pulled by track-mounted tractors operating at average speeds of 4 to 5 mph. They are also much more difficult to turn and they cannot operate in reverse easily.

To obtain:	Given by:
Compactor production, Ecy/hr	$$\dfrac{\left(\begin{array}{c}\text{Thickness of}\\ \text{compacted}\\ \text{layer, ft}\end{array}\right)\left(\begin{array}{c}\text{Compactor}\\ \text{speed, mph}\end{array}\right)\left(\dfrac{88}{\text{ft/min/mph}}\right)\left(\begin{array}{c}\text{Width of}\\ \text{one pass,}\\ \text{ft}\end{array}\right)\left(\dfrac{60}{\text{min}}\right)\left(\dfrac{\text{Efficiency \%}}{100}\right)}{(\text{No. of passes required})(27\ \text{cf/cy})}$$
Thickness of compacted, layer, ft	$$\dfrac{\left(\begin{array}{c}\text{Thickness of layer}\\ \text{before compaction, ft}\end{array}\right)}{\left(1 + \dfrac{\text{swell \%}}{100}\right)}\left(1 - \dfrac{\text{shrink \%}}{100}\right)$$
Compactor production, Bcy/hr	$$\dfrac{(\text{Production in Ecy/hr})}{\left(1 - \dfrac{\text{shrink \%}}{100}\right)}$$

FIGURE 9-13 Compaction production relationships.

Width of One Pass and Number of Required Passes

Compactor manufacturers' literature lists the effective width in feet of the compaction rolling unit in contact with the top of the layer being compacted during one traverse, or pass, of the compactor. If there are two or more rolling units that are fixed in tandem on the compactor, the latter rolling unit will follow in the path of the first during one pass of the compactor. This fact must be taken into account when determining the number of passes of the compactor that will be required to achieve the specified degree of compaction.

Occasionally, construction specifications will state not only the layer thickness that the material must be placed in before compaction but will also specify that each layer is to be compacted by a minimum number of passes of a designated type of compactor with a specified weight per compactor drum (or wheel) per foot of width or by some other method of defining the magnitude of compaction contact loading. Sometimes, specifications will state the minimum number of required passes of a particular compactor make and model that is required to be used.

For instance, a Caterpillar 825C compactor is a self-powered, two-axle, articulated unit with two 3-ft-8-in.-wide tamping foot wheels on each axle. The rear two wheels track the path of the lead wheels. The unit has an operating weight of 71,429 lb. The load per foot of wheel width is (71,429 lb)/(4 × 3.67 ft) = 4,866 lb, or 2.43 ton/ft. If four passes *of the compactor* had been specified, the compacted layer would have received the 2.4 tons/ft load eight separate times. Under this type of specification, this level of compactive effort would have been determined by the project designer to be sufficient (probably based on compaction tests made during the project geotechnical investigation).

More commonly, the compaction specifications will state that the material must be placed at a moisture content within narrow limits of the optimum moisture content[4] for the particular material in layers of a specified uncompacted thickness and then must be compacted to a specified dry density in pounds per cubic foot. In this case, a bidding contractor is free to choose the type of compaction equipment but, to estimate the probable cost of compaction, must make the determination of the number of passes that

---

[4]Optimum moisture content of fine-grained soils such as silts and clays is the moisture content that has been determined by soil tests to be that at which maximum soil density can be achieved by application of a standard compaction effort. The achievable density of embankments constructed of naturally broken or blasted rock is relatively insensitive to moisture content and the term has no meaning in this case.

will be required to achieve the specified degree of compaction. This would be done on the basis of prior experience, consultation with hired geotechnical experts, or prebid field tests.

General Efficiency of Compaction Operation

Like all construction operations, compaction spreads will never achieve production levels calculated on theoretical assumptions of no delay or other inefficiencies. Load-and-haul spread production is discounted by most estimators to a 50 min hour (83% efficiency) or a 45 min hour (75% efficiency). The accompanying compaction operation should be discounted similarly.

Example Compactor Production Calculations

Assume soil was being compacted by means of the previously mentioned Caterpillar 825C compactor traveling an average speed of 6 mph. Four passes *of the compactor* were believed to be required.[5] Further assume that the thickness of the compacted layer had been determined to be 5.6 in. Taking an operating efficiency of 75%,

$$\text{Production} = \frac{(5.6 \text{ in.}/12)(6.0 \text{ mph})(88)(2 \times 3.67 \text{ ft})(60 \text{ min})(0.75)}{(4)(27)}$$

$$= 754 \text{ Ecy/hr}$$

If the shrink factor had been 15%, this is equivalent to $754/(1 - 0.15) = 887$ Bcy/hr.

Similarly, if a 50 ton, 12-ft-wide single-axle pneumatic roller towed by a Caterpillar D8 tractor moving at an average travel speed of 3 mph was compacting blasted rock to a 2 ft compacted layer thickness requiring four passes of the roller, the compaction production would be

$$\frac{(2 \text{ ft})(3 \text{ mph})(88)(12 \text{ ft})(60 \text{ min})(0.75)}{(4)(27)} = 2640 \text{ Ecy/hr}$$

and if 1 Bcy of the rock prior to blasting swelled to the extent that it made 1.15 Ecy in the compacted rock fill (15% negative shrink), the compaction production in terms of bank cubic yards of rock compacted in the fill would be $2640/1.15 = 2296$ Bcy/hr.

COMPREHENSIVE EXAMPLE PROBLEM ILLUSTRATING LOAD, HAUL, AND COMPACT PRINCIPLES

To illustrate the foregoing principles in this chapter consider a hypothetical earthfill dam project illustrated by Figures 9-14 and 9-15. The dam is to be constructed on a foundation composed of river bed alluvium underlain by an impervious rock formation of substantial depth. The dam embankment consists of two separate zones—an impervious core to be obtained from a clay borrow area located about 1 mile upstream from the dam

---

[5]This is equivalent to eight passes of a single compaction wheel loaded at 2.43 tons $\times$ 3.67 ft = 8.92 tons.

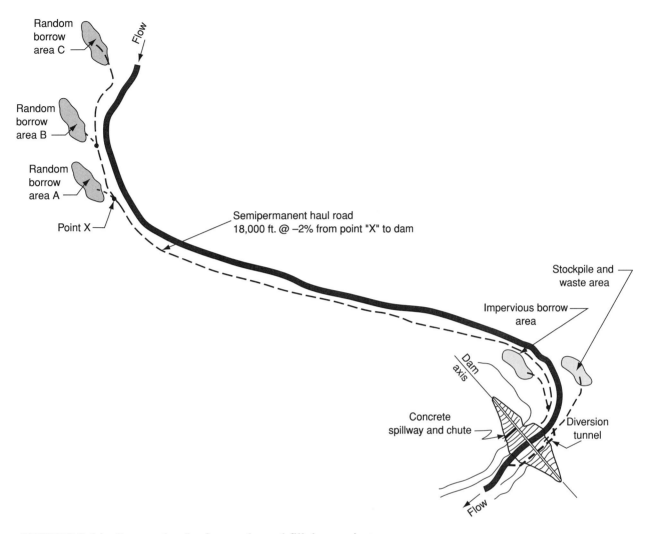

FIGURE 9-14   Layout sketch of example earthfill dam project.

and random fill made up of silty clays, sandy gravel, and small cobbles obtained from required foundation excavation and from borrow areas approximately $3\frac{1}{2}$ miles upstream from the dam. The dam embankment contains a reinforced concrete gated spillway and chute section and contains filter and drainage blankets consisting of processed graded material protecting the impervious core from migration of fines due to seepage.

The general working season during which embankment may be placed extends from March 1st through November 30th each year. The river high water flow period occurs during March and April. Ten work days can be expected to be lost each working season due to rain.

After construction of the diversion tunnel, the plan for the excavation and embankment of the dam consists of the following major steps.

First Construction Season

- Divert the river through the tunnel on April 1 by placing the upstream closure cofferdam. Isolate the stream bed construction area by placing the downstream closure cofferdam

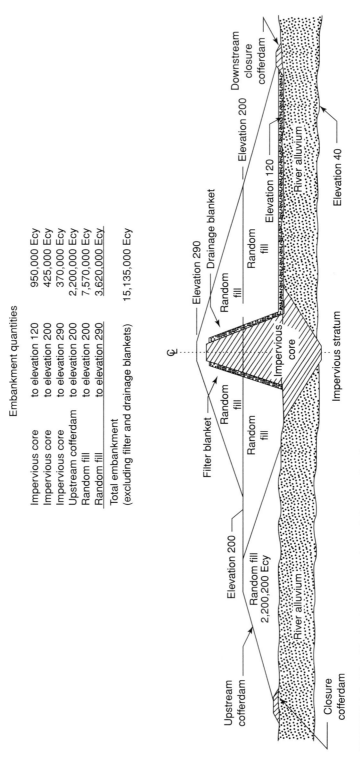

Embankment quantities

Impervious core	to elevation 120	950,000 Ecy
Impervious core	to elevation 200	425,000 Ecy
Impervious core	to elevation 290	370,000 Ecy
Upstream cofferdam	to elevation 200	2,200,000 Ecy
Random fill	to elevation 200	7,570,000 Ecy
Random fill	to elevation 290	3,620,000 Ecy
Total embankment (excluding filter and drainage blankets)		15,135,000 Ecy

FIGURE 9-15 Typical cross section of dam embankment.

PHOTO 9-8 These photos illustrate embankment zones in an earthfill dam that are typical for the type of project illustrated by the example problem featured in this chapter.

- Install deep well dewatering system, dewater, and excavate core trench to impervious rock formation
- Construct upstream protection cofferdam to elevation 200 and place impervious embankment in core trench to original stream bed

Second Construction Season

- Construct concrete spillway to a height sufficient to permit placement of embankment to elevation 200
- Place impervious core and random embankment

Third Construction Season

- Complete all dam embankment by November 30
- Complete concrete spillway

The bid form in the bidding documents contains the following three bid items for payment to the contractor for the major excavation and embankment work:

BI 10 Core trench and   1,150,000 Bcy at $_____/Bcy = $_____
       abutment excavation

BI 11 Impervious borrow   2,181,250 Bcy at $_____/Bcy = $_____

BI 12 Random borrow   14,870,500 Bcy at $_____/Bcy = $_____

BI 13 Impervious embankment   1,745,000 Ecy at $_____/Ecy = $_____

BI 14 Random embankment   13,390,000 Ecy at $_____/Ecy = $_____

The major issues confronting the construction contractor in planning and estimating the cost of the excavation and embankment work for this type of project consist of the following:

- What is the plan for the required movement of material?
- How are the various required hauls for each required movement of material to be characterized?
- What load-and-haul equipment choices are reasonable for accomplishment of the work?
- How many major load-and-haul equipment units must be employed to meet the time requirements of the project?
- What type and amount of compaction equipment and other support equipment will be required?
- What crew compositions are appropriate for the work?
- What are the estimated costs?

Each of these issues will be dealt with in turn.

Load-and-Haul Plan

Assume that, after the necessary takeoff work had been completed, the contractor concluded that the necessary major embankment quantities could be broken down into the quantities shown in Figure 9-15 to model the required load-and-haul operations. The relatively small quantity required for the closure cofferdams is not included because this work is usually included as part of a separate bid item for care and diversion of the river. Also, the filter and drainage blankets are covered under separate bid items and are excluded because they will be composed of sand and gravel material recovered from the stream bed and processed through a screening plant. Both operations are distinctly different from the major excavation and embankment operations and, for this reason, are not considered in the illustrative sample problem.

Based on the embankment quantities shown in Figure 9-15, the bid quantities for impervious and random borrow excavation are consistent with a shrink factor of 20% for the impervious material and 15% for the random if one makes the assumption that 750,000 Ecy of random embankment can be made from a portion of the 1,150,000 Bcy of abutment and core trench excavation. This is a reasonable assumption and on this basis the load-and-haul plan for the excavation bid items is reflected in Table 9-3. Eight separate hauls (hauls 1 through 8) are involved.

The intended use of the material from the load-and-haul operations to construct the various embankment bid items shown in Table 9-3 is shown in Table 9-4.

Haul Road Characterization

The abutment and core trench excavation to the upstream stockpile and waste area is generally an uphill haul (haul 1) on a ramp out of the core trench and then on a slight uphill haul on a haul road along the river to the dump area. The return haul for the reclaimed material for part of the random fill (haul 2) is generally the reverse of haul 1 and then ramping up on to the random fill at the dam.

The impervious borrow hauls (hauls 3, 4, and 5) are generally slightly downhill hauls to the dam and then either down or up on a ramp to the fill surface depending on the height of the impervious fill (haul 3 ramps down, and hauls 4 and 5 ramp up).

TABLE 9-3 Load-and-Haul Plan for Excavation Bid Items

Bid item	Quantity, Bcy	Use of material and fill quantities produced	Haul
10	882,400	To stockpile at waste area for use in random fill (882,400 Bcy $\times$ 0.85 = 750,000 Ecy)[1]	1
	267,600	To be wasted at waste area	1
	$\overline{}$	Recover 882,400 Bcy from stockpile to random fill to elevation 200	2
	1,150,000		
11	1,187,500	Impervious borrow area to core embankment to original stream bed (1,187,500 Bcy $\times$ 0.80 = 950,000 Ecy)[2]	3
	531,250	Impervious borrow area to core embankment to elevation 200 (531,250 Bcy $\times$ 0.80 = 425,000 Ecy)[2]	4
	462,500	Impervious borrow area to core embankment to elevation 290 (462,500 Bcy $\times$ 0.80 = 370,000 Ecy)[2]	5
	$\overline{}$		
	2,181,250		
12	2,588,200	Random borrow area A to upstream cofferdam (2,588,200 Bcy $\times$ 0.85 = 2,200,000 Ecy)[1]	6
	8,023,500	Random borrow area B to random embankment to elevation 200 (8,023,500 Bcy $\times$ 0.85 = 6,820,000 Ecy)[1]	7
	4,258,800	Random borrow area C to random embankment to elevation 290 (4,258,800 Bcy $\times$ 0.85 = 3,620,000 Ecy)[1]	8
	$\overline{}$		
	14,870,500		

[1]Based on assumed shrinkage factor of 15%.

[2]Based on assumed shrinkage factor of 20%.

TABLE 9-4   Source of Material for Embankment Construction

Bid item	Quantity, Ecy	Quantity and source of material	Haul notes
13	950,000	1,187,500 Bcy from impervious borrow	Delivered under haul 3
	425,000	531,250 Bcy from impervious borrow	Delivered under haul 4
	370,000	462,500 Bcy from impervious borrow	Delivered under haul 5
	1,745,000		
14	750,000	882,400 Bcy from stockpile	Delivered under haul 2
	2,200,000	2,588,200 Bcy from random borrow area A	Delivered under haul 6
	6,820,000	8,023,500 Bcy from random borrow area C	Delivered under haul 7
	3,620,000	4,258,800 Bcy from random borrow area B	Delivered under haul 8
	13,390,000		

The random borrow hauls (hauls 6, 7, and 8) are all extremely long ($+18,000$ ft) on a 2% favorable grade ramping up to the fill surface at the dam.

Reasonable rolling resistance values would be 200 lb/ton in the cuts, 150 lb/ton on the fills, 40 lb/ton on the semipermanent haul roads, and 100 lb/ton on the ramps.

Based on the layout shown in Figure 9-14 and the typical embankment cross sections shown in Figure 9-15 and the above discussion, the eight separate hauls can be characterized as shown in Table 9-5.

Equipment Selection

The abutment and core trench excavation and the impervious borrow hauls (hauls 1 through 5) are well suited to scrapers. The adverse grades are modest ($+6\%$), so twin-engine scrapers would not be necessary. On the other hand, the loaded haul lengths (5,300 to 10,300 ft) and the relatively open areas in which to operate favor the larger-sized scrapers. For this exercise, assume that the contractor selected Caterpillar 631E (21 cy) units push-loaded by two D9 pushcats. Caterpillar 651E (32 cy) scrapers push-loaded with two D9 pushcats would also be an acceptable choice.

The balance of the hauls for moving the random borrow to the dam are all very long hauls (up to 26,800 ft) with very little adverse grade. Large bottom-dump wagons pulled by a separate prime mover are designed for these kinds of haul conditions. Assume that the contractor intends to utilize 130 ton DART haul units loaded by a 15 cy DART front-end loader assisted by a Caterpillar D9 track-mounted bulldozer. Large (50 to 85 ton) end-dump trucks loaded by large front-end loaders would also be a possible choice. Also, large hydraulic backhoes or front shovels could be substituted for the front-end loaders.

Amount of Equipment Required to Meet Construction Schedule

The time requirements for the excavation and embankment operations are defined by the simple arrow diagram schedule shown in Figure 9-16. The contractor plans to work two 10 hr shifts per day, 5 days per week (10.0/2/5). Because 10 working days will be lost to rain each working season plus five holidays (Easter, Memorial Day, Independence Day, Labor Day, and Thanksgiving), the available working hours for each major operation shown above can be calculated as shown in Table 9-6.

TABLE 9-5   Haul Road Characterization for Hauls 1 through 8

Haul no.	Travel in cut (200 lb/ton)	Ramp (100 lb/ton)	Haul road (40 lb/ton)	Ramp (100 lb/ton)	Travel on fill (150 lb/ton)	Total haul
1	1,000 ft at 0%	1,500 ft at 6%	2,500 ft at 2%	500 ft at 6%	800 ft at 0%	6,300 ft
2	800 ft at 0%	500 ft at −6%	2,500 ft at −2%	500 ft at 6%	1,000 ft at 0%	5,300 ft
3	800 ft at 0%		3,500 ft at −2%	1,500 ft at −6%	1,000 ft at 0%	6,800 ft
4	800 ft at 0%		4,500 ft at −2%	700 ft at 6%	1,500 ft at 0%	7,500 ft
5	800 ft at 0%		5,500 ft at −2%	2,200 ft at 6%	1,800 ft at 0%	10,300 ft
6	800 ft at 0%		18,000 ft at −2%	700 ft at 6%	1,000 ft at 0%	20,500 ft
7	800 ft at 0%		20,000 ft at −2%	700 ft at 6%	1,500 ft at 0%	23,000 ft
8	800 ft at 0%		22,000 ft at −2%	2,200 ft at 6%	1,800 ft at 0%	26,800 ft

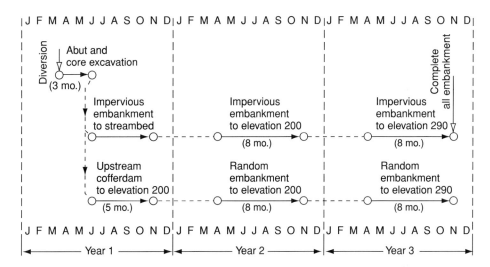

FIGURE 9-16   Time requirements for excavation and embankment operations.

TABLE 9-6   Calculation of Available Hours for Load-and-Haul Operations

Abutment and core excavation	3 month × 4.33 weeks/month × 5 days per week − 5 days
	= 65 − 5
	= 60 days
	= 60 × 20 hours per day = 1,200 hr
Impervious embankment to stream bed	5 month × 4.33 weeks per month × 5 days per week − 10 days
	= 108 − 10
	= 98
	= 98 × 20 hours per day = 1,960 hr
Upstream cofferdam	1,960 hr
Each of the remaining operations	8 month × 4.33 weeks per month × 5 days per week − 15 days
	= 173 − 15
	= 158 days
	= 158 × 20 hours per day = 3,160 hr

The following constants are assumed for required input into the pushcat/scraper and truck/loader load-and-haul computer programs:

All operating efficiencies	75%
Scraper load factor, excavation	0.85
Scraper load factor, borrow	0.90
Scraper load time	0.70 min
Pushcat return time	0.28 min
Pushcat maneuver time	0.15 min
Scraper turn-and-dump time	0.50 min
Random in situ unit weight	3200 lb/Bcy
Swell factor	30%
Loader bucket factor	0.90
Loader cycle time	0.40 min
Truck spot and maneuver time	0.50 min
Truck dump time	0.50 min

On the basis of the above assumptions, the key productivity figures from the load-and-haul analysis printouts (not included here) are shown in Table 9-7.

TABLE 9-7   Key Productivity Figures from Load-and-Haul Analysis Computer Printouts

Haul	Scraper haul unit productivity, Bcy/hr	Truck haul unit productivity, Bcy/hr	Scrapers or trucks per load unit,[1] each
1	97		11
2	133		8
3	131		8
4	117		9
5	84		13
6		160	7
7		144	7
8		116	9

[1]Low side of balance number.

Based on these load-and-haul analysis results, the number of load-and-haul units required to meet schedule requirements for each critical construction period shown in Figure 9-16 are developed in Table 9-8. The maximum scraper requirement (10 units) occurs during the critical 3-month period following the river diversion when the abutment and core trench excavation is carried out. If it were possible to phase this work[6] or find a stockpile area nearer the dam for all or part of the 882,400 Bcy of material to be used later as part of the random fill embankment, this requirement could be reduced.[7] Note that even if these steps are not taken, only one loading unit (consisting of two D9 pushcats) is required.

---

[6]Bottoming out part of the foundation area first, permitting the impervious embankment operation to begin as scheduled on July 1, and then overlapping completion of the core trench excavation with continuation of the impervious embankment operation.

[7]Another possibility would be to employ a truck/loader spread (composed of equipment used later in that season) to move part of the core-trench excavation.

TABLE 9-8 Determination of Number of Units of Each Type of Load-and-Haul Equipment Required to Meet Time Requirements of Construction Schedule

(1) Construction period	(2) Operation description	(3) Haul Quantity,[1] Bcy	(4) Haul no.[1]	(5) Haul equipment	(6) Load equipment	(7) Haul unit productivity,[2] Bcy/hr	(8) Required haul unit,[3] hr	(9) Haul unit hours required for period[4]	(10) Available, hr[5]	(11) Haul units required, each[6]	(12) Load units required, each[7]
First year, April through June	Abutment/core excavation	1,150,000	1	631E Scraper	2 D9 P.C.	97	11,856	11,856	1,200	10	1
First year, July through November	Impervious to streambed	1,187,500	3	631E Scraper	2 D9 P.C.	131	9,065	9,065	1,960	5	1
	Upstream cofferdam	2,588,200	6	Dart 5130	15 cy F.E.L.	160	16,177	16,177	1,960	9	2
Second and third year, April through November	Impervious to elevation 200	531,250	4	631E Scraper	2 D9 P.C.	117	4,541				
	Random to elevation 200	882,400	2	631E Scraper	2 D9 P.C.	133	6,635				
	Impervious to elevation 290	462,500	5	631E Scraper	2 D9 P.C.	84	5,506	16,682	6,320	3	1
	Random to elevation 200	8,023,500	7	Dart 5130	15 cy F.E.L.	144	55,719				
	Random to elevation 290	4,258,800	8	Dart 5130	15 cy F.E.L.	116	36,714	92,433	6,320	15	2

[1] From Table 9-3.
[2] From Table 9-7.
[3] Col. (3)/Col. 7.
[4] The cumulative total haul unit hours for the particular type of equipment required for the period.
[5] Available work hours for the period from Table 9-6.
[6] Col. (9)/Col. (10).
[7] From Table 9-7.

A maximum of nine 130 ton bottom-dump haul units and two 15 cy front-end loaders are required during the placement of the upstream cofferdam during the latter part of the first season. For the balance of the random fill over the final two construction seasons, 15 haul units and two loaders will be needed.

Additional equipment will be required for spreading and compacting the embankments and for support of the load-and-haul/compact operations. The development of the numbers of units and operated hours for this supplementary equipment is keyed to the now completed load-and-haul analysis and can be determined quite logically as the detailed crews are set up. This process will now be explained.

Compaction Equipment Requirements

Assume that (1) the contractor elected to compact the impervious embankment with Caterpillar 815B compactors, (2) the project specifications required the material to be spread in 8 in. lifts before compaction, and (3) the best geotechnical information available dictated that six passes of the compactor would be required to achieve the specified compaction density. Further assume that material swell = 30% and shrink = 20%.

Additionally, the random embankment is to be laid down in 12 in. layers and to be compacted to required density with four passes of a two-drum vibratory compactor, each drum 60 in. wide. Swell = 30%, and shrink = 15%. Based on these assumptions, compactor requirements can be determined as shown in Tables 9-9 and 9-10

Work Crew Compositions for Pricing

The full complement of information necessary to price out each bid item can now be developed. Because all of this work will be performed on two 10 hr shifts per day, both

TABLE 9-9   Compaction Requirements for Impervious Embankment

Thickness of compacted layer $= \dfrac{(0.67 \text{ ft})(1 - 0.20)}{(1 + 0.30)} = 0.41$ ft

Efficiency = 75%

Speed = 6 mph

Width of compactor wheel = 38.5 in.

Total compactor width $= (4)\dfrac{(38.5)}{12} = 12.8$ ft

Production $= \dfrac{(0.41 \text{ ft})(12.8 \text{ ft})(6 \text{ mph})(88)(60 \text{ min})(0.75)}{(27)(6)}$

$= 770$ Ecy/hr

$= \dfrac{770}{(1 - 0.20)} = 963$ Bcy/hr

Construction period	Rate of material delivery, Bcy/hr	Required compactors, each
First year	(5)(131) = 655	1
Second year	(3)(117) = 351	1
Third year	(3)(84) = 252	1

TABLE 9-10 Compaction Requirements for Random Embankment

Thickness of compacted layer $= \dfrac{(1.0\ \text{ft})(1 - 0.15)}{(1 + 0.30)} = 0.65\ \text{ft}$

Efficiency $= 75\%$
Speed $= 4$ mph
Width of compactor drum $= 60$ in. $= 5$ ft
Total compactor width $= (2)\,(5) = 10$ ft

Production $= \dfrac{(0.65\ \text{ft})(10\ \text{ft})(4\ \text{mph})(88)(60\ \text{min})(0.75)}{(27)(4)}$

$\quad = 953\ \text{Ecy/hr}$

$\quad = \dfrac{953}{(1 - 0.15)} = 1{,}121\ \text{Bcy/hr}$

Construction period	Rate of material delivery, Bcy/hr	Required compactors, each
First year	$9 \times 160 = 1{,}440$	2
Second and third years	$15 \times 144 = 2{,}160$	2

overtime and shift work premiums will be involved. For overtime payment provisions of 1.5/1.5/2.0 (see Chapter 2), the labor premium for pricing purposes (see Chapter 6) can be calculated as follows:

Shortest repeating time period $= 1$ work day

Shift	Hours paid	Hours worked
Day	11.00	10.00
Swing	11.75	10.00
Totals for day	22.75	20.00

Labor premium $= \left(\dfrac{22.75 - 20.00}{20.0}\right)(100)$

$\quad = 13.75\%$

Also, for this type of work, a small tools and supplies (ST&S) allowance (see Chapter 6) of 5% should be adequate.

Table 9-11 lists the details of all relevant information for bid item pricing.

Note the addition of the following units of support equipment to fill out the load-and-haul crews and to provide crews on the fills to spread, compact, and finish slopes:

- D8 dozer on the waste dump to spread dumped material and maintain the fill
- D8 dozer to finish exposed embankment slopes
- Caterpiller 14 motor graders on the fills to spread delivered material in controlled thickness layers prior to compaction
- Caterpiller 815B compactor and two-drum vibrator compactors on the impervious and random embankments, respectively, to compact delivered material to the required density
- Caterpiller 14 motor graders and 10,000 gal water tankers for haul road maintenance
- Three-quarter ton pickups for crew foremen
- Portable 12 kW light towers with generators for night lighting

TABLE 9-11 Work Crew Composition and Listing of All Relevant Information for Pricing

Bid item no.	Bid item information	Work operation		Crew details			Required crew hours
		No.	Description and work quantity	Description	No.	Crew productivity	
10	Core trench and abutment excavation	1	Load and haul dam to waste area (Haul 1)	D9 pushcat	2.00		
	Bid quantity = 1,150,000 Bcy			631E scraper	10.00		
	T.O. quantity = 1,150,000 Bcy			D8 dozer on fill	1.00		
			1,150,000 Bcy	Dozer operators	3.00	(10)(97) = 970 Bcy/hr	1,186
			10.0/2/5	Scraper operators	10.00		
				Grade checker	1.00		
				Dumpman on fill	1.00		
				ST&S, %	5.00		
				Labor premium, %	13.75		
		2	Recover from stockpile and haul to dam (Haul 2)	D9 pushcat	2.00		
				631E scraper	3.00		
			882,400 Bcy	Dozer operators	2.00	(3)(133) = 399 Bcy/hr	2,211
			10.0/2/5	Scraper operators	3.00		
				ST&S, %	5.00		
				Labor premium, %	13.75		
		3	Supervision and haul road maintenance	3/4 ton pickup	1.00		
				Cat. 14 motor grader	1.00		
				10,000 gal water tanker	1.00		
			1,186 + 2,211 = 3,397 hr	Operator foreman	1.00		
			10.0/2/5	Motor grader operator	1.00	1 hr/hr	3,397
				Teamster	1.00		
				ST&S, %	5.00		
				Labor premium, %	13.75		
		4	Night lighting	12 kW light tower with generator	3.00	1 hr/hr	1,698
			3,397 × 1/2 = 1,698 hr				
11	Impervious borrow	1	Load and haul borrow to stream bed elevation (Haul 3)	D9 pushcat	2.00		
	Bid quantity = 2,181,250 Bcy			631E scraper	5.00		
	T.O. quantity = 2,181,250 Bcy			Dozer operators	2.00	(5)(131) = 655/Bcy/hr	1,813
			1,187,500 Bcy	Scraper operators	5.00		
			10.0/2/5	ST&S, %	5.00		
				Labor premium, %	13.75		

TABLE 9-11 (Page 2 of 3) Work Crew Composition and Listing of All Relevant Information for Pricing

Bid item no.	Bid item information		Work operation — Description and work quantity	Crew details — Description	No.	Crew productivity	Required crew hours
		2	Load and haul borrow to elevation 200 (Haul 4)	D9 pushcat	2.00	(3)(117) = 351 Bcy/hr	1,514
				631E scrapers	3.00		
				Dozer operators	2.00		
			531,250 Bcy	Scraper operators	3.00		
			10.0/2/5	ST&S, %	5.00		
				Labor premium, %	13.75		
		3	Load and haul borrow to elevation 290 (Haul 5)	Same as operation 2	–	(3)(84) = 252 Bcy/yr	1,835
			462,500 Bcy		–		
			10.0/2/5				
		4	Supervision and haul road maintenance	Same as bid item 10 operation 3		1 hr/hr	5,162
			1,813 + 1,514 + 1,835 + 5,162 hr				
		5	Night lighting	12 kW light tower with generator	1.00	1 hr/hr	2,581
			5,162 × 1/2 = 2,581 hr				
12	Random borrow	1	Random borrow to upstream cofferdam (Haul 6)	15 cy Dart F.E.L.	2.00	(9)(160) = 1,440 Bcy/hr	1,798
				Dart 5130 bottom-dump trucks	9.00		
	Bid quantity = 14,870,500 Bcy			D9 dozer	2.00		
	Takeoff quantity = 14,870,500 Bcy		2,588,200 Bcy	Loader operators	2.00		
			10.0/2/5	Dozer operators	2.00		
				Teamsters	9.00		
				ST&S, %	5.00		
				Labor premium, %	13.75		
		2	Random borrow to elevation 200	15 cy Dart F.E.L.	2.00	(15)(144) = 2,160 Bcy/hr	3,715
				Dart 5130 bottom-dump trucks	15.00		
			8,023,500 Bcy	D9 dozer	2.00		
			10.0/2/5/	Loader operators	2.00		
				Dozer operators	2.00		
				Teamsters	15.00		
				ST&S, %	5.00		
				Labor premium, %	13.75		
		3	Random borrow to elevation 290	Same as operation 2	–	(15)(116) = 1,740 Bcy/hr	2,448
			4,258,800 Bcy				
			10.0/2/5				

TABLE 9-11 (Page 3 of 3) Work Crew Composition and Listing of All Relevant Information for Pricing

Bid item no.	Bid item information	Work operation No.	Description and work quantity	Crew details Description	No.	Crew productivity	Required crew hours
		4	Supervision and haul road maintenance	3/4 ton pickup	1.00		
				Motor graders	2.00		
				10,000 gal water tankers	2.00		
			1,798 + 3,715 + 2,448 = 7,961 hr	Operator foreman	1.00	1 hr/hr	7,961
				Grader operators	2.00		
				Teamsters	2.00		
				ST&S, %	5.00		
				Labor premium, %	13.75		
		5	Night lighting	12 kW light tower with generator	1.00	1 hr/hr	3,981
			7,961 × 1/2 = 3,981 hr				
13	Impervious embankment	1	Spread and compact	815B compactor	1.00		
				Cat. 14 motor grader	1.00		
	Bid quantity = 1,745,000 Ecy		1,813 + 1,514 + 1,835 = 5,162 hr	Compactor operator	1.00	1 hr/hr	5,162
	T.O. quantity = 1,745,000 Ecy		10.0/2/5	Motor grader operator	1.00		
				Dumpman	1.00		
				ST&S, %	5.00		
				Labor premium, %	13.75		
		2	Night lighting	12 kW light tower with generator	1.00	1 hr/hr	2,581
			5,162 × 1/2 = 2,581 hr				
14	Random embankment	1	Spread and compact	2 drum vibratory compactor	2.00		
			First year	Cat 14 motor grader	1.00		
	Bid quantity = 13,390,000 Ecy		1,798 hr	Compactor operators	2.00		
	T.O. quantity = 13,390,000 Ecy		10.0/2/5	Motor grader operator	1.00	1 hr/hr	1,798
				Dumpman	1.00		
				ST&S, %	5.00		
				Labor premium, %	13.75		
		2	Spread and compact	Same as operation 1		1 hr/hr	6,163
			Second and third years				
			3,715 + 2,448 = 6,163 hr				
		3	Night lighting	12 kW light tower with generator	2.00	1 hr/hr	3,981
			(1,798 + 6,163) × 1/2 = 3,981 hr				
		4	Finish slopes and miscellaneous, day shift only	D8 dozer	1.00	1 hr/hr	3,981
			(1,798 + 6,163) × 1/2 = 3,981	Dozer operator	1.00		

Similarly, additional added personnel include

- Grade checker in foundation excavation crew when cutting to line and grade
- Dumpmen on waste fills and/or embankments
- Operators for motor graders
- Teamsters for water tankers
- Crew foremen where required

Additionally, the crew compositions and the required crew-hours shown in Table 9-11 are predicated on the basis that

- Costs for all heavy equipment service and repair units such as mechanics trucks, fuel trucks, grease units, and tire changing equipment are provided for in the production equipment operating expense (EOE) and rental (R) hourly rates.
- All repair labor is provided for in the production equipment repair labor (RL) hourly rates.
- Required crew-hours for the spread and compact crews are determined on the basis that these crews will be required on the embankments to match the time durations that material is being delivered.
- Supervision and haul road maintenance crew hours are determined on the basis that these crews will be required for the full time durations that haul operations are conducted over the major haul roads.
- Night lighting hours are determined as 50% of the hours work is performed in any cut or fill area on a two 10-hour-shift-per-day basis.

By carefully studying Figure 9-11 and the analytical procedures leading up to it, the reader should be able to fully understand the principles involved in planning and pricing the load, haul, and compact work reflected in the five bid items for this hypothetical dam construction project.

Estimated Costs

By use of the computer-aided cost estimating program developed at California State University, Chico, the input information shown in Table 9-11 can easily be input and converted into dollars and cents costs. The California State University program is one in which discrete labor, equipment, material, and subcontract libraries can be input for a particular project as a preliminary step, when setting up the estimate as described in Chapter 5.

Examples of completed direct cost estimates for two of these bid items are shown as Figure 9-17 (for bid item 10) and Figure 9-18 (for bid item 13). The libraries that were preloaded into the computer for these estimates are included in the Appendix to this book. All other input information is shown in Table 9-11. Direct cost estimates for the remaining three bid items could have been similarly generated and included herein, but Figures 9-17 and 9-18 should be sufficient to illustrate the format of the direct costs generated by the estimating program for a typical bid item. Costs generated by the manual procedures explained in Chapter 6 using the libraries in the Appendix would be the same.

				DATE: 11/09/98 ESTIMATOR:	BID ITEM: JOB							10 CORE TRENCH AND ABUTMENT EXCAVATION			TAKEOFF QUANTITY: 1,150,000 BCY BID QUANTITY: 1,150,000 BCY				
				— UNIT-COSTS —							REQD.							SUB	
QTY	CODE	DESCRIPTION	UNIT	E.O.E	R-L	RENTAL	LABOR	E.M.	P.M.	SUBS	HOURS	E.O.E	R-L	RENTAL	LABOR	E.M.	P.M.	CONTRACTS	TOTAL
OPRATN: 1		L&H DAM FON. EX TO WASTE AREA																	
1150000 BCY			970 BCY	/ Hour															
2	EQ33	D9 Push Cat	ea	36.050	8.510	31.340					1186	$85,479	$20,178	$74,311					$179,969
10	EQ42	CAT 631E Scraper (21CY)	ea	71.370	12.390	23.010						846,139	146,892	272,799					1,265,830
1	EQ28	D8 Dozer (Dirt)	ea	33.310	10.500	32.040						39,491	12,448	37,986					89,925
3	OE4	Dozer Operator	ea				44.09								156,815				156,815
10	OE8	Scraper Operator	ea				43.280								513,113				513,113
1	OE20	Grade Checker/Setter	ea				41.840								49,604				49,604
1	LAB2	Dumpman	ea				28.010								33,208				33,206
5.00 %		ST & S	LL													37,637			37,637
13.75 %		PREMIUM PAY	LL												103,502				103,502
OPRATN: 2		RECOVER FROM STOCKPILE & HAUL TO DAM																	
882400 BCY			399 BCY	/ Hour															
2	EQ33	D9 Push Cat	ea	36.050	8.510	31.340					2212	$159,451	$37,640	$138,619					$335,710
3	EQ42	CAT 631E Scraper (21CY)	ea	71.370	12.390	23.010						473,510	82,203	152,662					708,375
3	OE4	Dozer Operator	ea				44.090								195,013				195,013
3	OE8	Scraper Operator	ea				43.280								287,145				287,145
5.00 %		ST & S	LL													24,108			24,108
13.75 %		PREMIUM PAY	LL												66,297				66,297
OPRATN: 3		SUPERVISION & HAUL ROAD MAINTENANCE																	
3397 HRS			1 HRS	/ Hour															
1	EQ80	3/4 Ton Pickup Truck	ea	5.000	0.500	1.910					3397	$16,985	$1,699	$6,488					$25,172
1	EQ24	CAT 14 Motor Grader	ea	21.360	6.920	25.980						72,560	23,507	88,254					184,321
1	EQ53	10,000 gal Water Truck	ea	70.220	7.930	68.570						238,537	26,938	232,932					498,408
1	OE17	Operator Foreman	ea				50.510								171,582				171,582
1	OE10	Motor Grader Operator	ea				44.090								149,774				149,774
1	TM7	Water Truck Driver	ea				32.910								111,795				111,795
5.00 %		ST & S	LL													21,658			21,658
13.75 %		PREMIUM PAY	LL												59,558				59,558

FIGURE 9-17 (PAGE 1 OF 2)   Computer-generated cost estimate for bid item 10.

| DATE: 11/09/98 | | BID ITEM: | | | | | 10 | | | | TAKEOFF QUANTITY: | 1,150,000 BCY | | | | | | | |
| ESTIMATOR: | | JOB | | | | | CORE TRENCH AND ABUTMENT EXCAVATION | | | | BID QUANTITY: | 1,150,000 BCY | | | | | | | |

QTY	CODE	DESCRIPTION	UNIT	E.O.E	R-L	RENTAL	LABOR	E.M.	P.M.	SUBS	REQD. HOURS	E.O.E	R-L	RENTAL	LABOR	E.M.	P.M.	SUB CONTRACTS	TOTAL
						— UNIT-COSTS —		S											
OPRATN:	**4**	NIGHT LIGHTING																	
1698	HRS		1 HRS	/Hour															
3	EQ68	12 KW Light Plant	ea	1.920	11.660	2.240					1698	$9,780	$59,396	$11,411					$80,587
		% ST & S	LL																
		% PREMIUM PAY	LL																
		TAKEOFF TOTALS:										$1,941,934	$410,901	$1,015,462	$1,897,406	$83,402			$5,349,105
		PRORATED BID TOTALS:										1,941,934	410,901	1,015,462	1,897,406	83,402			5,349,105
		UNIT PRICES /	BCY									1.69	0.36	0.88	1.65	0.07			4.65

FIGURE 9-17 (PAGE 2 OF 2) Computer-generated cost estimate for bid item 10.

235

DATE: 08/06/98
ESTIMATOR: JQB

BID ITEM: 13 IMPERVIOUS EMBANKMENT

TAKEOFF QUANTITY: 1,745,000 ECY
BID QUANTITY: 1,745,000 ECY

QTY	CODE	DESCRIPTION	UNIT	E.O.E	R-L	RENTAL	LABOR	E.M	P.M	SUBS	REQD. HOURS	E.O.E	R-L	RENTAL	LABOR	E.M	P.M	SUB CONTRACTS	TOTAL
							UNIT-COSTS												
OPRATN:	1	SPREAD & COMPACT																	
5162 HRS			1 HRS	/Hour															
1	EQ36	Cat 815 Compactor	ea	42.880	9.710	26.140					5162	$221,347	$50,123	$134,935					$406,404
1	EQ24	CAT 14 Motor Grader	ea	21.360	6.920	25.980						110,260	35,721	134,109					280,090
1	OE9	Compactor Operator	ea				44.090								227,593				227,593
1	OE10	Motor Grader Operator	ea				44.090								227,593				227,593
1	LAB2	Dumpman	ea				28.010								144,588				144,588
	5.00	% ST & S	LL													29,989			29,989
	13.75	% PREMIUM PAY	LL												82,469				82,469
OPRATN:	2	NIGHT LIGHTING																	
2581 HR			1 HR	/Hour															
1	EQ68	12 KW Light Plant	ea	1.920	11.660	2.240					2581	$4,956	$30,094	$5,781					$40,831
		% ST & S	LL																
		% PREMIUM PAY	LL																
		TAKEOFF TOTALS:										$336,562	$115,939	$274,825	$682,242	$29,989			$1,439,556
		PRORATED BID TOTALS:										336,562	115,939	274,825	682,242	29,989			1,439,556
		UNIT PRICES /	ECY									0.19	0.07	0.16	0.39	0.02			0.82

FIGURE 9-18 Computer-generated cost estimate for bid item 13.

CONCLUSION

This long chapter has explained the basic principles governing the operation of the common types of heavy earth and rock excavation and haulage equipment such as scrapers, various types of loaders, and trucks. Further, the chapter explained how large, complex projects can be broken down and analyzed, operation by operation, so that a plan for the accomplishment of the work can be developed. Finally, procedures were explained to enable the amount of equipment that must be furnished to meet a required construction schedule to be determined and to estimate the resulting dollars and cents costs.

The following chapter considers heavy concrete operations—a class of labor-intensive expensive work that will be found on most heavy construction projects.

QUESTIONS AND PROBLEMS

1. What three unions are involved for load, haul, and compact operations when the work is performed under collective bargaining agreements? What type of equipment is operated by the separate union workers and/or what other specific duties does each union worker perform?

2. What are the two basic categories of loading equipment discussed in this chapter? What are the eight separate categories of top-loading equipment that were mentioned? What are the present-day commonly available bucket sizes for front-end loaders, front shovels, and backhoes? Answer the same questions for cable-operated clamshells and draglines.

3. What two basic categories of haul equipment are commonly used in present-day heavy construction?

4. What are two basic types of scrapers discussed in this chapter? How are scraper payload capacities expressed? What bowl capacity ranges are commonly available for push-loaded scrapers? What type of use (on-highway or off-highway) are scrapers intended for and what haul speeds are attainable?

5. What two common types of trucks were discussed? State whether each type is intended for on- or off-highway use. How is their payload capacity expressed and what typical payload ranges are commonly available for end- and bottom-dump trucks? What range of haul speed can be attained?

6. What two broad classes of compaction equipment are commonly used? What four separate subtypes for each were mentioned in this chapter? Why are the larger self-powered compactors often articulated? What size range of towed pneumatic compactors is available?

7. What five classes of support equipment were discussed in this chapter?

8. What is the first step required to analyze load-and-haul operations for either a lump sum estimate for an entire project or for a single bid item for a schedule-of-bid-items estimate?

9. Define or explain the following terms:
   - **(a)** Schedule of haul quantities
   - **(b)** Average haul
   - **(c)** Haul length and profile
   - **(d)** Switchback
   - **(e)** Percent grade
   - **(f)** Rolling resistance

(g)	Total haul road resistance	**(h)**	Sustained travel speed
(i)	Average travel speed	**(j)**	Haul road segment
(k)	Segment travel time	**(l)**	Total travel time

10. Sustained travel speed depends on what four factors?

11. Once the sustained speed is known, on what three additional factors does the average speed for a haul road segment depend?

12. Do the answers to questions 8 through 11 above depend on whether the load-and-haul operation uses scrapers or trucks, or are the answers equally applicable to both?

13. Define the following with regard to scraper operations and indicate the units of measure for each:

(a)	Load-and-boost time	**(b)**	Loaded travel time
(c)	Turn-and-dump time	**(d)**	Return travel time
(e)	Pushcat maneuver time	**(f)**	Pushcat return time
(g)	Scraper cycle time	**(h)**	Pushcat cycle time
(i)	Balance number	**(j)**	Scraper struck capacity
(k)	Scraper load factor	**(l)**	Efficiency factor
(m)	Scraper unit productivity	**(n)**	Scraper spread productivity

14. Define the following with regard to truck/loader operations and indicate the units of measure for each:

(a)	Truck rated capacity	**(b)**	In situ material weight
(c)	Truck payload	**(d)**	Rated bucket size
(e)	Swell percent	**(f)**	Bucket fill factor
(g)	Loader cycle time	**(h)**	Truck spot-and-maneuver time
(i)	Truck load time	**(j)**	Truck-loaded travel time
(k)	Truck turn-and-dump time	**(l)**	Truck return travel time
(m)	Balance number	**(n)**	Truck unit productivity
(o)	Truck spread productivity		

15. Define the following with regard to compaction operations and indicate the units of measurement for each:

(a)	Swell percentage	**(b)**	Shrink percentage
(c)	Thickness of the uncompacted layer	**(d)**	Thickness of the compacted layer
(e)	Compactor speed	**(f)**	Width of one pass
(g)	Number of required passes	**(h)**	Compactor unit productivity

16. Revise the data given for the comprehensive example problem in the text as follows and complete the stated tasks.

Figure 9-14

Change the 18,000 ft semipermanent haul road length and loaded-haul grade to 14,000 ft and -3%, respectively.

Figure 9-15

- Change impervious core to elevation 120 from 950,000 to 760,000 Ecy
- Change impervious core to elevation 200 from 425,000 to 340,000 Ecy
- Change impervious core to elevation 290 from 370,000 to 296,000 Ecy
- Change upstream cofferdam to elevation 200 from 2,200,000 to 1,760,000 Ecy
- Change random fill to elevation 200 from 7,570,000 to 6,056,000 Ecy
- Change random fill to elevation 290 from 3,620,000 to 2,896,000 Ecy
- Change total embankment from 15,135,000 to 12,108,000 Ecy

Bid Form

- Change bid item 10 quantity from 1,150,000 to 920,100 Bcy
- Change bid item 11 quantity from 2,181,250 to 1,745,000 Bcy
- Change bid item 12 quantity from 14,870,500 to 11,896,300 Bcy
- Change bid item 13 quantity from 1,745,000 to 1,396,000 Ecy
- Change bid item 14 quantity from 13, 390,000 to 10,712,100 Ecy

Table 9-3

Maintaining the same shrink factors, revise to read as follows:

Bid item	Quantity, Bcy	Use of material and fill quantities produced	Haul
10	706,000	To stockpile at waste area for use in random fill (706,000 Bcy $\times$ 0.85 = 600,100 Ecy)[1]	1
	214,100	To be wasted at waste area	1
		Recover 706,000 Bcy from stockpile to random fill to elevation 200	2
	920,100		
11	950,000	Impervious borrow area to core embankment to original streambed (950,000 Bcy $\times$ 0.80 = 760,000 Ecy)[2]	3
	425,000	Impervious borrow area to core embankment to elevation 200 (425,000 Bcy $\times$ 0.80 = 340,000 Ecy)[2]	4
	370,000	Impervious borrow area to core embankment to elevation 290 (370,000 Bcy $\times$ 0.80 = 296,000 Ecy)[2]	5
	1,745,000		
12	2,070,400	Random borrow area A to upstream cofferdam (2,070,400 Bcy $\times$ 0.85 = 1,760,000 Ecy)[1]	6
	6,418,800	Random borrow area B to random embankment to elevation 200 (6,418,800 Bcy $\times$ 0.85 = 5,456,000 Ecy)[1]	7
	3,407,100	Random borrow area C to random embankment to elevation 290 (3,407,100 Bcy $\times$ 0.85 = 2,896,000 Ecy)[1]	8
	11,896,300		

[1]Based on assumed shrinkage factor of 15%.
[2]Based on assumed shrinkage factor of 20%.

Table 9-4

Revise the table to read as follows:

Bid item	Quantity, Ecy	Quantity and source of material	Haul notes
13	760,000	950,000 Bcy from impervious borrow	Delivered under haul 3
	340,000	425,000 Bcy from impervious borrow	Delivered under haul 4
	296,000	370,000 Bcy from impervious borrow	Delivered under haul 5
	1,396,000		
14	600,100	760,000 Bcy from stockpile	Delivered under haul 2
	1,760,000	2,070,600 Bcy from random borrow area A	Delivered under haul 6
	5,456,000	6,418,800 Bcy from random borrow area B	Delivered under haul 7
	2,896,000	3,407,000 Bcy from random borrow area C	Delivered under haul 8
	10,712,100		

Table 9-5

Revise the table to read as follows:

Haul no.	Travel in cut (200 lb/ton)	Ramp (100 lb/ton)	Haul road (40 lb/ton)	Ramp (100 lb/ton)	Travel on fill (150 lb/ton)	Total haul
1	800 ft at 0%	1,200 ft at 6%	1,850 ft at 3%	600 ft at 6%	500 ft at 0%	4,950 ft
2	500 ft at 0%	600 ft at 6%	1,850 ft at −3%	800 ft at 6%	800 ft at 0%	4,350 ft
3	600 ft at 0%		3,200 ft at −3%	1,200 ft at −6%	800 ft at 0%	5,800 ft
4	600 ft at 0%		4,200 ft at −3%	600 ft at 6%	1,200 ft at 0%	6,600 ft
5	600 ft at 0%		5,200 ft at −3%	1,900 ft at 6%	1,500 ft at 0%	9,200 ft
6	600 ft at 0%		14,000 ft at −3%	600 ft at 6%	800 ft at 0%	16,000 ft
7	600 ft at 0%		16,000 ft at −3%	600 ft at 6%	1,300 ft at 0%	18,500 ft
8	600 ft at 0%		20,000 ft at −3%	2,000 ft at 6%	1,600 ft at 0%	24,200 ft

Equipment Selection

Change the scraper/pushcat equipment from 631E scrapers/2D9 pushcats to 651E scrapers/2D9 pushcats.

Work Conditions

Change 10.0/2/5 to 7.5/3/5.

All other data and/or assumptions stated in the text remain unchanged. Based on the above,

(a)   Perform all analyses shown in the text. Use the California State University, Chico, load-and-haul-analysis computer programs included in the appendix for analysis of the eight required hauls.

(b)   Using the computer cost-estimating program included in the appendix, prepare direct cost estimates for the five bid items.

10

Concrete Operations

Key Words and Concepts

Cost of concrete, f.o.b. jobsite
Cement
Fine and coarse aggregate
Concrete admixtures
Reinforcing steel
Miscellaneous metal
Premolded joint filler
Waterstops
Waste and overbreak concrete
Waste on other permanent material items
Dimension lumber
Form plywood
Form purchase/rental
Form ties and hardware
Sandblast sand
Curing compound/paper mats
Cement for dry finish
Quantities taken-off from the construction
   drawings
Preservation of bid-item organization
Strength classes of concrete
Concrete structure elements
Classes of formwork
Form complexity code
Shoring volume

Foundation contact areas
Horizontal construction joint areas
Wet finish areas
Cure contact areas
Supplementary quantities developed from primary
   quantities
Ready-mix concrete quantities
Cement and aggregate quantities
Form ratio
Form reuse factors
Dimension lumber and form plywood use factors
Quantities for "allowance" pricing
General pricing approach using labor-hour factors
Place and vibrate labor-hour factors
Miscellaneous concrete operation labor-hour
   factors
Form work labor-hour factors
Form erect and strips labor-hour factors
Crew productivity
Consolidated approach to reduce work operations
   for pricing
Weighted-average labor-hour factor
Typical work crew compositions
Alternate method to provide crane service

241

There is hardly a major heavy construction project that does not involve concrete operations to some extent. Some projects consists almost entirely of concrete work, at least as far as the major dollars represented in the cost of the project are concerned. Classic examples include concrete dams and hydroelectric powerhouses, pumping plants, navigation locks, dry docks, and the like. Other projects where concrete operations are also important include substructures of bridges over major rivers, cut-and-cover and driven tunnel subways, freeway overpass and bridge construction, and water, sewage, and other utility tunnels.

The field of concrete construction is too broad for this text to cover in detail. The material that follows is intended to apply generally to mass concrete and structural concrete operations. Airport and freeway paving, slipform construction, precast work, and similar highly specialized concrete operations are not included.

CONSTRUCTION CRAFTS AND EQUIPMENT

Construction Crafts

Six construction crafts are usually involved in concrete work. **Carpenters** are required to erect and strip formwork and to set items in the forms that will be embedded in the concrete. **Laborers** place and vibrate the concrete, perform joint and other concrete cleanup tasks preparatory to placing concrete, apply curing compound, perform water curing, and assist carpenter crews. **Cement masons** perform wet finish operations, plug tie holes and repair misaligned joints and honeycomb, and carry out dry finish treatments on exposed formed surfaces. **Ironworkers** tie reinforcing steel in the forms prior to concrete placement as well as install miscellaneous metalwork that fastens onto finished concrete surfaces.

Teamsters drive transit-mix and dumpcrete concrete haul vehicles and drive interjob flat rack and boom-trucks used for miscellaneous tool and minor material deliveries throughout the work. Finally, **operating engineers** operate cranes, concrete batch and mix plants, large compressor plants, and concrete pumps.

Batching and Mixing Equipment

On projects where the concrete is not furnished by an off-site supplier, some type of **batching and mixing plant** must be erected and operated on site. These plants must have the capability of accurately batching the precise quantities of cement, aggregates, and admixtures required for the various mixes used on the project and then to properly mix these ingredients into wet concrete for delivery to the forms. A number of years ago these plants for large projects such as concrete dams were ponderous affairs, erected on fixed foundations, often containing four to six 4 cy mixer bowls and capable of producing over 300 cy of concrete per hour. Today, batching and mixing plants are usually portable and highly automated and can be quickly erected and dismantled at the jobsite. Typically, the larger present day plants will be equipped with one or more 4 cy mixer bowls and are capable of supplying several hundred cubic yards per hour. Automated recorders maintain accurate records of quantities of ingredients batched for each charge of the mixers. Fully automated installations permit the entire operation to be controlled by a single operator.

Transport Equipment

If the concrete for the project is furnished by an off-site supplier, the concrete will be delivered in 8 to 12 cy on-highway **transit-mix trucks.** Usually, the concrete is batched, mixed, and discharged into the truck at the concrete supplier's off-site plant. During transit, the drum continually rotates to prevent the wet concrete from segregating. Alternately, the concrete may be batched and discharged at the plant into transit-mix trucks without water being added. Water for the batch is carried on board the truck and added during transit or, sometimes, not until the jobsite is reached. At that point, after a short mixing period, the concrete is ready to be discharged into the forms. Such equipment is advantageous when the haul to the jobsite is very long and time consuming, which might permit the concrete to take an initial set after the water is added but before the concrete is discharged from the truck.

When on-site batch-and-mix plants are employed, the contractor must provide some sort of transport equipment to haul the concrete from the batch-and-mix plant to the forms. This will ordinarily be done with transit-mix or **dumpcrete trucks.**[1]

Hoisting Equipment

Various types of **cranes** are usually required for handling formwork and reinforcing steel and for placing concrete when bottom-dump buckets are used. Crane sizes vary from 15 or 18 ton single-operator center-mount cranes to 300 ton electrically powered revolvers that are usually mounted on high gantries running on grade level rail or on high trestles. In the past, spectacular high-speed cableways erected across canyons were used in dam construction, each end of the cableway supported by either a fixed or moving tower running in trackways on the dam abutments. Other than the previously mentioned small center-mount cranes, today's hoisting equipment usually consists of moderate- to large-sized crawler mounted or motor-cranes with boom lengths up to 200 ft (and occasionally longer).

Placing Methods

For on-grade work, concrete can frequently be discharged from a transit-mix truck or dumpcrete directly into the forms. When the concrete must be placed at some height above grade or at some distance from the point at which it is delivered by the transport equipment, the traditional method of placement has been by crane and **bottom-dump bucket,** the bucket size ranging from $\frac{1}{2}$ cy to 4 cy.[2] Although crane and bucket placement is still common, the most popular means of concrete placement today is by means of high volume **concrete pumps.** Such pumps have capacities of 60 to 80 cy/hr (or more) and are capable of pumping concrete several hundred feet vertically or horizontally underground in tunnels.

Another popular method utilizes **conveyor belts.** The concrete is delivered to a central point over the placement area where it is chuted down onto a 360° rotating swinger belt that also has the ability to rack in or out. A swinger belt with a 50 ft racking ability would thus be able to place concrete anywhere within a 100 ft circle centered on the point of central concrete delivery. This method is particularly effective when the

---

[1] Open-bed rear-dump trucks with a narrow discharge gate, sometimes fitted with agitator screws to keep the concrete fresh during transit for long on-site hauls.

[2] Eight cubic yard buckets were not uncommon when cableway placement was used for large dams.

PHOTO 10-1 Use of crane and bucket and long portable conveyor truss to place concrete in a bridge pier cofferdam. Note transit-mix concrete delivery trucks upper left.

PHOTO 10-2   Further examples of concrete placing methods. Upper photo shows crew placing and vibrating concrete for a supported slab using concrete pump with placing boom. Lower photo illustrates placement and vibration of concrete for horizontal and sloping slabs on grade.

PHOTO 10-3   Use of conveyor belts to deliver concrete to inaccessible locations (in this case to a cofferdam in the middle of the Missouri River). Note the concrete placement swinger in the lower photograph.

PHOTO 10-4 Upper photo, portable concrete batch-and-mix plant loading a dumpcrete truck (note control trailer on left). Lower photo, dumpcrete truck discharging concrete for a lift of roller-compacted concrete slope protection blanket for an earthfill dam.

concrete is transported horizontally some distance by conveyor belts to reach the area of placement, as is frequently the case for projects with difficult access problems such as hard to reach areas in dam construction or when constructing bridge piers in a river.[3]

Miscellaneous Equipment

Additional equipment required for concrete work includes **air compressors, pumps, waterblast or sandblast equipment, vibrators,** and **welding machines.**

Compressed air is required for blowing out forms prior to placing concrete, for joint cleanup by the greencut[4] or sandblast[5] process, or for air operated bottom-dump buckets and/or air-operated vibrators. Waterblast equipment is utilized in a relatively new method of horizontal joint cleanup for large open lift joints encountered in mass concrete dam construction. An extremely high-pressure stream of water (+400 psi) is directed against the hardened concrete surface, removing the weaker surface material and exposing the coarse aggregate.

When the sandblast method is used, **sandblast pots** are required to hold a supply of sand and to feed it into a compressed air stream through a hose to the sandblast nozzle.

Vibrators are required for consolidating wet concrete after placement in the forms. For surface work, such vibrators are of the **immersion type** usually consisting of a $1\frac{1}{2}$- to 3-in.-diameter circular head connected by a 10 to 12 ft "wiggle tail" that houses a high-speed rotating drive. The drive is either electrically actuated or driven by compressed air. Electrically driven vibrators require special high-cycle electric generators. For mass concrete work, immersion vibrators are larger and hand-held when in use. **Form vibrators,** for use in underground work, clamp onto the outside of tunnel forms where access to the space enclosed by the forms is extremely limited. The vibrators are spaced at uniform intervals on the forms so that, when actuated, the entire form vibrates, transmitting the vibration to the concrete.

Portable welding machines (300 to 400 A), either electric or gasoline/diesel engine powered, are a common tool-of-the-trade used primarily by the carpenter crews erecting formwork and by ironworker crews installing templates and/or supports for reinforcing steel.

MATERIALS REQUIRED FOR CONCRETE CONSTRUCTION

Permanent Materials

If the concrete is obtained from an off-site concrete supplier, an important permanent material item in the estimate will be the **cost of the concrete delivered f.o.b. jobsite.** Otherwise, the contractor must purchase and stockpile on site the individual constituents, such as **cement** and the required **fine and coarse aggregate.**[6] Various **concrete admixtures** must also be purchased and handled at the jobsite.

---

[3]In this latter case, the horizontal conveyors would be supported on a trestle built out into the river.

[4]A process for removing laitance from freshly placed concrete by directing a stream of compressed air and water against horizontal lift joints just after the initial concrete set.

[5]A joint cleanup method where a stream of compressed air containing particles of sharp angular sand is directed against horizontal lift joint surfaces after the concrete has taken its final set.

[6]On large mass concrete projects such locks and dams, fine and coarse aggregate is sometimes produced by the contractor from naturally occurring deposits on site.

Reinforcing steel is also a major permanent material item that will almost always be required. It generally is purchased from a reinforcing steel fabricator, cut, bent, and delivered f.o.b. jobsite. Sometimes, reinforcing steel is obtained on an installed basis through a subcontract. In this event, it will appear in the contractor's estimate as a subcontract rather than a permanent material item.

Other permanent materials include **miscellaneous metal** embedded items consisting of anchor bolts and various fabricated items such as edge angles, weld plates, pipe sleeves, stair treads, and so on. Various types of expansion joint fabrications and **premolded expansion joint fillers** and **waterstops** are also often required.

Expendable Materials

In addition to small tools and supplies (ST&S), a number of expendable material items will always be required. As explained in Chapter 2, one prominent expendable material category consists of **waste and overbreak concrete** and **waste on other permanent material items.**

If formwork is to be job-built, **dimension lumber, form plywood,** and miscellaneous minor form materials such as nails, bolts, and form oil must be purchased and delivered to the jobsite. If it is expected that manufactured modular form panels or custom-built forms are to be used, they must be purchased from an off-site form supplier and delivered to the job. The **purchase cost or a portion of the purchase cost**[7] (or the **supplier's rental charges** in the event the forms are rented) must be included in the estimate as an expendable material item. Regardless of the source of the form panels, the cost of **form ties** and **associated hardware** must also be provided in the estimate as an expendable material item.

A number of additional minor expendable material items consisting of **sand blast sand,** spray-on **curing compound, curing paper or mats,** and **cement for dry finish work** must also be included.

QUANTITIES THAT MUST BE TAKEN OFF FROM THE CONSTRUCTION DRAWINGS

Certain kinds of quantities needed for pricing concrete operations must be computed directly from the dimensional and other information appearing on construction drawings. The following kinds of quantities are usually involved:

- Concrete volumes
- Formwork areas and shoring volumes
- Foundation contact areas
- Horizontal construction joint areas
- Wet finish areas
- Cure contact areas
- Dry finish areas
- Reinforcing steel weights
- Miscellaneous embedded items

---

[7]Form "write off" consisting of purchase cost less salvage value at the end of the job.

For schedule-of-bid-item bids, it is usual to **preserve the organization of bid items established by the owner in the bid form.** Accordingly, any applicable items in the above list are taken off separately for the work of each separate bid item.

Concrete Volumes

Each separate class of concrete, measured in cubic yards (i.e., 3,000, 4,000, 5,000 psi, etc.), must be separately determined because the cost of the concrete materials involved varies with the concrete strength. Also, it is necessary to break down the quantities of concrete required for each of the various structural elements involved in the project because the labor required for placing and vibrating the concrete depends heavily on the size and shape of the element involved. This general breakdown by **structural element** applying to most concrete structures would include the following:

- Mass concrete
- Structural footings, pile caps, and grade beams
- Heavy slabs on grade
- Thin slabs on grade
- Heavy walls
- Thin walls
- Columns
- Heavy supported slabs
- Thin supported slabs

Special structures such as bridge piers, by their nature, require the following breakdown:

- Abutment concrete
- Tremie seals[8]
- Footings
- Pier pedestals
- Piers and connecting web walls
- Pier caps

Cut and cover structures and tunnels logically break down into:

- Inverts
- Walls and roof slabs
- Tunnel inverts
- Tunnel arch and/or full-round concrete placements

Similar breakdowns can be logically developed for other types of structures.

Formwork Areas

Because the amount of labor required for fabricating, erecting, and striping concrete formwork greatly depends on the type and configuration of the particular form involved, the **various classes of formwork** required for the project should be segregated.

---

[8]Concrete placed underwater inside of sheet pile cofferdams prior to dewatering to resist hydrostatic seepage pressures and to prevent the cofferdam from floating.

A general breakdown applicable to most concrete structures includes the following form types:

- Built in place (BIP) off of irregular rock foundations
- Other BIP forms
- Slab-on-grade (SOG) edge forms
- SOG bulkheads (vertical construction joint forms)
- Single-face walls
- Double-face walls
- Wall bulkheads
- Supported slab soffits
- Supported slab edge forms
- Supported slab bulkheads
- Beam soffits
- Beam sides
- Beam bulkheads
- Curb forms
- Hanging forms[9]
- Stair forms
- Transition forms
- Gallery forms[10]
- Blockouts[11]

Similarly, formwork types required for special structures might include the following:

- Large bridge piers
- Large bridge caps
- Large girder and beam forms
- Large multiple-box culvert and/or subway invert forms
- Multiple-box wall, haunch, and roof forms
- Tunnel invert forms
- Tunnel arch forms
- Tunnel full-round forms
- Circular shaft forms
- Cantilever forms[12]

Some of the above forms occur in various degrees of complexity. For instance, columns and bridge piers can consist of flat, curved, or warped surfaces or combinations thereof, which affect the amount of labor required to fabricate (build) the forms. Consequently, many estimators classify form types yet further by assigning a **form complexity code** as follows:

---

[9]Forms suspended and braced from above supports.

[10]Side and arch forms for small formed tunnels in mass concrete structures.

[11]Forms fastened to walls or slab soffits to form recesses in the flat formed surface.

[12]Vertical or inclined single-face forms used for upstream or downstream faces or monolith line joints on dams and/or other similar structures.

F	Flat	Large flat areas
S	Straight	Flat but with one right-angle dimension much larger than the other
B	Broken	Flat faces but with short dimensions in each right-angle direction
C	Curved	Straight in one right-angle direction but curved in the other
W	Warped	Curved in both right-angle directions

Shoring Volumes

Supported slab, girder, and beam soffit forms must be carried on some sort of falsework or shoring. This vertical support is often provided by sectional patent scaffold frames (or by individual wood or steel shores) when the soffit forms to be supported are not too high above the supporting surface. Some contractors price these types of support systems on a volumetric basis, the **shoring volume** in cubic feet being considered the area of the forms to be supported in square feet times the height of the shoring systems in feet. When this system is used, costs are then estimated for installing and later removing scaffolding or shoring occupying this "shoring volume." In these cases, the shoring volumes required are taken off the drawings as part of the formwork takeoff.

When the forms to be supported are very high above the supporting surface, an engineered support system must be devised for which the component elements must be separately taken off for pricing.

Foundation Contact Areas

Invariably, concrete structures are founded upon some sort of excavated surface. Structures **resting on soil or granular alluvial materials** usually will be preexcavated to grade tolerances of ±0.1 ft. Even so, it is usually necessary to smooth and firm-up the excavated surfaces to erect supports for the lower mats of reinforcing steel. This operation requires a labor expenditure, so the **contact area involved,** measured in square feet, must be taken off.

If the structure is founded on horizontal or inclined areas **in rock excavation** removed by heavy ripping or blasting operations, the excavated surface will usually be irregular and somewhat broken. All loose or "drummy" rock must be removed prior to placing concrete. This operation is very labor intensive, and it is important that such areas be taken off so that the work can be properly priced.

When the structure is supported on driven piles, the preparation for footing concrete placement can be very costly due to the necessity to remove excess foundation material resulting from **ground swell or heaving** from the in situ earth volumes being displaced by the solid volume of the piles. Also, when structural footings are placed on **top of tremie seal pours** in sheet pile cofferdams, the top surface of the tremie seal will usually be of poor quality and will require removal of mud and laitance. In both of these situations, the cost of preparing the underlying surface for placement of the footing concrete can approach or equal the cost of foundation preparation on rock surfaces.

Horizontal Construction Joint Areas

Horizontal construction joint areas, measured in square feet, at the top of each previously placed lift of concrete must be **cleaned by removing laitance and roughening the lift joint** to ensure proper bond of the following lift of concrete. The work is usually accomplished by the **green cut, waterblast,** or **sandblast process,** and the areas requiring such treatment must be taken off to properly price the work required. The takeoff should segregate large open areas (that might be encountered in the horizontal joints of a large concrete dam, for instance) from the more normal smaller congested areas occurring in highly reinforced structural concrete work.

Wet Finish Areas

Freshly placed concrete requires a **wet finish at the top of the final lift,** which is measured in square feet. This operation is more or less labor intensive, depending on the type of wet finish required. The following types of wet finish should be segregated in the takeoff:

- *Strike-off only:* Leveling off the concrete at a specified elevation
- *Wood float finish:* Rubbing a hard coarse-grain finish into the concrete with a wood-base finishers float
- *Steel trowel finish:* Following wood floating, finishing with a smooth steel trowel
- *Wall tops and other small, isolated areas:* Regardless of the type of finish

Cure Contact Areas

After placement, concrete **must be kept wet or moist during a curing period** to prevent evaporation of water that is necessary to hydrate the cement. Another effective method to prevent water loss for vertical surfaces is to spray the surfaces with a curing compound that forms a waterproof membrane on the surface of the concrete. This method is also effective for horizontal areas that are not to receive an additional placement of concrete (such as finished floors or roof slabs), the underside of slab soffits, and beam and girder sides and soffits. The cure contact areas, measured in square feet, that must be taken off include all vertical and horizontal form surfaces, horizontal construction joint areas, and all wet finish areas.

Dry Finish Areas

After vertical and/or horizontal form work is stripped, the resulting concrete surface will contain tie holes, joint off-sets and leakage fins, and occasionally, honeycomb areas where the cement mortar has not completely flowed into the voids between aggregate particles. These imperfections usually require some sort of **dry finish treatment.** Dry finish operations can be quite labor intensive, depending on the finish requirements for the particular formed area. A common takeoff breakdown for dry finish areas, measured in square feet, includes the following:

- *Non-exposed areas:* Areas that are either backfilled or otherwise not exposed to public view. These areas require only repair of honeycomb and plugging of tie holes.
- *Exposed areas:* Areas exposed to public view that, in addition to plugging tie holes and repairing honeycomb, require all leakage fins and irregular or offset joints to be ground smooth and the concrete surface generally be made visually attractive.

- *Sack rub finish:* Highly important exposed areas are often required to be sack rubbed, a process wherein the concrete surface is rubbed with rough burlap sacking material containing a moist fine sand/cement mixture. This further smoothes the surface by filling air bubble holes and results in a highly attractive finish.
- *Other special finishes:* Occasionally, special architectural finishes such as sandblasting or bush hammering will be specified for selected areas.

Reinforcing Steel Weight

Most concrete structures are reinforced with **steel reinforcing bars,** some very heavily reinforced. For schedule-of-bid-item bids, reinforcing steel is usually paid as a separate bid item, although sometimes it is included within the bid price for concrete containing the reinforcing steel. In any case, the weights, in pounds, of the various reinforcing steel bars required must be known so that the material and labor costs may be priced. Usually, reinforcing steel suppliers furnish and fabricate the required reinforcing steel, cut and bent ready for installation. They normally will independently make the required material takeoff, which is a highly specialized task. The takeoff information usually will be made available to the bidding general contractors. Occasionally, however, prime contractors may wish or be required to take off reinforcing steel quantities. Detailed takeoffs are segregated by reinforcing bar size and by quantities of bars that are straight and quantities that must be bent to particular shapes.

Miscellaneous Embedded Items

Examples of such items include **waterstops,**[13] **premolded expansion joint filler,** and **miscellaneous metal work items.** These items are sometimes separately paid for in other bid items, but are often included in the work of the concrete bid item. In either case, the quantities must be determined for pricing.

Quantities for prefabricated items such as miscellaneous metalwork are often taken off by the suppliers who traditionally fabricate and supply these items. The quantity information is made available to bidding general contractors through the material supply quotation of such suppliers. However, the bidding general contractors often wish to, or (if supplier quotations are not expected) must, take off these materials themselves.

Waterstops are usually taken off in units of linear feet for the various widths required, joint filler by the square feet, and miscellaneous metalwork by the number of pieces or by the cumulative weight in pounds for similar kinds of items.

DETERMINATION OF SUPPLEMENTARY QUANTITIES THAT ARE DEVELOPED FROM PRIMARY QUANTITIES TAKEN OFF THE DRAWINGS

In addition to the primary quantities taken directly off of the construction drawings (discussed above), a further category of quantities required for pricing must be developed. Examples of these kinds of quantities include the following:

---

[13]Metal or polyvinylchloride strips with a central circular bulb, generally 6 to 9 in. wide, placed across construction and expansion joints to prevent the passage of water through the joint.

- Cement and aggregate quantities (when on-site batch plants are expected to be used)
- Quantity of formwork panels to be fabricated or purchased and, when job-built forms are to be used, dimension lumber and form plywood quantities
- Quantities for "allowance" pricing

Cement and Aggregate Quantities

When concrete is to be batched and mixed by the contractor in an on-site plant, the necessary **cement and aggregate quantities** must be determined so that the supply and delivery of these materials to the jobsite or, alternately, the production of these materials from on-site natural deposits may be properly priced. Bidding specifications usually define the following for each particular strength class of concrete required:

- 28 day strength (i.e., 2,000, 3,000, 4,000 psi, etc.)
- Maximum size aggregate permitted (i.e., $1\frac{1}{2}$ in.)
- Minimum cement content in sacks of cement per cubic yard of concrete, or in pounds of cement per cubic yard of concrete (i.e., 5.0 sacks per cubic yard, or 470 lb/cy)
- Maximum water/cement ratio by weight or maximum water content in gallons per sack of cement (i.e., water/cement = 0.62 or 7.0 gal/per sack)

Based on the above specification information, estimates can be made (satisfactorily for pricing purposes) for the required quantities of cement, coarse aggregate, and fine aggregate. In making these estimates, the following physical facts are utilized:

- One sack of cement weighs 94 lb.
- One barrel of cement contains four sacks.
- The specific gravity of cement is 3.15.
- The specific gravity of coarse and fine aggregate is 2.65.
- Water weighs 62.4 lb/cf.
- One cubic foot of water contains 7.48 U.S. gal.
- One gallon of water weighs 8.33 lb.
- The usual proportion of fine aggregate varies between 25 and 45% of total aggregate.
- The absolute volume of any ingredient in cubic feet equals the weight of the ingredient in pounds ÷ (specific gravity of the ingredient × 62.4 lb/cf).

For instance, assume it is desired to compute the constituent material quantities required for 75,000 cy of concrete with the following specification requirements:

Maximum size aggregate: $1\frac{1}{2}$ in.
Minimum cement content: 6.5 sacks per cubic yard
Maximum water/cement ratio: 0.65 (by weight)

Cement/cy	6.0 sacks = (6.0)(94 lb/sack) = 564 lb
Absolute volume of cement	564 lb/(3.15 × 62.4) = 2.87 cf
Water/cy	(564 lb)(0.65) = 366.6 lb
Absolute volume of water	366.6 lbs/(1.0 × 62.4) = 5.88 cf
Absolute volume of 1 cy of concrete	27.0 cf
Assume 6% air voids	(27.0)(0.06) = 1.62 cf

Absolute volume of aggregate	$27.00 - 1.62 - 2.87 - 5.88 = 16.63$ cf
Assume % sand (fine aggregate)	35%
Absolute volume of sand	$(0.35)(16.63) = 5.82$ cf
Absolute volume of coarse aggregate	$16.63 - 5.82 = 10.81$ cf
Pounds fine aggregate/cy	$(5.82)(2.65)(62.4) = 962.39$ lb/cy
Pounds coarse aggregate/cy	$(10.81)(2.65)(62.4) = 1,787.54$ lb/cy
Total quantity cement	$(75,000)(6.0)(94)/2,000 = 21,150$ tons
Total quantity fine aggregate	$(75,000)(962.39)/2,000 = 36,090$ tons
Total quantity coarse aggregate	$(75,000)(1,787.54)/2,000 = 67,033$ tons

These quantities would have to be increased to take waste and overbreak concrete into account. In other words, if waste and overbreak concrete were expected to be 6% of the neat concrete quantities shown on the drawings, each of the quantities for cement, sand, and fine aggregate shown above would need to be increased by 6%.

Form Ratio and Quantity of Formwork Panels to Be Fabricated or Purchased

Once the contact areas to be formed are taken off, the form ratio and the quantity of form panels required for each separate form element in the structure can be determined. The **form ratio** for a particular element is the square feet of form contact area (sfca) for that element divided by the neat cubic yards quantity of concrete in the element. It expresses directly the ratio of the contact area of formwork per cubic yard of concrete, thus indicating the relative formwork quantity requirement for the element formed. For instance, the form ratio for a 54-ft-wide monolith in a mass concrete dam is 1.0 sfca/cy, whereas for a 6-in.-thick-wall in a pumping plant it is 108.0 sfca/cy. The average form ratio for an entire structure is the total square feet of form contact area divided by the total neat volume of concrete in cubic yards.

Because the cost of formwork is usually the major contributor to the cost of concrete in place, the form ratio is one important indicator of the relative cost of structural concrete.

The **reuse factor** is simply the number of times a form panel will be used. It can be thought of as the total contact area for a particular element in square feet divided by the area in square feet of form panels that will be required on hand to form the entire contact area for the element. The form reuse factor has some influence on the probable total cost of the concrete but not to the same extent as the form ratio.

These concepts are illustrated in Figure 10-1. Case 1 shows a 1-ft-6-in.-thick, 20-ft-0-in.-high concrete wall, 325 ft long, that is uniform in height from one end to the other. The form ratio is 37.2 sfca/cy, and the reuse factor is 13.0. In case 2, the wall thickness is the same but the height diminishes from 20 ft 0 in. at one end of the wall to 4 ft. 0 in. at the other end. For this case, the form ratio is also 37.2 sfca/cy, even though computed by use of different dimensions. The reuse factor decreases to 7.8.

The form ratio serves in cost estimating only as an indicator of relative formwork quantity requirements for a given quantity of concrete, whereas the reuse factor determines the quantity of form panels (in square feet) that will be required to form a given quantity of form contact area (in square feet of contact area). Establishing the reuse factor is a judgment call on the part of the estimator and is highly related to the type of formwork involved and to the project time schedule. In traveling form systems for tunnel linings and cut-and-cover box structures, form reuse factors as high as 100 (or more) are not uncommon.

Once the reuse factor has been determined for each class of formwork required, the quantities of required form panels for each class can be calculated by dividing the total contact area of each form class by the appropriate reuse factor.

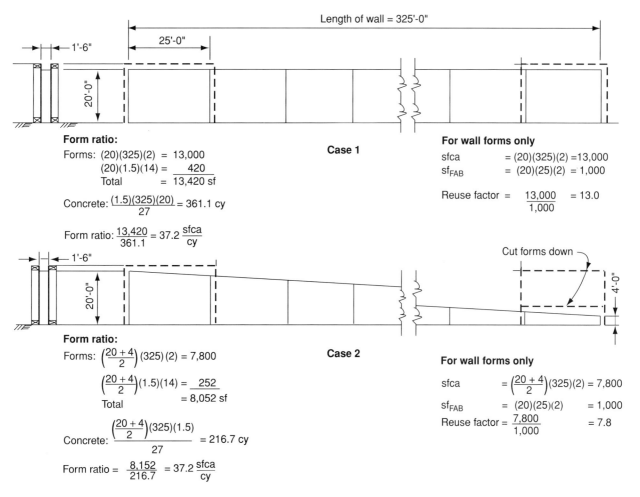

FIGURE 10-1   Form ratios and reuse factors.

Dimension Lumber and Form Plywood Requirements

Once the reuse factor and quantity of form panels have been determined for each class of form that is to be job-built, the required quantities of **dimension lumber** and **plywood** can be calculated by means of the **use factors** shown in Table 10-1. The calculation procedure can best be explained by means of the following two examples. The first is a flat single-face wall:

Contact area	22,000 sfca
Reuse factor	15
Fabrication quantity required	$22,000/15 = 1,467$ sf
Dimension lumber use factor, (F)	$3.0 + (15 - 1)(5.0)/(100)(3.0)$
	$3.0 + 2.10 = 5.10$ FBM/sf
Plywood use factor, (F)	$1.1 + (15 - 1)(2.50/100)(1.1)$
	$1.1 + 0.39 = 1.49$ sf/sf
Total dimension lumber	$(1,467)(5.10) = 7,482$ FBM
Total plywood	$(1,467)(1.49) = 2,186$ sf

TABLE 10-1 Dimension Lumber and Plywood Use Factors for Job-Built Wood Forms

Description of form	Dimension lumber factors, BF/SF FAB						Form plywood factors,[9] SF/SF FAB					
	Complexity of form					% additional uses over one	Complexity of form					% additional uses over one
	F[1]	S[2]	B[3]	C[4]	W[5]		F[1]	S[2]	B[3]	C[4]	W[5]	
BIP off rock	3.0	3.0	3.5	4.0	8.0	(7)						
Other BIP forms	3.0	3.0	3.2	3.5	8.0	(7)						
SOG edge	3.0	3.0	3.2	3.5	8.0	5.0	1.1	1.1	1.2			2.5
SOG bulkhead	3.5	3.5	3.5	4.0		(7)						
Single-faced walls	3.0	3.0	3.2	3.5	8.0	5.0	1.1	1.1	1.2	1.5	1.5	2.5
Double-faced walls	3.0	3.0	3.2	3.5		5.0	1.1	1.1	1.2	1.5	1.5	2.5
Columns	3.0	3.0		4.0	8.0	10.0	1.1	1.1		1.5(8)	1.5(8)	2.5
Wall bulkheads	3.0	3.0	3.5			(7)	1.1	1.1	1.1			2.5
Supported slab soffit[6]	2.6	2.6	2.6			5.0	1.1	1.5	1.5			2.5
Supported slab edge		2.2	2.5	3.0		10.0		1.5	1.5	1.5		2.5
Support slab bulkhead	3.0	3.0	3.5			(7)						
Beam soffit[6]		3.0	3.5	4.0		10.0		1.5	1.5	1.5		5.0
Beam side		3.0	3.5	4.0		10.0		1.5	1.5	1.5		5.0
Beam bulkhead		3.0	3.5			(7)						
Curb forms		2.5	2.5	3.0		10.0						
Hanging forms		3.0	3.5	4.0		10.0		1.1	1.5	1.5(8)		5.0
Stair forms		6.0	6.0			(7)		1.5	1.5			
Transition forms			4.0	8.0	8.0	10.0		1.1	1.5(8)	1.5(8)	1.5(8)	10.0
Gallery forms		3.5	3.5	4.0		10.0		1.1	1.5	1.5		5.0
Blockouts			2.5			(7)			1.5			5.0

(1)Flat: Large flat areas.
(2)Straight: Flat with one right-angle dimension much larger than the other.
(3)Broken: Flat with short dimensions in each right-angle direction.
(4)Curved: Straight in one right-angle direction but curved in the other.
(5)Warped: Curved in both right-angle directions.
(6)Soffit form factors do not include shoring.
(7)Normally a one-use form (i.e., no reuse).
(8)When plywood used, it is normally only a thin overlay (say 1/4 in.).
(9)All plywood requirements are in addition to dimension lumber requirements.

258

The second example is an inlet form for water discharge conduits:

Contact area	30,000 sfca
$\frac{1}{3}$ broken (B)	10,000 sfca
$\frac{2}{3}$ warped (W)	20,000 sfca
Reuse factor	3
Dimension lumber:	
Broken portion factor (B)	$4.0 + (3 - 1)(10.0/100)(4.0)$
	$4.0 + 0.80 = 4.80$ FBM/sf
Broken portion quantity	$(10,000/3)(4.80) = 16,000$ FBM
Warped portion factor, (W)	$8.0 + (3 - 1)(10.0/100)(8.0)$
	$8.0 + 1.60 = 9.60$ FBM/sf
Warped portion quantity	$(20,000/3)(9.60) = 64,000$ FBM
Total dimension lumber	$16,000 + 64,000 = 80,000$ FBM
Plywood:	
Factor for both broken and warped portion	$1.5 + (3 - 1)(10.0/100)(1.5)$
	$1.5 + 0.30 = 1.80$ sf/sf
Total plywood	$(30,000/3)(1.80) = 18,000$ sf

Dimension lumber and plywood quantities for the forms for other structural elements can be computed similarly. When the number of form elements involved is large, these calculations are more easily performed in tabular form using spreadsheet software.

Quantities for "Allowance" Pricing

A number of expendable material items are required that are so small and varied that it is not practical to determine quantities for them as individual items. These kinds of material requirements are usually priced by providing a **monetary allowance** against some key quantity figure to generate sufficient money in the estimate to buy the entire class of small miscellaneous materials involved. The monetary "allowance" is a figure determined by past experience obtained from prior job records. Examples of key quantity figures for these types of expendable materials include the following:

- **Form ties and miscellaneous form materials:** The materials involved include all required form ties and accompanying hardware as well as nails, bolts, form oil, and similar materials consumed to provide project formwork. The monetary allowance is applied against the total square feet of contact area for all formed elements required for each particular concrete bid item. The key quantity therefore is the total square feet of contact area for all of the forms included in the bid item.
- **Sandblast sand:** The monetary allowance in this case is applied to the total cubic yards of horizontal construction joints required for a particular bid item to be cleaned by the sandblast process. The key quantity figure is therefore this total horizontal joint contact area.
- **Cure materials:** The monetary allowance is applied against the total area for all concrete elements in the bid item that must be cured. The key quantity figure is this total cure area measured in square feet.
- **Finish materials:** Here the monetary allowance is applied against the total area requiring any type of dry finish. The key quantity figure therefore is the sum of

all dry finish areas measured in square feet for all concrete elements included in the bid item.
- **Shoring materials:** The allowance figure is applied against the shoring volume takeoff figure in cubic feet.

LABOR REQUIREMENTS FOR VARIOUS CONCRETE OPERATIONS

General Pricing Approach Using Labor-Hour Factors

One reliable way to price structural concrete operations is by utilizing **labor-hour factors** obtained from past experience on similar work. Determination of work crew productivities for the various required operations then follows, depending on the labor-hour factor and the number of workers in the crew. When this system is used, it must be recognized that the calculated required labor-hours ostensibly required for the central work or purpose of the crew will include labor-hours for many other related subsidiary tasks required to be performed as well as labor-hours that may be lost due to delays and/or other causes of crew inefficiency.

For instance, the use of a historic labor-hour factor for a crew placing concrete for a particular kind of structural element might result in a calculated crew productivity of, say, 15 cy per crew-hour, a figure far less than the productivity of the crew when it is actually placing and vibrating the concrete in the forms. That is because the time required for moving between pours, cleaning out and otherwise preparing the forms for concrete, moving and cleaning tools and chutes, and similar tasks are all built into the historic labor-hour factor, as is time lost for any one of a number of different reasons. Thus, although use of the labor-hour factor method results in reliable pricing when everything that has to be done and overall crew efficiency is taken into consideration, the crew productivities appearing on the estimate sheets may seem to the uninitiated to be unrealistically low.

The following explanations and discussion pertain to the use of labor-hour factor estimating. The specific labor-hour factor information presented herein reflects a consolidation of job records obtained by the writer, and others, for a wide variety of structures. It must be understood that these values are general representations only of labor productivity that can be expected for well-managed concrete work. Individual contractors, at times, have achieved more, or less, favorable results on their individual projects.

Place and Vibrate Concrete Labor-Hour Factors

Representative labor-hour factors for the work of **form cleanup and placing and vibrating concrete** are shown in Table 10-2 for various types of concrete structural elements. Based on Table 10-2, a concrete crew consisting of a crane operator and oiler, a concrete foreman and six concrete laborers preparing for and placing and vibrating mass concrete on a dam would have a crew productivity of 9.0 labor-hours per crew-hour/0.20 labor-hour per cubic yard = 45 cy per crew-hour, whereas the same crew performing the same work for moderate to small pier caps would have a crew productivity of 9.0 labor-hour per crew-hour/0.90 labor-hour per cubic yard = 10.0 cy per crew-hour. As mentioned above, the concrete placing rates attained when these crews are actually placing concrete would be considerably higher.

Crew productivities for any other size crew performing the work of form cleanup and placing and vibrating concrete for any of the structural elements shown in Table 10-2 can be similarly obtained, as long as the crew size is reasonable.

TABLE 10-2 Place and Vibrate Concrete Labor-Hour Factors for Various Kinds of Structures

Type of structure	Labor-hour per cubic yard
Dams, locks, and similar massive structures	
Mass concrete, 4–8 cy bottom-dump buckets	0.20
Heavy structural walls, columns, and slabs	0.45
Moderately heavy structures	
Pile caps, grade and perimeter beams, footings	0.50
Slab on grade	0.40
Heavy walls	0.60
Thin walls	1.50
Columns, beams, and girders	1.00
Supported slabs	0.80
Bridge substructures	
Large piers and caps	0.75
Moderate to small piers and caps	0.90
Underground concrete	
Tunnel inverts, 100–300 ft placements, traveling form system	0.60
Arch or full-round, 100–300 ft placements, traveling forms	0.75
Miscellaneous structural concrete	1.65

TABLE 10-3 Miscellaneous Concrete Operations Labor-Hour Factors

Description of work	Labor-hours per square foot
Fine grade on earth foundations (not supported by driven piling)	0.02
Foundation preparation on rock foundations, on top of tremie seals, or on earth foundations with extensive pile heave	0.15
Sandblast construction joint cleanup: Large open areas	0.06
Sandblast construction joint cleanup: Small congested areas	0.08
Wet finish: Strike-off only	0.005
Wet finish: Wood float finish	0.015
Wet finish: Wood float/steel trowel finish	0.030
Wet finish: Wall tops/other small isolated areas	0.050
Cure: Spray-on curing compound	0.001
Cure: Water spray/curing paper or mats	0.005
Point and patch: Nonexposed areas	0.005
Point and patch: Exposed areas	0.015
Point and patch/sack rub	0.020
Shoring: In and out	0.005[1]

[1]This value is expressed in labor-hours per cubic foot of shoring volume.

Miscellaneous Concrete Operations Labor-Hour Factors

Labor-hour factors for miscellaneous required operations for concrete work are shown in Table 10-3. The various kinds of work operations reflected have been discussed earlier in this chapter.

Crew productivities can be determined in the same manner as for the place and vibrate concrete operation discussed above. For the results obtained by use of Table 10-3 to be realistic, the crew composition established for performing the various work operations reflected in the table must be reasonable.

Reasonable crew sizes will vary considerably. For instance, the "crew" for the work of applying spray-on curing compound after the forms have been stripped would consist of one or two laborers only, whereas the crew for foundation preparation on a blasted rock foundation could easily consist of a crane operator and oiler, a labor foreman, and six to eight laborers.

Formwork Labor-Hour Factors

The most labor-intensive work required for structural concrete is the fabrication, erecting, and stripping of formwork. Labor-hour factors for **the fabrication and erect and strip of job-built wood forms** are shown in Table 10-4, and **erect and strip labor-**

TABLE 10-4   Fabrication and Erect and Strip Labor-Hour Factors for Job-Built Wood Forms

Formwork description	F[1]	S[2]	B[3]	C[4]	W[5]
Fabricate only,[6] all wood forms	0.09[8]	0.12	0.20	0.30	0.60
E&S[7] BIP off rock	0.35[9]	0.35	0.40	0.50	
E&S Other BIP	0.25	0.25	0.35	0.40	
E&S SOG edges		0.30	0.35	0.45	
E&S SOG bulkheads		0.35			
E&S Single-faced walls	0.15	0.15	0.20	0.30	0.40
E&S Double-faced walls	0.12	0.12	0.15	0.20	
E&S Wall bulkheads		0.45			
E&S Supported slab soffits	0.12	0.15	0.20		
E&S Supported slab edges		0.35	0.45	0.50	
E&S Supported slab bulkheads		0.40			
E&S Columns		0.20		0.20	
E&S Beam soffits		0.25	0.25	0.30	
E&S Beam sides		0.20	0.25	0.30	
E&S Beam bulkheads		0.40			
E&S Curb forms		0.35			
E&S Hanging forms		0.40	0.45	0.50	
E&S Stair forms		0.45			
E&S Transition forms		0.30		0.35	0.45
E&S Gallery forms		0.20			
E&S Blockouts		0.20	0.25	0.30	

[1]Flat: Large flat areas.

[2]Straight: Flat with one right-angle dimension much longer than the other.

[3]Broken: Flat with short dimensions in each right-angle direction.

[4]Curved: Straight in one right-angle direction but curved in the other.

[5]Warped: Curved in both right-angle directions.

[6]Job-built form panels to be used one or more times.

[7]Erect, strip, and repair form panels.

[8]Fabricate values are labor-hours per square foot of form panels built.

[9]E&S values are labor-hours per square foot of form contact area.

hour factors for purchased, steel or combination wood-steel, factory fabricated forms are shown in Table 10-5.

Examples of **form fabrication crew productivities** for several different varieties of job-built wood-forms determined by the use of Table 10-4 for a crew consisting of a carpenter foreman, four carpenters, and a center-mount crane operator (one-half time) are reflected below:

Flat forms (F)

Crew productivity = 5.5 labor-hours per crew-hour/0.09 labor-hours per square foot = 61.1 sf per crew-hour

Curb wood forms (C)

Crew productivity = 5.5 labor-hours per crew-hour/0.30 labor-hours per square foot = 18.3 sf per crew-hour

TABLE 10-5 Erect and Strip Labor-Hour Factors for Purchased, Steel or Combination Wood-Steel, Factory-Fabricated Forms

Formwork description	F[1]	S[2]	B[3]	C[4]	W[5]
Large bridge piers[6]		0.22		0.22	0.22
Large bridge caps[7]		0.38			
Large girder and beam forms		0.15			
Single-faced walls	0.12	0.15		0.25	
Double faced walls	0.10	0.12		0.15	
Gallery forms		0.15			
Large double-box culvert or subway invert forms[8]		0.15			
Double-box wall, haunch, and roof forms[9]		0.025			
Tunnel invert forms[10]		0.15			
Tunnel arch forms[11]				0.025	
Tunnel full-round forms[12]				0.03	
Circular shaft forms[13]				0.15	
Cantilever forms[14]		0.10			

[1]Flat: Large flat areas.

[2]Straight: Flat with one right-angle dimension much longer than the other.

[3]Broken: Flat with short dimensions in each right-angle direction.

[4]Curved: Straight in one right-angle direction but curved in the other.

[5]Warped: Curved in both right-angle directions.

[6]No external bracing, interior ties where required. Custom built for a particular project. Concrete placed in 15 to 25 ft vertical lifts.

[7]Including vertical supports. Concrete placed monolithically.

[8]Forms moved, supported, and braced by traveling carrier. Concrete placed in 50 ft formed sections. Labor-hour values in this case are per linear foot of individual suspended form surface.

[9]Forms moved, supported, and braced by traveling carrier. Concrete placed in 50 ft formed sections.

[10]Forms moved, supported, and braced by traveling carrier. Concrete placed monolithically in 100 to 300 ft formed lengths. Labor-hour values in this case are per linear foot of individual suspended form surface.

[11]Telescopic forms moved by traveling carrier, supported and blocked off of preplaced invert and excavated tunnel walls and arch. Concrete placed monolithically in 100 to 300 ft formed lengths.

[12]Same as for tunnel arch forms in invert-first, arch-second method, except full-round tunnel forms blocked and supported off of full-round excavated surfaces. Monolithic concrete placement in 100 to 300 ft lengths.

[13]Forms supported off of previous lift. Blocked off of excavated shaft surface. Concrete placed in 10 to 20 ft vertical lifts.

[14]Vertical of inclined single face forms. Concrete placed in $7\frac{1}{2}$ ft vertical lifts.

PHOTO 10-5   Examples of two different formwork systems for bridge pier construction. Upper photo illustrates job-built wood form panels using steel channel wales and strong backs. Lower photo shows crew assembling steel factory fabricated form panels (note job-built wood form panel on lower right).

PHOTO 10-6   Example of patented shoring and curved formed
concrete surfaces (upper photo), beam sides, and soffets and
supported slab soffit forms being stripped after concrete placed and
cured (lower photo).

PHOTO 10-7   These phots illustrate custom-made cantilever forms for the faces and monolith joints for a tall thin-arch dam. These forms are designed for a $7\frac{1}{2}$ ft vertical lift. Note the high-quality exposed concrete surface on the dam upstream face (no dry finish required) and the horizontal lift joint (requiring either a greencut or sandblast joint cleanup before placing the next lift).

PHOTO 10-8 These photos illustrate typical heavy formwork required for cut-and-cover subway construction. Upper photo shows a heavy job fabricated single-face form blocked at bottom and secured with a single steel tie-rod welded to soldier pile at top. Lower photo shows custom-built steel double box forms for side-by-side subway tunnels.

PHOTO 10-9   Use of plastic spacer wheels to ensure adequate concrete cover before setting column forms (upper photo). Example of excellent formwork line and grade and high-quality dry finish typical of first-class concrete work.

PHOTO 10-10 A 300-ft-long section of telescopic tunnel forms being concreted at a rate of over 300 ft per day. Note concrete pumpline discharge into port at crown of tunnel (invert had been previously placed at a rate of 300 ft per day with separate set of invert traveling forms). Note hydraulic ram in upper photo forcing concrete discharge "stinger" into port at crown.

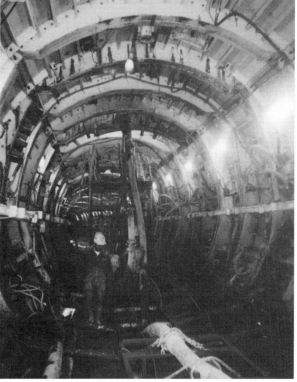

Warped forms (W)

Crew productivity = 5.5 labor-hours per crew-hour/0.60 labor-hours per square foot = 9.2 square feet per crew-hour

Similarly, **erect and strip crew productivities** for a crew consisting of a carpenter foreman, six carpenters, two laborers, and a crane operator and oiler (half-time) erecting and stripping various types of forms would be:

Flat double-faced job-built wood forms (F)

Crew productivity = 10.0 labor-hours per crew-hour/0.12 labor-hours per square foot of contact area = 83.3 square feet of contact area per crew-hour (Table 10-4)

Straight beam soffit wood forms (S)

Crew productivity = 10.0 labor-hours per crew-hour/0.25 labor-hours per square foot of contact area = 40.0 square feet of contact area per crew-hour (Table 10-4)

Steel cantilever forms (S)

Crew productivity = 10.0 labor-hours per crew-hour/0.10 labor-hours per square foot of contact area = 100.0 square feet of contact area per crew-hour (Table 10-5)

Similarly, a tunnel labor crew consisting of one shifter, a form carrier operator, and 15 tunnel laborers erecting and moving tunnel arch forms would achieve a crew productivity of

17.0 labor-hours per crew-hour/0.025 labor-hours per square foot of contact area = 680 square feet of contact area per crew-hour (Table 10-5)

This latter crew productivity is typical for highly mechanized form operations where sophisticated form carriers are utilized.

Consolidated Approach to Reduce Number of Work Operations to Be Individually Priced

Depending on the complexity of the structure, many of the required work operation categories will have so many individual takeoff quantities for a variety of individual operations under the general category heading that if a work operation were set up on the estimate sheet for each of them, the resulting estimate would be very long and difficult to read and comprehend. For example, assume that the takeoff quantities for placing and vibrating the concrete for a structure were

Description	Quantity to place, cy
Pile caps and grade beams	250
SOG	3,575
Heavy walls	2,540
Columns, beams, and girders	1,210
Supported slabs	1,875
Parapet walls (thin walls)	75
Total	9,525

If the concrete was intended to be placed by the same general method, the six separate work operations could be **consolidated into one work operation to be performed by one "typical" concrete crew** as follows:

Description	Quantity, cy	Labor-hour factor, labor-hour per cubic yard	Labor-hour
Pile caps and grade beams	250	0.50	125
SOG	3,575	0.40	1,430
Heavy walls	2,540	0.60	1,524
Columns, beams, and girders	1,210	1.00	1,210
Supported slabs	1,875	0.80	1,500
Parapet walls	75	1.50	113
Totals	9,525	0.62[1]	5,902

[1]5,902 labor-hours / 9,525 cy = 0.62 labor-hours per cubic yard.

Thus, only one work operation with a quantity of 9,525 cy would appear in the estimate that would be priced on the basis of **a weighted average labor-hour factor** of 0.62 labor-hours per cubic yard.

Detailed breakdowns of quantities for detailed work operations for all other kinds of required work under general work categories such as formwork, wet finish, dry finish, and so on can be consolidated similarly for pricing by means of composite, weighed average labor-hour productivity factors.

COMPOSITION OF TYPICAL WORK CREWS FOR CONCRETE OPERATIONS

Typical Work Crews

The **composition of work crews for the various required work operations** will vary depending on the kind of structure involved and the work practices of the particular contractor. The crews shown in Table 10-6 are typical for the normal work operations required for general structural concrete work and most such work can be reliably priced on the basis of these crews.

Alternate Methods for Providing Crane Service

Crane hoisting equipment and the required operators, and (where required) oilers, are included in the crew makeups shown in Table 10-6. A common **alternate method to provide crane service** in the estimate is to carry the required cranes (with their operators and oilers) as separate work operations. If this choice is made, an additional crew would be set up for each different class of crane intended to be utilized for the bid item consisting of the crane equipment and the required operating personnel. In this case, all the crane operators and oilers shown in Table 10-6 would not be included in those crews.

Required crew-hours for concrete operation work crews are determined by the operation work quantity and the work crew productivity. The work crew productivity

TABLE 10-6   Typical Crews for General Structural Concrete Work Operations

Work description	Crew composition	
1. Batch, mix, haul concrete	Batch and mix plant[1]	1.00
	4 cy front-end loader	0.25
	Dumpcrete trucks	3.00[2]
	Batch plant operator	1.00
	Common laborer	1.00
	Loader operator	0.25
	Teamsters	3.00[2]
	ST&S, %	5.00
2. Fine grade	Labor foreman	1.00
	Common laborers	6.00
	ST&S, %	5.00
3. Foundation preparation	3 cy hydraulic backhoe	1.00
	600 cfm compressor	1.00
	25 HP diesel contractor's pump	1.00
	Paving breakers	6.00
	Labor foreman	1.00
	Laborer-air tool	6.00
	Common laborer	2.00
	Backhoe operator	1.00
	ST&S, %	10.00
4. Fabricate job-built forms	18 ton center-mount crane	0.50
	Carpenter foreman	1.00
	Carpenters	4.00
	Crane operator	0.50
	ST&S, %	5.00
5. E&S forms (wood or steel)	100 ton crawler or motor crane	0.50
	400 A welding machine	1.00
	Carpenter foreman	1.00
	Carpenters	6.00
	Common laborer	2.00
	Crane operator	0.50
	Oiler	0.50
	ST&S, %	5.00
6. Shoring in and out	Carpenter foreman	1.00
	Carpenters	4.00
	Common laborer	2.00
	ST&S, %	5.00
7. Place and vibrate concrete (crane and bucket)	100 ton crawler or motor crane	1.00
	4 cy concrete bucket	2.00
	185 cfm compressor	1.00
	High-cycle generator with vibrators, (lot)	1.00
	Labor foreman	1.00
	Laborers, concrete	6.00
	Common laborer	1.00
	Crane operator	1.00
	Oiler	1.00
	ST&S, %	5.00

8. Place and vibrate concrete (concrete pump)	Truck-mounted concrete pump with boom	1.00
	185 cfm compressor	1.00
	High-cycle generator with vibrators (lot)	1.00
	Labor foreman	1.00
	Laborers, concrete	6.00
	Common laborer	1.00
	Pump operator	1.00
	ST&S, %	5.00
9. Wet finish	Concrete finisher foreman	1.00
	Concrete finishers	3.00
	ST&S, %	5.00
10. Sandblast joints	Sandblast pot	1.00
	185 cfm compressor	1.00
	Laborers	2.00
	ST&S, %	5.00
11. Greencut joints	185 cfm compressor	1.00
	Laborers	3.00
	ST&S, %	5.00
12. Cure	Laborers	2.00
	ST&S, %	5.00
13. Dry finish	185 cfm compressor concrete finisher foreman	1.00
	Concrete finishers	2.00
	ST&S, %	5.00

[1]Productive capacity depends on project size. 100–200 cy/hr is typical.

[2]Number depends on concrete on-site haul length; two or three haul units is typical.

is first determined by the crew size and the labor-hour production factor for the particular kind of work involved. When required crew-hours have been determined for all operations in the bid item, the number of crew-hours of each operation requiring cranes is known. The number of crew-hours for each separately provided class of crane service would then be the sum of the crane hours for all of the work operations requiring that particular class of crane.

A similar method is often utilized for providing **compressed air service** on large jobs where compressed air is provided by a centrally located plant, as well as for other **centrally provided services** such as underground ventilation and/or shaft-service or portal-service crews on tunnel projects.

EXAMPLE OF CONCRETE ANALYSIS AND PRICING

Description of Project and Takeoff Data from Drawings

Consider the concrete substructure for a hydroelectric powerhouse. The substructure concrete and reinforcing steel are to be paid under bid items 12 and 13, which are described on the bid form as:

BI 12	Substructure concrete	10,150 cy at $_____/cy = $_____
BI 21	Reinforcing steel	1,675,800 lb at $_____/lb = $_____

PHOTO 10-11   This small diversion dam (although not a hydro-electric powerhouse) illustrates the type of formwork and concrete joint configuration typically involved for the example concrete estimate in this chapter.

The contractor's takeoff quantities from the bidding drawings were as follows:

Foundation preparation		20,000 sf	
Concrete volumes		Neat Line, cy	Overbreak cy
SOG	3,000 psi	2,890	650
Heavy walls	3,000 psi	2,596	467
Mass concrete	3,000 psi	2,790	944
Columns, beams, and girders	4,000 psi	1,160	
Supported slabs	4,000 psi	260	
E&S forms			
Single-faced walls (F)		13,920 sfca	
Block outs (B)		4,624 sfca	
Supported slabs (F)		19,938 sfca	
Wall bulkheads (S)		4,939 sfca	
Double-face walls (F)		76,343 sfca	
Single-face curved walls (C)		1,235 sfca	
SOG edge (S)		6,817 sfca	
Columns (S)		1,820 sfca	
Beam sides and soffits (S)		11,488 sfca	
Shoring in and out		223,302 cf	
Wet finish			
Steel trowel		26,893 sf	
Wood float		7,845 sf	
Sandblast joints		12,518 sf	
Cure		151,510 sf	
Dry finish			
Exposed		67,514 sf	
Unexposed		34,700 sf	

Reinforcing steel
  No takeoff was performed for bid item 21, reinforcing steel, but the potential supplier advised that the bid quantity was reasonably accurate.

Contractor's Construction Plans for Bid Items 12 and 21

The powerhouse is founded on a rock foundation that must be blasted prior to excavation to design foundation grade. This will result in both horizontal, vertical, and inclined rock contact surfaces.

The configuration of the site is such that satisfactory hoisting equipment reach can be achieved by use of 100 ton crawler cranes operating on an on-grade surface along the downstream face of the powerhouse. The crane runway surface is on the upstream part of the powerhouse tailrace excavation, which is left in place until completion of the powerhouse structural, mechanical, and electrical work. This remaining excavation will then be taken out near the end of the job following completion of the powerhouse.

The contractor has chosen to utilize a purchased patented steel forming system for 60% of the wall forms and job-built wood forms for the balance of the formwork.

Concrete is available from a nearby concrete supplier using 8 cy transit-mix trucks to deliver concrete on an f.o.b. jobsite basis.

Reinforcing steel and all important embedded items, including waterstops, pre-molded joint filler, and miscellaneous metal work, are paid for under separate bid items. The contractor will purchase reinforcing steel for bid item 21, cut and bent, on an f.o.b. jobsite basis and plans to engage a subcontractor to install the steel. Under this arrangement, it has been agreed that the prime contractor must furnish unloading and hoisting service to the reinforcing placement subcontractor free of charge and must furnish the fabricated reinforcing steel and pay the subcontractor for installation of all templates and other required supports.

All work is intended to be performed on a straight-time, two-shift-per-day basis, 5 days per week.

Concrete Analysis for Place and Vibrate and Concrete Purchase Quantities for Bid Item 12

Working from the takeoff information from the drawings, all required concrete purchase and place and vibrate quantities can be easily developed as shown in Figure 10-2. The significant developed quantities required for pricing include

3,000 psi PM purchase	8,276 cy
3,000 psi waste and overbreak purchase	$2,061 + 310 = 2,371$ cy
4,000 psi PM purchase	1,420 cy
4,000 psi waste and overbreak purchase	43 cy
SOG place and vibrate	3,646 cy
Heavy walls place and vibrate	3,155 cy
Mass concrete place and vibrate	3,846 cy
Columns, beams, and girders place and vibrate	1,195 cy
Supported slab place and vibrate	268 cy

In addition, as a check of the bid quantity, the total takeoff figure for pay concrete = 8,276 cy + 1,420 cy = 9,696 cy. Thus, on the basis of the contractor's takeoff, the bid quantity of 10,150 cy is considerably overstated.

Formwork Analysis for Form Fabrication, Form Buy, Dimension Lumber, and Plywood Quantities for Bid Item 12

Supplemental required quantities for formwork fabrication, steel form buy, and the materials required for job-built forms can be developed from the basic take off information obtained from the bidding drawings in a similar manner to that employed for developing quantities for concrete material purchase and for the concrete placing quantities. This work for the example hydroelectric powerhouse is shown as Figure 10-3. The significant supplemental quantities developed in Figure 10-3 are

Total quantity of job-built forms to be fabricated for each form type totaling 43,793 sf

Total quantity of dimension lumber to be purchased = 132,973 FBM = 133.0 MBF

Concrete purchase quantities

Description	3,000 psi concrete, cy				4,000 psi concrete, cy				Place/vibrate quantity,[2] cy
	Paylines	Overbreak	Waste[1]	Total	Paylines	Overbreak	Waste[1]	Total	
SOG	2,890	650	106	3,646	–	–	–	–	3,646
Heavy walls	2,596	467	92	3,155	–	–	–	–	3,155
Mass concrete	2,790	944	112	3,846	–	–	–	–	3,846
Columns, beams, & girders	–	–	–	–	1,160	–	35	1,195	1,195
Supported slabs	–	–	–	–	260	–	8	268	268
Totals	8,276	2,061	310	10,647	1,420	–	43	1,463	12,110

[1] Waste taken as 3% of payline + overbreak quantity.
[2] Place/vibrate quantity taken as including waste.

FIGURE 10-2   Concrete analysis for place and vibrate and concrete purchase quantities for bid item 12.

Form description	Code	sfca	No. of uses	Total quantity to fab	Dimension lumber Use factor, fbm/sf[1]	Dimension lumber Total quantity, fbm	Plywood Use factor, fbm/sf[1]	Plywood Total quantity, sf	Steel forms Quantity to buy, sf
Single-face walls: wood	F	5,568[2]	2	2,784	3.15	8,770	1.13	3,146	–
Single-face walls: steel	F	8,352[3]	6	–	–	–	–	–	1,392
Blockouts	B	4,624	1	4,624	2.50	11,560	1.50	6,936	–
Supported slab soffits	F	19,938	2	9,969	2.73	27,215	1.13	11,265	–
Wall bulkheads	S	4,939	1	4,939	3.00	14,817	–	–	–
Double-face walls: wood	F	30,537[4]	3	10,179	3.30	33,591	1.16	11,808	–
Double-face walls: steel	F	45,806[5]	6	–	–	–	–	–	7,634
Single-face curved walls	C	1,235	1	1,235	3.50	4,323	1.50	1,853	–
SOG edge	S	6,817	2	3,409	3.15	10,739	1.13	3,852	–
Columns	S	1,820	2	910	3.30	3,003	1.16	1,056	–
Beam sides and soffits	S	11,488	2	5,744	3.30	18,955	1.58	9,076	–
Totals		141,124	–	43,793[6]	–	132,973	–	48,992	9,026
		Flat		22,932					
		Straight		15,002					
		Broken		4,624					
		Curved		1,235					

[1] From Table 10-1.
[2] 13,920 sfca − 8,352 sfca = 5,568 sfca.
[3] 60% of 13,920 sfca = 8,352 sfca.
[4] 76,343 sfca − 45,806 sfca = 30,537 sfca.
[5] 60% of 76,343 sfca = 45,806 sfca.
[6] See complexity breakdown.

FIGURE 10-3 Formwork analysis for form fabrication, form buy, dimension lumber, and plywood quantities for bid item 12.

Total quantity of plywood to be purchased = 48,992 sf = 49.0 MSF

Total quantity of modular steel form panels to be purchased for the planned patented forming system = 9,026 sf

Note that information from Figures 10-2 and 10-3 also yield the average form ratio and average form reuse factor for the powerhouse:

$$\text{Average form ratio} = \frac{141,124 \text{ sfca}}{8,276 \text{ cy} + 1,420 \text{ cy}} \quad \begin{array}{l}\text{(Figure 10–3)}\\\text{(Figure 10–2)}\end{array}$$

$$= \frac{141,124 \text{ sfca}}{9,696 \text{ cy}}$$

$$= 14.6 \text{ sfca/cy}$$

$$\text{Average reuse Factor} = \frac{141,124 \text{ sfca}}{(43,793 + 9,026) \text{ sf fab}} \quad \begin{array}{l}\text{(Figure 10–3)}\\\text{(Figure 10–3)}\end{array}$$

$$= \frac{141,124}{52,819}$$

$$= 2.67$$

Development of Quantities for Allowance Pricing

The previously discussed significant quantities for allowance pricing in the case of the example hydroelectric powerhouse and the source of the quantity information are

Form ties and miscellaneous form materials	141,124 sfca (Figure 10-3)
Sandblast sand	12,518 sf (takeoff from plans)
Cure materials	151,510 sf (takeoff from plans)
Dry finish materials	67,514 sf + 34,700 sf = 102,214 sf (takeoff from plans)
Furnish shoring materials	223,302 cf (takeoff from plans)

Development of Consolidated Operations Productivity Factors

For this example, the foundation preparation, shoring in and out, sandblast joints, and cure work operations are associated with only a single quantity taken off from the drawings. The other work operations, however, have one or more subdivisions for which quantities have been taken off that can be consolidated into a single total quantity for which a weighted labor-hour productivity figure can be developed. These general work operations include

Place and vibrate concrete
Fabricate job-built forms
Erect and strip forms
Wet finish
Dry finish

Computations to develop the weighted labor-hour factor for each of these consolidated work operations follows:

Place and Vibrate Concrete

Description	Quantity, cy[1]	Labor-hour per cubic yard[2]	Required labor-hours
SOG	3,646	0.40	1,458
Heavy walls	3,155	0.60	1,893
Mass concrete	3,846	0.40[3]	1,538
Columns, beams, girders	1,195	1.00	1,195
Supported slabs	268	0.80	214
Total	12,110	0.52	6,298

[1]From Figure 10-2.
[2]From Table 10-2.
[3]Not a dam. Use SOG factor.

Fabricate Forms

Description	Quantity, sf[1]	Labor-hours per square foot[2]	Required labor-hours
Flat forms	22,932	0.09	2,064
Straight forms	15,002	0.12	1,800
Broken forms	4,624	0.20	925
Curved forms	1,235	0.30	371
Total	43,793	0.118	5,160

[1]From Figure 10-3.
[2]From Table 10-4.

E&S Forms

Description	Quantity, sfca[1]	Labor-hours per square foot of contact area[2]	Required labor-hours
F: Single-face wall, wood	5,568	0.15	835
F: Single-face wall, steel	8,352	0.12	1,002
B: Blockouts	4,624	0.25	1,156
F: Supported slab soffits	19,938	0.12	2,393
S: Wall bulkheads	4,939	0.45	2,223
F: Double-face walls, wood	30,537	0.12	3,664
F: Double-face walls, steel	45,806	0.10	4,581
C: Single-face curved	1,235	0.30	371
S: SOG edge	6,817	0.30	2,045
S: Columns	1,820	0.20	364
S: Beam sides and soffits	11,488	0.25	2,872
Total	141,124	0.152	21,506

[1]From Figure 10-3.
[2]From Tables 10-4 and 10-5.

Wet Finish

Description	Quantity, sf[1]	Labor-hours per square foot[2]	Required labor-hours
Steel trowel	26,893	0.030	807
Wood float	7,845	0.015	118
Total	34,738	0.027	925

[1]From takeoff from drawings.
[2]From Table 10-3.

Dry Finish

Description	Quantity, sf[1]	Labor-hours per square foot[2]	Required labor-hours
Exposed	67,514	0.015	1,013
Nonexposed	34,700	0.005	174
Total	102,214	0.012	1,187

[1]From takeoff from drawings.
[2]From Table 10-3.

In summary, the consolidated work operations and their corresponding weighted labor-hour productivity factors are as follows:

Place and vibrate concrete	12,110 cy	0.52 labor-hours per cubic yard
Fabricate forms	43,793 sf	0.118 labor-hours per square foot
Erect and strip forms	141,124 sfca	0.152 labor-hours per square foot of contact area
Wet finish	34,738 sf	0.027 labor-hours per square foot
Dry finish	102,214 sf	0.012 labor-hours per square foot

Typical Work Crews and Required Crew-Hours

Referring to the typical work crews shown in Table 10-6, the crew productivities for the various required work operations discussed earlier in this chapter and the corresponding required number of crew hours can now be developed as shown in Figure 10-4.

Estimate Pricing for Bid Item 12

Utilizing the California State University, Chico, computer estimating program and all of the previously developed input information, the direct cost estimate for bid item 12 is easily developed. It appears as Figure 10-5. It should be noted that because of the overstatement of the bid quantity in the schedule of bid items, a considerable upward proration is necessary for costs developed against the takeoff quantities to be raised to the level of the bid quantity. The unit prices for each cost category column remain the same, whether at the takeoff level or at the bid quantity level.

Work crew	Crew no.[1]	Work quantity	Labor hr/crew hr[1]	Labor hr productivity factor	Crew productivity	Required crew hr
Foundation preparation	3	20,000 sf[2]	10.0	0.15 mh/sf[3]	66.7 sf/crew/hr	300
Place/vibrate concrete	7	12,110 cy[4]	10.0	0.52 mh/cy[4]	19.2 cy/crew/hr	631
Fabricate forms	4	43,793 sf[4]	5.5	0.118 mh/sf[4]	46.6 sf/crew/hr	940
Erect and strip forms	5	141,124 sfca[4]	10.0	0.152 mh/sfca[4]	65.8 sfca/crew/hr	2,145
Shoring in and out	6	223,302 cf[2]	7.0	0.005 mh/cf[3]	1,400.0 cf/crew/hr	160
Wet finish	9	34,738 sf[4]	4.0	0.027 mh/sf[4]	148.1 sf/crew/hr	235
Sandblast joints	10	12,518 sf[2]	2.0	0.08 mh/sf[3]	25.0 sf/crew/hr	501
Cure	12	151,510 sf[2]	2.0	0.001 mh/sf[3]	2,000.0 sf/crew/hr	76
Dry finish	13	102,214 sf[4]	3.0	0.012 mh/sf[4]	250.0 sf/crew/hr	409

[1]See Table 10.6.
[2]Take off quantity from drawings.
[3]Table 10-3.
[4]Consolidated operations computations.

FIGURE 10-4 Determination of work crew productivities and required crew-hours.

DATE: 11/09/98
ESTIMATOR:
BID ITEM: JQB
TAKEOFF QUANTITY: 9,696 CY
BID QUANTITY: 10,150 CY

12

SUBSTRUCTURE CONCRETE

QTY	CODE	DESCRIPTION	UNIT	E.O.E	R-L	RENTAL	LABOR	S E.M.	P.M.	SUBS	REQD. HOURS	E.O.E	R-L	RENTAL	LABOR	E.M.	P.M.	SUB ONTRACT	TOTAL
		1 FOUNDATION PREPARATION																	
20000	SF		SF	66.7		/Hour					300								
1	EQ19	3 CY Hyd Backhoe	ea	50.020	5.000	29.720						$14,999	$1,499	$8,912					$25,409
1	EQ62	600 CFM Diesel Comp.	ea	15.300	1.610	7.430						4,588	483	2,228					7,298
1	EQ87	25HP Diesel Contr. Pump	ea	5.000	1.750	5.000						1,499	525	1,499					3,523
1	LAB8	Labor Foreman	ea				33.070								9,916				9,916
6	LAB4	Air Tool Operator	ea				30.080								54,117				54,117
2	LAB1	Common Laborer	ea				29.740								17,835				17,835
1	OE6	Backhoe Operator	ea				44.780								13,427				13,427
6	EQ90	Paving Breaker	ea	0.1	0.43	0.26						180	774	468					1,421
10.00	%	ST & S	LL													9,530			9,530
3.23	%	PREMIUM PAY	LL												3,078				3,078
		2 PLACE/VIBRATE CONCRETE																	
12110	CY		CY	19.2		/Hour					631								
1	EQ4	100 Ton Crawler Crane	ea	23.250	8.940	27.950						$14,664	$5,639	$17,629					$37,932
2	EQ77	4 CY Concrete Bucket	ea	0.130	0.400	1.040						164	505	1,312					1,980
1	EQ78	Vibrator w/ Generator	ea	1.650	7.210	1.460						1,041	4,548	921					6,509
1	EQ61	185 CFM Diesel Comp.	ea	5.460	1.610	2.430						3,444	1,015	1,533					5,992
1	LAB8	Labor Foreman	ea				33.070								20,858				20,858
7	LAB3	Concrete Laborer	ea				30.080								132,806				132,806
1	OE1	Crane Operator>150 Ft.	ea				50.160								31,637				31,637
1	OE3	Oiler	ea				39.890								25,160				25,160
5.00	%	ST & S	LL													10,523			10,523
3.23	%	PREMIUM PAY	LL												6,798				6,798
		3 FABRICATE FORMS																	
43793	SF		SF	46.6		/Hour					940								
0.5	EQ9	18 Ton Ctr. Mt. Crane	ea	13.200	7.020	14.330						$6,202	$3,299	$6,733					$16,234
1	CARP2	Carpenter Foreman	ea				40.290								37,863				37,863
0.5	OE2	Crane Operator<150 Ft.	ea				48.170								22,634				22,634
4	CARP1	Journeyman Carpenter	ea				38.270								143,859				143,859

FIGURE 10-5 (PAGE 1 OF 4) Direct cost estimate for hydroelectric powerhouse bid item 12 (cranes with work operations).

DATE: 11/09/98
ESTIMATOR:
BID ITEM:
JOB

SUBSTRUCTURE CONCRETE 12

TAKEOFF QUANTITY: 9,696 CY
BID QUANTITY: 10,150 CY

QTY	CODE	DESCRIPTION	UNIT	E.O.E	R-L	RENTAL	LABOR	S	E.M	P.M	SUBS	REQD. HOURS	E.O.E	R-L	RENTAL	LABOR	E.M	P.M	SUB ONTRACT	TOTAL
		5.00 % ST & S	LL																	
		% PREMIUM PAY	LL														10,218			10,218
OPRATN:		4 ERECT/STRIP FORMS																		
141124	SFCA		65.8 SFCA	/ Hour																
1	EQ65	400 AMP Diesel Welder	ea	3.200	1.030	1.000						2145	$6,863	$2,209	$2,145					$11,217
1	CARP2	Carpenter Foreman	ea				40.290									86,412				86,412
6	CARP1	Journeyman Carpenter	ea				38.270									492,476				492,476
2	LAB1	Common Laborer	ea				29.740									127,569				127,569
0.5	OE1	Crane Operator>150 Ft	ea				50.160									53,790				53,790
0.5	OE3	Oiler	ea				39.890									42,777				42,777
0.5	EQ4	Ton Crawler Crane	ea	23.25	8.94	27.95							24,933	9,587	29,973					64,492
		5.00 % ST & S	LL														40,151			40,151
		3.23 % PREMIUM PAY	LL													25,938				25,938
OPRATN:		5 SHORING IN & OUT																		
223302	CF		1400 CF	/ Hour																
1	CARP2	Carpenter Foreman	ea				40.290					160				$6,426				$6,426
4	CARP1	Journeyman Carpenter	ea				38.270									24,416				24,416
2	LAB1	Common Laborer	ea				29.740									9,487				9,487
		5.00 % ST & S	LL														2,016			2,016
		3.23 % PREMIUM PAY	LL													1,303				1,303
OPRATN:		6 WET FINISH																		
34738	SF		148.1 SF	/ Hour																
3	CM1	Concrete Finisher	ea				33.890					235				$23,847				$23,847
1	CM2	Conc. Finisher Foreman	ea				35.240									8,266				8,266

FIGURE 10-5 (PAGE 2 OF 4) Direct cost estimate for hydroelectric powerhouse bid item 12 (cranes with work operations).

284

DATE: 11/09/98
ESTIMATOR: JQB
BID ITEM: 12

TAKEOFF QUANTITY: 9,696 CY
BID QUANTITY: 10,150 CY

SUBSTRUCTURE CONCRETE

QTY	CODE	DESCRIPTION	UNIT	UNIT-COSTS E.O.E	R-L	RENTAL	LABOR	E.M.	P.M.	SUBS	REQD. HOURS	E.O.E	R-L	RENTAL	LABOR	E.M.	P.M.	SUB CONTRACT	TOTAL
	5.00	% ST & S	LL													1,606			1,606
	3.23	% PREMIUM PAY	LL												1,037				1,037
OPRATN:	7	SANDBLAST JOINTS																	
12518	SF		25 SF / Hour																
1	EQ61	185 CFM Diesel Comp.	ea	5.460	1.610	2.430					501	$2,734	$806	$1,217					$4,757
2	LAB4	Air Tool Operator	ea				30.080								30,123				30,123
1		Sandblast Pot (allow)	ea		1.5	1.5							751	751					1,502
	5.00	% ST & S	LL													1,506			1,506
	3.23	% PREMIUM PAY	LL												973				973
OPRATN:	8	CURE																	
151510	SF		2000 SF / Hour																
2	LAB1	Common Laborer	ea				29.740				76				$4,506				$4,506
	5.00	% ST & S	LL													225			225
	3.23	% PREMIUM PAY	LL												146				146
OPRATN:	9	DRY FINISH																	
102214	SF		250 SF / Hour																
1	EQ61	185 CFM Diesel Comp.	ea	5.460	1.610	2.430					409	$2,232	$658	$994					$3,884
2	CM1	Concrete Finisher	ea				33.890								27,712				27,712
1	CM2	Conc. Finisher Foreman	ea				35.240								14,408				14,408

FIGURE 10-5 (PAGE 3 OF 4) Direct cost estimate for hydroelectric powerhouse bid item 12 (cranes with work operations).

| DATE: 02/2299 | | BID ITEM: | | 12 | | | | | | TAKEOFF QUANTITY: | | 9,696 CY | | | | | | |
| ESTIMATOR: JQB | | | | SUBSTRUCTURE CONCRETE | | | | | | BID QUANTITY: | | 10,150 CY | | | | | | |

QTY	CODE	DESCRIPTION	UNIT	E.O.E	R-L	UNIT-COSTS RENTAL	LABOR	S E.M.	P.M.	SUBS	REQD. HOURS	E.O.E	R-L	RENTAL	LABOR	E.M.	P.M.	SUB ONTRACT	TOTAL
5.00 %		ST & S	LL													2,106			2,106
3.23 %		PREMIUM PAY	LL												1,360				1,360
OPRATN:	10	BUY FORM MATERIALS																	
133	EM3	Dimension Lumber	MBF					400.00								53,200			$53,200
49	EM4	3/4" Form Plywood	MSF					600.00								29,400			29,400
9026	EM5	BUY Steel Forms (Flat)	SF					30.00								270,780			270,780
9026	EM7	Sal. Steel Forms (Flat)	SF					-15.00								(135,390)			(135,390)
223302	EM9	Shoring Rental (Allow)	CF					0.038								8,485			8,485
141124	EM10	Form Ties & Mis (Allow)	SFCA					0.25								35,281			35,281
		% ST & S	LL																
		% PREMIUM PAY	LL																
OPRATN:	11	BUY CONCRETE MATERIALS																	
8276	PM1	3000 PSI Concrete	CY						45.00								$372,420		$372,420
1420	PM2	4000 PSI Concrete	CY						50.00								71,000		71,000
2371	EM1	3000 PSI Waste/OB Conc.	CY					45.00								106,695			106,695
43	EM2	4000 PSI Waste/OB Conc.	CY					50.00								2,150			2,150
12518	EM11	Sandblast M & S (Allow)	SF					0.15								1,878			1,878
151510	EM13	Cure M & S (Allow)	SF					0.01								1,515			1,515
102214	EM12	Finish M & S (Allow)	SF					0.02								2,044			2,044
		% ST & S	LL																
		% PREMIUM PAY	LL																
		TAKEOFF TOTALS:										$83,543	$32,297	$76,313	$1,502,962	$453,920	$443,420		$2,592,455
		PRORATED BID TOTALS										87,455	33,809	79,886	1,573,336	475,174	464,182		2,713,842
		UNIT PRICES /	CY									8.62	3.33	7.87	155.01	46.82	45.73		267.37

FIGURE 10-5 (PAGE 4 OF 4) Direct cost estimate for hydroelectric powerhouse bid item 12 (cranes with work operations).

Alternate Format for Bid Item 12 Estimate Showing Separate Crane Service

If the estimator preferred to show crane service as separate work operations, the direct cost estimate for bid item 12 would appear as shown in Figure 10-6. Note that because two classes of cranes were utilized, two separate crane service operations appear in the estimate (operations 12 and 13).

Estimate Pricing for Bid Item 21

According to the agreed terms of the planned subcontract, the prime contractor must provide crane service to the placing subcontractor. A work operation will have to be set up in the estimate for this work. Also, the contractor will be required to purchase the material for templates and supports and for nonpay splices (for which no payment will be received from the owner). An expendable material purchase item must therefore be included in the estimate for this material in the amount of, say, 5% of the total weight of reinforcing steel for which payment will be received.

Assuming that the full bid quantity of 1,675,800 lb will develop, the weight of nonpay steel that the prime contractor must purchase would be $(0.05)(1,675,800) = 83,790$ lb. The installation quantity for which the subcontractor would be paid would then be $1,675,800 + 83,790 = 1,759,590$ lb.

A reasonable labor-hour productivity factor for placing powerhouse reinforcing steel would be about 15 labor-hours per ton, or $15/2,000 = 0.008$ labor-hours/lb. On this basis, a crew of eight reinforcing ironworkers would place an average of 8.0 labor-hours per crew-hour/0.008 labor-hours per pound $= 1,000$ lb per crew-hour. Thus, a total of 1,759,590 lb would require $1,759,590/1000$ lb per crew-hour $= 1,760$ subcontractor crew-hours. It would be reasonable for the general contractor to anticipate that crane service would be required for 25% of the duration of the subcontract crew-hours, or $(1,760)(0.25) = 440$ crane crew-hours.

On the basis of the above analysis, the key quantities for pricing bid item 21 would be

Purchase permanent material	1,675,800 lb
Purchase support and nonpay splices	83,790 lb
Subcontract pay quantity	1,759,590 lb
100 ton crawler crane service	440 crew-hours

The resulting cost estimate for bid item 21 is shown as Figure 10-7.

ALTERNATE ESTIMATE FOR BID ITEM 12 ON BASIS OF UTILIZATION OF ON-SITE BATCH AND MIX PLANT

To illustrate the format of estimate involved, assume that the preceding example project was in an isolated location, compelling the contractor to batch, mix, and haul the concrete on-site rather than purchase the concrete ready-mixed from a nearby off-site supplier. The only part of the direct cost estimate shown in Figures 10-5 or 10-6 that would be affected would be operation 11 (where the permanent material purchase costs for 3,000 and 4,000 psi ready-mix concrete and the expendable material costs for

DATE: 02/22/99
ESTIMATOR:
BID ITEM: 12
JOB:
TAKEOFF QUANTITY: 9,696 CY
BID QUANTITY: 10,150 CY

SUBSTRUCTURE CONCRETE

QTY	CODE	DESCRIPTION	UNIT	E.O.E	R-L	RENTAL	LABOR	E.M.	S P.M.	SUES	REQD. HOURS	E.O.E	R-L	RENTAL	LABOR	E.M.	P.M.	SUB CONTRACTS	TOTAL
		OPRATN: 1 FOUNDATION PREPARATION																	
20000	SF		66.7 SF / Hour																
1	EQ19	3 CY Hyd. Backhoe	ea	50.020	5.000	29.720					300	$14,999	$1,499	$8,912					$25,409
1	EQ62	600 CFM Diesel Comp.	ea	15.300	1.610	7.430						4,588	483	2,228					7,298
1	EQ87	25HP Diesel Contr. Pump	ea	5.000	1.750	5.000						1,499	525	1,499					3,523
1	LAB8	Labor Foreman	ea				33.070								9,916				9,916
6	LAB4	Air Tool Operator	ea				30.080								54,117				54,117
2	LAB1	Common Laborer	ea				29.740								17,835				17,835
6	OE6	Backhoe Operator	ea				44.780								13,427				13,427
6	EQ90	Paving Breaker	ea	0.1	0.43	0.26						180	774	468					1,421
10.00		% ST & S	LL													9,530			9,530
3.23		% PREMIUM PAY	LL												3,078				3,078
		OPRATN: 2 PLACE/VIBRATE CONCRETE																	
12110	CY		19.2 CY / Hour								631								
2	EQ77	4 CY Concrete Bucket	ea	0.130	0.400	1.040						164	505	1,312					1,980
1	EQ78	Vibrator w/ Generator	ea	1.650	7.210	1.460						1,041	4,548	921					6,509
1	EQ61	185 CFM Diesel Comp.	ea	5.460	1.610	2.430						3,444	1,015	1,533					5,992
1	LAB8	Labor Foreman	ea				33.070								20,858				20,858
7	LAB3	Concrete Laborer	ea				30.080								132,806				132,806
5.00		% ST & S	LL													7,683			7,683
3.23		% PREMIUM PAY	LL												4,963				4,963
		OPRATN: 3 FABRICATE FORMS																	
43793	SF		46.6 SF / Hour								940								
1	CARP2	Carpenter Foreman	ea				40.290								37,863				37,863
4	CARP1	Journeyman Carpenter	ea				38.270								143,859				143,859
5.00		% ST & S	LL													9,086			9,086
		% PREMIUM PAY	LL																

FIGURE 10-6 (PAGE 1 OF 4) Direct cost estimate for hydroelectric powerhouse bid item 12 (with crane service isolated).

DATE: 02/22/99
ESTIMATOR:

BID ITEM: 12
JOB:

SUBSTRUCTURE CONCRETE

TAKEOFF QUANTITY: 9.696 CY
BID QUANTITY: 10.150 CY

QTY	CODE	DESCRIPTION	UNIT	— UNIT-COSTS — E.O.E	R-L	RENTAL	LABOR	S E.M.	P.M.	SUBS	REQD. HOURS	E.O.E	R-L	RENTAL	LABOR	E.M.	P.M.	SUB CONTRACTS	TOTAL
OPRATN:	4	ERECT/STRIP FORMS																	
141124	SFCA		65.8 SFCA	/Hour															
1	EQ65	400 AMP Diesel Welder	ea	3.200	1.030	1.000					2145	$6,863	$2,209	$2,145					$11,217
1	CARP2	Carpenter Foreman	ea				40.290								86,412				86,412
6	CARP1	Journeyman Carpenter	ea				38.270								492,476				492,476
2	LAB1	Common Laborer	ea				29.740								127,569				127,569
	5.00 %	ST & S	LL													35.323			35.323
	3.23 %	PREMIUM PAY	LL												22,819				22,819
OPRATN:	5	SHORING IN & OUT																	
223302	CF		1400 CF	/Hour															
1	CARP2	Carpenter Foreman	ea				40.290				160				$6,426				$6,426
4	CARP1	Journeyman Carpenter	ea				38.270								24,416				24,416
2	LAB1	Common Laborer	ea				29.740								9,487				9,487
	5.00 %	ST & S	LL													2,016			2,016
	3.23 %	PREMIUM PAY	LL												1,303				1,303
OPRATN:	6	WET FINISH																	
34738	SF		148.1 SF	/Hour															
3	CM1	Concrete Finisher	ea				33.890				235				$23,847				$23,847
1	CM2	Conc. Finisher Foreman	ea				35.240								8,266				8,266
	5.00 %	ST & S	LL													1,606			1,606
	3.23 %	PREMIUM PAY	LL												1,037				1,037
OPRATN:	7	SANDBLAST JOINTS																	
12518	SF		25 SF	/Hour															
1	EQ61	185 CFM Diesel Comp.	ea	5.460	1.610	2.430					501	$2,734	$806	$1,217					$4,757
2	LAB4	Air Tool Operator	ea				30.080								30,123				30,123
1		Sandblast Pot (allow)	ea		1.5	1.5							751	751					1,502

FIGURE 10-6 (PAGE 2 OF 4) Direct cost estimate for hydroelectric powerhouse bid item 12 (with crane service isolated).

289

DATE: 02/22/99
ESTIMATOR:
BID ITEM JOB 12 SUBSTRUCTURE CONCRETE
TAKEOFF QUANTITY: 9,696 CY
BID QUANTITY: 10,150 CY

QTY	CODE	DESCRIPTION	UNIT	E.O.E	R-L	RENTAL	LABOR	S E.M.	P.M.	SUBS	REQD HOURS	E.O.E	R-L	RENTAL	LABOR	E.M.	P.M.	SUB CONTRACTS	TOTAL
		5.00 % ST & S														1,506			1,506
		3.23 % PREMIUM PAY													973				973
OPRATN:		8 CURE																	
151510 SF			2000 SF	/Hour															
2	LAB1	Common Laborer	ea				29.740				76				4,506				$4,506
		5.00 % ST & S														225			225
		3.23 % PREMIUM PAY													146				146
OPRATN:		9 DRY FINISH																	
102214 SF			250 SF	/Hour															
1	EQ61	185 CFM Diesel Comp.	ea	5.460	1.610	2.430					409	$2,232	$658	$994					$3,884
2	CM1	Concrete Finisher	ea				33.890								27,712				27,712
1	CM2	Conc. Finisher Foreman	ea				35.240								14,408				14,408
		5.00 % ST & S														2,106			2,106
		3.23 % PREMIUM PAY													1,360				1,360
OPRATN:		10 BUY FORM MATERIALS																	
133	EM3	Dimension Lumber	MBF					400.00								$53,200			$53,200
49	EM4	3/4" Form Plywood	MSF					600.00								29,400			29,400
9026	EM5	BUY Steel Forms (Flat)	SF					30.00								270,780			270,780
9026	EM7	Sal. Steel Forms (Flat)	SF					-15.00								(135,390)			(135,390)
223302	EM9	Shoring Rental (Alow)	CF					0.038								8,485			8,485
141124	EM10	Form Ties & Mis (Alow)	SFCA					0.25								35,281			35,281
		% ST & S																	
		% PREMIUM PAY																	
OPRATN:		11 BUY CONCRETE MATERIALS																	

FIGURE 10-6 (PAGE 3 OF 4) Direct cost estimate for hydroelectric powerhouse bid item 12 (with crane service isolated).

DATE: 02/22/99
ESTIMATOR:
BID ITEM: 12
JOB: SUBSTRUCTURE CONCRETE
TAKEOFF QUANTITY: 9,696 CY
BID QUANTITY: 10,150 CY

| QTY | CODE | DESCRIPTION | UNIT | UNIT-COSTS | | | | S | | SUBS | REQD. HOURS | | | | | | | SUB CONTRACTS | TOTAL |
				E.O.E	R-L	RENTAL	LABOR	E.M.	P.M			E.O.E	R-L	RENTAL	LABOR	E.M.	P.M.		
8276	PM1	3000 PSI Concrete	CY						45.00								$372,420		$372,420
1420	PM2	4000 PSI Concrete	CY						50.00								71,000		71,000
2371	EM1	3000 PSI WasteOB Conc.	CY					45.00								106,695			106,695
43	EM2	4000 PSI WasteOB Conc.	CY					50.00								2,150			2,150
12518	EM11	Sandblast M & S (Allow)	SF					0.15								1,878			1,878
151510	EM13	Cure M & S (Allow)	SF					0.01								1,515			1,515
102214	EM12	Finish M & S (Allow)	SF					0.02								2,044			2,044
		% ST & S	LL																
		% PREMIUM PAY	LL																
OPRA.TN	12	TON CRANE SERVICE																	
1704	HR	1 HR	/Hour																
1	EQ4	100 Ton Crawler Crane	ea	23.250	8.940	27.950					1704	$39,618	$15,234	$47,627					$102,479
1	OE1	Crane Operator>150 Ft.	ea				50.160								85,473				85,473
1	OE3	Oiler	ea				39.890								67,973				67,973
5.00		% ST & S	LL													7,672			7,672
3.23		% PREMIUM PAY	LL												4,956				4,956
OPRATN.	13	TON CRANE SERV																	
470	HR	1 HR	/Hour																
1	EQ9	18 Ton Ctr. Mt. Crane	ea	13.200	7.020	14.330					470	$6,204	$3,299	$6,735					$16,239
1	OE2	Crane Operator<150 Ft.	ea				48.170								22,640				22,640
5.00		% ST & S	LL													1,132			1,132
		% PREMIUM PAY	LL																
		TAKEOFF TOTALS:										$83,565	$32,306	$76,340	$1,503,052	$453,924	$443,420		$2,592,607
		PRORATED BID TOTALS										87,478	33,818	79,914	1,573,430	475,178	464,182		2,714,001
		UNIT PRICES /	CY									8.62	3.33	7.87	155.02	46.82	45.73		267.39

FIGURE 10-6 (PAGE 4 OF 4) Direct cost estimate for hydroelectric powerhouse bid item 12 (with crane service isolated).

TE:
4/21/99
ESTIMATOR:

BID ITEM:
JOB   FURNISH & INSTALL STEEL

TAKEOFF QUANTITY: 1,675,800 LB
BID QUANTITY: 1,675,800 LB

QTY	CODE	DESCRIPTION	UNIT	E.O.E	R-L	RENTAL	LABOR	E.M.	P.M.	SUBS	REQD. HOURS	E.O.E	R-L	RENTAL	LABOR	E.M.	P.M.	SUB CONTRACTS	TOTAL
RATN:	1	PERMANENT MATERIALS & SUBCONTRACTS																	
675800	PM6	Fabricated Rebar	LB						0.25								$418,950		$418,950
83790	EM42	Non Pay Rebar	LB					0.25								20,948			20,948
759590	SUB1	Install Reinforcing	LB							0.20								351,918	351,918
		% ST & S	LL																
		% PREMIUM PAY	LL																
RATN:	2	TON CRAWLER SERVICE																	
440	HR	1 HR /Hour																	
1	EQ4	100 Ton Crawler Crane	ea	23.250	8.940	27.950					440	$10,230	$3,934	$12,298					$26,462
1	OE1	Crane Operator>150 Ft.	ea				50.180								22,070				22,070
1	OE3	Oiler	ea				39.890								17,552				17,552
	5.00	% ST & S	LL													1,981			1,981
	3.23	% PREMIUM PAY	LL												1,280				1,280
		TAKEOFF TOTALS:										$10,230	$3,934	$12,298	$40,902	$22,929	$418,950	$351,918	$861,160
		PRORATED BID TOTALS:										10,230	3,934	12,298	40,902	22,929	418,950	351,918	861,160
		UNIT PRICES /	LB :									0.01	0.00	0.01	0.02	0.01	0.25	0.21	0.51

FIGURE 10-7 Direct cost estimate for hydroelectric powerhouse bid item 21.

ready-mix waste and overbreak concrete were priced). Also an additional operation 13 would have to be added to price the on-site batch, mix, and haul operation.[14]

Cement and Aggregate Purchase Quantities

Development of the quantities of cement, fine aggregate, and coarse aggregate that must be purchased for the quantities of concrete required for bid item 12 (See Figure 10-2) is shown on Figures 10-8 and 10-9. Assume that the bidding documents contained the following specification information:

	3,000 psi concrete	4,000 psi concrete
Maximum size aggregate	3 in.	3 in.
Minimum cement content	5.0 sacks per cubic yard	6.0 sacks per cubic yard
Maximum water/cement ratio (by weight)	0.65	0.65

Although several mixes with different maximum size aggregate sizes would be required, the various gradations of coarse aggregate that would have to be purchased can be ignored because the purchase costs from an off-site aggregate plant source would not vary greatly due to variations in the coarse aggregate maximum size. Also, for cost-estimating purposes for a bid, the percentage of sand can be assumed to be constant (although it will vary with the maximum size aggregate used).

The key quantity figures for pricing the alternate batch and mix plant operation for bid item 12 are then:

Permanent materials
Cement   2,346 tons
Fine aggregate   5,016 tons
Coarse aggregate   9,317 tons

Expendable materials
Cement   571 tons
Fine aggregate   1,262 tons
Coarse aggregate   2,345 tons

It should be noted that at the level of precision required for a bid estimate, the small cost for the purchase of concrete admixtures can be neglected.

Batch, Mix, and Haul Work Operation

A work operation by the contractor's own forces will now become necessary for batching, mixing, and hauling concrete on-site from the batch-and-mix plant to the forms. Crew 1 from Table 10-6 would be appropriate for this work (utilizing two dumpcrete trucks only).

---

[14]Indirect costs would also be affected because the contractor would have to provide for the erection and later removal of the on-site batch and mix plant in the general plant in and out section of the indirect cost estimate (see Chapter 11).

Concrete strength, psi	Cement			Water			Total aggregate Absolute volume, cf[5]	Fine aggregate		Coarse aggregate	
	Sacks/cy	Absolute volume, cf[2]	lb/cy[1]	W/C	lb/cy[3]	Absolute volume, cf[4]		Absolute volume, cf[6]	lb/cy[7]	Absolute volume, cf[8]	lb/cy[9]
3,000	5.0	2.39	470	0.65	305.5	4.90	18.09	6.33	1046.7	11.76	1,944.6
4,000	6.0	2.87	564	0.65	366.6	5.88	16.63	5.82	962.4	10.81	1,787.5

[1] (94 lb/sack) (sacks/cy).
[2] lb/cy ÷ (3.15 × 62.4 lb/cf).
[3] (lb cement/cy) (W/C).
[4] lb/cy ÷ 62.4 lb/cf.
[5] 6% air voids: 27.0 cf/cy (1 − 0.06) − absolute volume cement (cf) − absolute volume water (cf).
[6] On basis 35% sand: (absolute volume total aggregate cf) (0.35).
[7] (Absolute volume sand, cf) (2.65) (62.4 lb/cf).
[8] (Absolute volume total aggregate, cf) − (absolute volume sand, cf).
[9] (Absolute volume coarse aggregate, cf) (2.65) (62.4 lb/cf).

FIGURE 10-8 Quantities of cement, fine aggregate, and coarse aggregate per cubic yard of concrete.

Description	Total concrete (cy)[1]	Cement		Fine aggregate		Coarse aggregate	
		lb/cy	Total tons	lb/cy	Total tons	lb/cy	Total tons
PM: 3,000 paylines	8,276	470	1,945	1,046.7	4,332	1,944.6	8,047
PM: 4,000 paylines	1,420	564	401	962.4	684	1,787.5	1,270
Total permanent material	–	–	2,346	–	5,016	–	9,317
EM: 3,000, waste and overbreak	2,371	470	558	1,046.7	1,241	1,944.6	2,306
EM: 4,000, waste and overbreak	43	564	13	962.4	21	1,787.5	39
Total expendable materials	–	–	571	–	1,262	–	2,345

[1] From Figure 10-2.

FIGURE 10-9 Total quantities of cement, fine aggregate, and coarse aggregate for bid item 12.

The determination of the required number of crew hours presents a problem that has not previously been discussed. The required crew work hours in this case are a time-related matter that depends on the duration of the period over which the total concrete quantity for bid item 12 will be batched, mixed, hauled, and placed. A crew productivity factor is more or less meaningless because it will vary greatly over the duration of the concrete work and, for many hours of this work duration, will be zero. Nonetheless, the crew will be present on the project and must be paid. They are not personnel that can easily be assigned to other work during work hours when concrete is not being placed.

In the case of this example problem, a total of 12,110 cy of concrete must be batched, mixed, hauled, and placed. Depending on the time requirements of the construction contract of which bid item 12 is a part, and the contractor's overall project work schedule, it would be reasonable to expect to perform the concrete work over a period of 8 months, on one straight-time shift per day.[15] Thus, the work quantity for the batch, mix and haul operations would be (8 months)(4.33 weeks per month)(5 days per week)(8.0 hr per day) = 1,386 hr.

Cost Estimate Difference for Bid Item 12 for On-Site Batch, Mix, and Haul Operation

The estimated direct costs for delivery of concrete to the forms, assuming purchase from an off-site ready-mix supplier, are shown in Figure 10-5, operation 11, as

Purchase 3,000 psi payline concrete	$372,420
Purchase 3,000 psi waste and overbreak concrete	$106,695
Purchase 4,000 psi payline concrete	$ 71,000
Purchase 4,000 psi waste and overbreak concrete	$ 2,150
Total concrete, f.o.b. forms at jobsite	$552,265

The comparable estimated direct costs for a contractor-provided batch, mix, and haul operation utilizing an on-site plant are shown in the substitute operations 11 and 13 of Figure 10-10 as follows

Purchase cement for payline concrete	$152,490
Purchase fine aggregate for payline concrete	$ 37,620
Purchase coarse aggregate for payline concrete	$ 93,170
Purchase cement for waste and overbreak concrete	$ 37,115
Purchase fine aggregate for waste and overbreak concrete	$ 9,465
Purchase coarse aggregate for waste and overbreak concrete	$ 23,450
Batch and mix plant equipment costs	$116,521
Front end loader equipment costs	$ 14,678
Dumpcrete trucks equipment costs	$ 90,894
Batch plant operation labor costs	$ 70,007
Common laborer labor costs	$ 41,220
Loader operator labor costs	$ 14,997
Teamsters labor costs	$ 97,103
ST&S	$ 11,166
Total concrete, f.o.b. forms at jobsite	$809,896

---

[15]Here, the assumption made is that although the concrete formwork and other operations are carried out on a two-shift-per-day basis, concrete will be batched, mixed, and hauled only on one shift per day.

DATE: 0807/98
ESTIMATOR:
BID ITEM:
JOB:
BID ITEM: 12
SUBSTRUCTURE CONCRETE
TAKEOFF QUANTITY: 9,696 CY
BID QUANTITY: 10,150 CY

QTY	CODE	DESCRIPTION	UNIT	E.O.E	R-L	RENTAL	LABOR	S E.M.	P.M.	SUBS	REQD. HOURS	E.O.E	R-L	RENTAL	LABOR	E.M.	P.M.	SUB CONTRACTS	TOTAL
OPRATN:	11	BUY CEMENT & CONCRETE AGGREGATES																	
2346	PM3	Type I Cement	TON						65.00								$152,490		$152,490
5016	PM4	Concrete Sand	TON						7.50								37,620		37,620
9317	PM5	Coarse Aggregate	TON						10.00								93,170		93,170
571	EM14	Cement	TON					65.00								37,115			37,115
1262	EM15	Concrete Sand	TON					7.50								9,465			9,465
2345	EM16	Concrete Aggregate	TON					10.00								23,450			23,450
		% ST & S	LL																
		% PREMIUM PAY	LL																
OPRATN:	12	BUY CONCRETE MISCELLANEOUS MATERIALS																	
12518	EM11	SandBlast M & S (Allow)	SF					0.15								$1,878			$1,878
102214	EM12	Finish M & S (Allow)	SF					0.02								2,044			2,044
151510	EM13	Cure M & S (Allow)	SF					0.01								1,515			1,515
		% ST & S	LL																
		% PREMIUM PAY	LL																
OPRATN:	13	BATCH/MIX/HAUL CONCRETE																	
1386 HR			1 HR	/ Hour							1386								
1	EQ100	ConcBatchMixPlt-100cy/h	ea	28.020	7.310	48.740						$38,836	$10,132	$67,554					$116,521
0.25	EQ14	4 CY Wheel Loader	ea	18.690	3.960	19.710						6,476	1,372	6,830					14,678
2	EQ102	8cy Dumpcrete Trk.	ea	18.880	3.560	10.350						52,335	9,868	28,690					90,894
1	OE17	Operator Foreman	ea				50.510								70,007				70,007
1	LAB1	Common Laborer	ea				29.740								41,220				41,220
0.25	OE5	Loader Operator	ea				43.280								14,997				14,997
2	TM5	12CY Transit Mix Driver	ea				35.030								97,103				97,103
5.00		% ST & S	LL													11,166			11,166
		% PREMIUM PAY	LL																

FIGURE 10-10 Batch, mix, and haul costs for contractor's on-site concrete batch-and-mix plant.

Thus, the estimated cost difference for provision of concrete by means of a contractor operation utilizing an on-site plant is $809,896 − $552,265 = $257,631. In considering this cost difference, it must be realized that its relative magnitude is dependent on the particular unit rates set up in the labor, equipment, permanent material, and expendable material libraries (see Chapter 5). Also, the permanent material and expendable unit rates, as well as the off-site concrete price, would be subject to adjustment just prior to bid to conform to final material supplier "best price" quotations. Even with this qualification, however, this comparison illustrates the fact that where the quantity of concrete is relatively small (in this case only 12,110 cy), it is usually not cost effective to batch, mix, and haul concrete by means of an on-site plant when a commercial ready-mix source is available nearby. In remote locations, there may be no choice and an on-site batch, mix, and haul operation must be utilized.

CONCLUSION

This chapter has surveyed the general construction methods and provided examples of the typical work crews and equipment commonly employed for heavy construction concrete work. Additionally, the important procedures involved in analyzing and pricing heavy construction mass and structural concrete operations have been explained and average labor-hour productivity data have been presented that are indicative of the results that can be expected to be achieved by well managed, experienced work crews. Example direct cost estimates have also been developed and explained for typical concrete bid items for a hydroelectric powerhouse project.

This chapter is the final chapter in this text dealing with construction operation analysis and direct cost estimate pricing associated with selected heavy construction topical areas. The following three chapters revert to general aspects of heavy construction estimates common to any type of heavy construction project. The first of these chapters deals with the import subject of the indirect cost portion of the estimate.

QUESTIONS AND PROBLEMS

1. What six separate construction crafts are usually involved in concrete operations?

2. Describe the two alternate concrete batch, mix, and delivery systems discussed in this chapter. What two types of concrete delivery equipment were discussed? Which of these would normally be restricted to on-site use?

3. Describe the smallest size and type and the largest size and type of hoisting equipment normally employed in concrete construction. What are the types and size ranges of cranes most often employed?

4. What three concrete placing methods and what five kinds of additional and auxiliary equipment were discussed in this chapter?

5. What seven examples of permanent materials and what eleven examples of expendable materials that may be required for concrete construction were discussed in this chapter?

6. Name and state the common units of measurement for the nine kinds, or categories, of quantities that must be computed directly from the dimensional and

other information appearing on the construction drawings separately for each individual bid item during the takeoff phase of a concrete cost estimate.

7. Why do the quantities of each strength class of concrete need to be segregated when the required concrete volumes are determined?

8. What are the nine typical separate structural elements required for general concrete construction, six structural elements for bridge piers, and four structural elements for cut and cover structures and tunnels that should be broken out separately during the takeoff phase of a concrete estimate?

9. What are the 19 classes of formwork typically involved in general concrete construction and 10 formwork classes that could be involved in special structure construction that should be broken out separately during the concrete takeoff phase?

10. Explain the five separate formwork complexity codes discussed in this chapter.

11. What two methods of obtaining takeoff information for pricing the placement and removal of formwork or shoring were discussed?

12. What four separate kinds of foundation contact area quantities should be segregated or broken out during concrete takeoff?

13. What two separate ways of removing laitance from horizontal construction joints were discussed?

14. What four separate categories of wet finish areas should be segregated during concrete takeoff?

15. What two general methods for curing concrete were discussed in this chapter?

16. What four separate categories for dry finish areas should be segregated during concrete takeoff?

17. What two methods of payment for reinforcing steel were mentioned in this chapter? How are reinforcing steel quantities typically segregated when a detailed quantity takeoff is made?

18. Name three categories of miscellaneous embedded items and their common units of measurement that were mentioned in this chapter.

19. Name three general categories of "supplemental" quantities that must be developed from the primary quantities taken off of the construction drawings.

20. If concrete is to be batched and mixed on the jobsite, what three "supplementary" quantities must be determined and what are the common units of measurement for each?

21. Define the terms "form ratio" and "reuse factor."

22. In addition to the quantity of form panels to be job fabricated, what additional "supplemental" formwork-related quantities must be determined and what are the common units of measurement for each?

23. Name five categories of quantities for allowance pricing discussed in this chapter and state the common units of measurement for each.

24. What are the four general categories of labor-hour factors tabulated in this chapter?

25. What is the reason or purpose of the "consolidated" approach when using the labor-hour factor method of pricing? What is a "weighted labor-hour factor"?

26. What are the generic names of the fifteen concrete work operations for which typical crews have been set up in this chapter?

27. What two separate methods for providing crane service were explained in this chapter? What other examples of centrally provided services were mentioned?

28. Revise the data for the example powerhouse problem in the text to read as stated below. Carry out the same analysis as made in the text and prepare revised cost estimates for bid item 12 and bid item 21. Use spreadsheets in the same format as in the text.

The substructure concrete and reinforcing steel are to be paid under bid items 12 and 21, which are described on the bid form as

BI 12 Sub-structure concrete   12,690 cy at $ _____/cy = $ _____

BI 21 Reinforcing steel   2,095,800 lbs at $ _____/lb = $ _____

The contractor's takeoff quantities from the bidding drawings are as follows:

		Neat Line	Overbreak
Foundation preparation		25,000 sf	
Concrete volumes		*Neat Line*	*Overbreak*
SOG	3,000 psi	3,610 cy	820 cy
Heavy walls	3,000 psi	3,250 cy	584 cy
Mass concrete	3,000 psi	3,490 cy	1,180 cy
Columns, beams, girders	4,000 psi	1,450 cy	
Supported slabs	4,000 psi	325 cy	
E&S forms			
Single-faced walls (F)		17,400 sfca	
Block outs (B)		5,780 sfca	
Supported slabs (F)		24,925 sfca	
Wall bulkheads (S)		6,173 sfca	
Double-face walls (F)		95,429 sfca	
Single-face curved walls (C)		1,544 sfca	
SOG edge (S)		8,522 sfca	
Columns (S)		2,275 sfca	
Beam sides and soffits (S)		14,360 sfca	
Shoring in and out		279,128 cf	
Wet finish			
Steel trowel		33,616 sf	
Wood float		9,806 sf	
Sandblast joints		15,648 sf	
Cure		189,388 sf	
Dry finish			
Exposed		84,393 sf	
Nonexposed		43,375 sf	

Reinforcing steel

No takeoff was performed for bid item 21, reinforcing steel, but the potential supplier advised that the bid quantity was reasonably accurate.

Contractor's Construction Plans for bid items 12 and 21

Change the size of the crawler crane intended to be used from 100 to 200 ton.

Change the percentage of the flat wall forms to be formed with purchased patented steel forms from 60 to 70.

Change the working conditions from 7.75/2/5 to 9.0/2/5.

29. On the basis that the project was in an isolated location requiring the use of an on-site batch plant, develop the quantities of cement, coarse aggregate, and fine aggregate that would be required and prepare a revised cost estimate for operations 11, 12, and 13 of bid item 12 in the format of Figure 10-10. Base your work on the following revised data:

	3,000 psi concrete	4,000 psi concrete
Maximum size aggregate	3 in.	3 in.
Minimum cement content	5.5 sacks per cubic yard	6.5 sacks per cubic yard
Maximum water/cement ration (by weight)	0.60	0.60

The Indirect Cost

Key Words and Concepts

Salaried payroll
Management
Supervision
Engineering
Job office personnel
Safety and Equal Employment Opportunity
 personnel
Time-related overhead expense
Survey
General service equipment
Automotive equipment
Warehouse operations
Supplies
Office utility bills
Miscellaneous services
Rentals
Maintenance expense
Distribute drinking water
Interim job cleanup
Job electrical energy consumption
Job water consumption
Protective clothing
Miscellaneous drayage
Miscellaneous personnel expense

Entertainment
Salaried personnel living allowance
Concrete testing
Non-time-related overhead expense
Automotive insurance
Equipment floater insurance
Marine insurance
Builder's risk insurance
Public liability and property damage insurance
Railroad protective insurance
Deductibles
Worker's compensation insurance
Gross receipts taxes
Property taxes
Permits and fees
Alternate method of costing insurance and payroll
 taxes
Construction plant in-and-out
Office and yard area site work
Furnish buildings
Install utilities
Install work access and protection
Remove plant facilities
Construction equipment in-and-out

The definition and concept of the indirect cost was broadly discussed in Chapter 3 and will not be repeated here. Rather, this chapter focuses on the manner in which contractors typically assemble an indirect cost estimate for a project following completion of the direct cost estimate. This will be illustrated by means of a typical example, and as this process is described, ways that the example project might differ from other projects that contractors typically encounter will be pointed out as well as the effect these differences would have on the entries made in the indirect cost estimate.

DESCRIPTION OF EXAMPLE PROJECT

The example project used for illustrative purposes is a major bridge substructure crossing a large river. The substructure consists of a series of bridge piers constructed on both banks of the river as well as in the water at intervals on each side of a navigation channel lying in the deepest part of the river. The work of the contract consists of first constructing sheet-pile cofferdams at each pier location. This operation is followed by underwater excavation of the in situ material within the cofferdam to the bottom elevation of concrete tremie seals that will eventually be placed by underwater concreting methods to permit unwatering the cofferdams for permanent pier construction. Following underwater cofferdam excavation, steel bearing pile are to be driven to underwater cutoff elevations, followed by placement of the tremie seals, unwatering of the cofferdams, construction of the reinforced concrete bridge piers on top of the bearing pile in the dry inside the protection of the cofferdams, and finally, removal of the cofferdams. The bridge superstructure that will eventually rest on the piers is not included in the contract.

The above described construction operations will be repeated throughout the period of the contract at a series of pier locations, some on the river banks utilizing land-based construction equipment, some in the shallower reaches of water adjacent to the banks utilizing land-based equipment operating from work access trestles built out into the water, and some in the deeper reaches of the river utilizing floating equipment, consisting of barges (serviced by tugboats) supporting the required construction equipment.

The project site is near a major city, existing highway access is available to each side of the river, and domestic water and electric power service are available on each bank. The owner is a state highway department that has obtained the necessary right of way on each river bank, affording access to the work. However, access roads from the state highway to the bridge abutment areas, lay down areas for construction materials, and areas for the contractor's job offices, shops, and change houses must be constructed by the contractor and removed upon completion of the work. Similarly, access trestles out into the water, electrical substations, and a distribution system for job electrical power must be provided by the contractor, maintained throughout the project, and removed upon completion of the work. The time span contractually provided for construction of the project is 24 months.

The contractor is a joint venture consisting of two partners, with one of the partners acting as project sponsor and providing all of the jobsite management and supervisory personnel. By agreement, the majority of the land-based construction equipment will be provided by the sponsoring joint-venture partner on an agreed rental rate basis, whereas the majority of the floating equipment (tugs and barges) will be rented from third parties. Some units of construction equipment and a large quantity of construction materials consisting of steel sheet piling and cofferdam bracing materials will be purchased directly by the joint venture and partly written off during the life of the project. Also, the joint venture will purchase custom-built steel formwork for the bridge piers, which will also be partly written off during the project duration.

The direct cost estimate and the expected construction schedule have been completed prior to making the indirect cost estimate. The direct cost line totals are

Equipment operating expense (EOE)	$ 898,500
Repair labor (RL)	489,800
Rental (R)	1,660,100
Labor (L)	5,363,400
Expendable materials (EM)	2,714,800
Permanent materials (PM)	4,746,200
Subcontracts (S)	37,300
Total	$15,910,100

Preprinted Indirect Cost Checklist and Estimate Form

The reader should recall from the discussion in Chapter 3 that the indirect cost estimate can conveniently be broken down into the following major sections:

- Salaried payroll
- Time-related overhead expense
- Non-time-related overhead expense
- Insurance and taxes other than payroll
- Construction plant in and out

Each of these major sections contains a myriad of individual line items, depending on the particular project. Most contractors utilize a preprinted **checklist** that is constructed in such a way that it also serves as an **estimating form** for each of the individual entries that might be required for a particular project. For any one project, it is not likely that all of these entries would be required.

In recent years, computer software programs have been developed for these checklist and estimate forms for use in computer-aided estimates. In the discussion that follows, a simple computer software program developed at California State University, Chico, has been utilized to illustrate the assembly of an indirect cost estimate for the above described example project.

SALARIED PAYROLL

Figure 11-1 shows the printout for the salaried payroll portion of the indirect cost estimate for the example project. Five major subsections of salaried payroll expense are shown, consisting of

- Management
- Supervision
- Engineering
- Job office personnel
- Safety and equal employment opportunity (EEO) personnel

DESCRIPTION	QUANTITY	UNIT	EOE U.C.	EOE AMOUNT	REPAIR LABOR U.C.	REPAIR LABOR AMOUNT	RENTAL U.C.	RENTAL AMOUNT	LABOR U.C.	LABOR AMOUNT	EM U.C.	EM AMOUNT	SUBCONTRACT U.C.	SUBCONTRACT AMOUNT	TOTAL AMOUNT
MANAGEMENT:															
Project Manager	24	MAN-MO							10961.00	263064					263064
Assist. Project Manager		MAN-MO													
SUPERVISION:															
General Superintendent	24	MAN-MO							10316.00	247584					247584
Excavation Superintendent		MAN-MO													
Pile Driver Superintendent	20	MAN-MO							9026.00	180520					180520
Carpenter Superintendent	14	MAN-MO							8204.00	114856					114856
Concrete Superintendent		MAN-MO													
Tunnel Superintendent		MAN-MO													
Tunnel Walker		MAN-MO													
Assist. Superintendent		MAN-MO													
ENGINEERING:															
Project Engineer	24	MAN-MO							8383.00	201192					201192
Field Engineer	33	MAN-MO							6448.00	212784					212784
Office Engineer		MAN-MO													
Materials Engineer		MAN-MO													
Draftsman		MAN-MO													

FIGURE 11-1 (PAGE 1 OF 2)   Salaried payroll.

DESCRIPTION	QUANTITY	UNIT	EOE U.C.	EOE AMOUNT	REPAIR LABOR U.C.	REPAIR LABOR AMOUNT	RENTAL U.C.	RENTAL AMOUNT	LABOR U.C.	LABOR AMOUNT	EM U.C.	EM AMOUNT	SUBCONTRACT U.C.	SUBCONTRACT AMOUNT	TOTAL AMOUNT
JOB OFFICE PERSONNEL:															
Office Manager	24	MAN-MO							5225.00	125400					125400
Accountant		MAN-MO													
Accounts Payable Clerk		MAN-MO													
Timekeeper		MAN-MO													
Buyer		MAN-MO													
Expeditor		MAN-MO													
Secretary	24	MAN-MO							3658.00	87792					87792
Typist		MAN-MO													
Reception/Phone		MAN-MO													
SAFETY & EEO:															
Safety Supervisor		MAN-MO													
First Aid Man		MAN-MO													
EEO Officer		MAN-MO													
A SUBTOTAL										1433192					1433192

FIGURE 11-1 (PAGE 2 OF 2)   Salaried payroll.

Management

Assume that the job schedule prepared during assembly of the direct cost estimate shows that the entire 24 month period allowed by the contract documents will be needed to construct the project. A joint venture project of this type would require a full-time on-site project manager, but the job is not so large that more than one competent individual in this position would be required. At the time the estimate for the project was set up, the sponsor would have established the scale of compensation (including all taxes, insurances, and company benefits) that would apply to the project for salaried personnel. The number of months and the salary rate for each salaried person required are the only computer input information needed.

On a larger job, it is possible that an assistant project manager would be required for all or part of the project period and, for very large projects, the joint-venture partners might provide money in the indirect cost estimate for a level of management above on-site project management. In those situations, the money for this additional salaried payroll need would be provided in the management subsection of the salaried payroll section of the indirect cost.

Supervision

Once the direct cost estimate and construction schedule have been completed, it is relatively easy to determine the numbers of each of the individual superintendents that will be required for the project work as well as the beginning and the ending dates for each individual person. In the case of the example bridge substructure project, it was determined that a general superintendent would be required for the entire 24 month duration of the project, a pile-driving superintendent for 20 months, and a carpenter/concrete superintendent for 14 months. This information, along with the loaded salary rates for these classifications determined when the estimate was set up, constitute the input information for supervision salaried payroll computation.

On larger projects, additional classifications of supervision and additional numbers of superintendents within a particular trade classification would be required, and the checklist template of Figure 11-1 indicates some of the additional possibilities and leaves blank spaces where others could be added. In all cases, the number of superintendent classifications required and the number of individuals within each classification would be driven by the complexity and the duration of the expected project work.

Engineering

For the example project, it was judged that a full-time project engineer would be required for the entire 24 month duration of the project, as indicated on Figure 11-1. The only other engineering need was the field engineer classification where it was judged necessary to provide one field engineer for the entire 24 month duration of the project and a second field engineer for a 9 month period of peak field activity within the project duration, for a total of 33 months.

Larger projects might require additional field engineers as well as one or more persons in the several additional classifications listed on Figure 11-1. For instance, tunneling projects, typically running three shifts a day, often are staffed with an engineer in each heading on each shift to check line and grade and to maintain accurate job records of the details of the tunneling operations. Also, in recent years owners have tended to contractually assign significant quality assurance/quality control (QA/QC) responsibilities to the contractor. In this case, additional engineers to perform the necessary QA/QC inspections and tests and to provide the required documentation would be necessary.

Job Office Personnel

For a bridge substructure project of the scope illustrated by the example project, only two persons would be required—an office manager and a secretary/typist/receptionist, each for the full 24 month duration of the project. Although the office manager would be required to perform the necessary joint-venture accounting tasks (which are more complex than if the project was not a joint venture), data processing assistance for payroll accounting, accounts receivable/accounts payable accounting, etc. is typically provided by the sponsoring contractor's home office administrative staff, which reduces the number of persons required at the jobsite. Also, for a project of this size, major purchasing activity would be performed at the onset of the job by the sponsor's home office purchasing department, leaving the more minor day by day purchasing activity to the project engineer and office manager at the jobsite. Larger projects could well require some, or all, of the additional classifications shown on Figure 11-1, in some cases multiple persons being required within each classification.

Safety and Equal Employment Opportunity Personnel

For the example project, none of the classifications shown on Figure 11-1 were judged to be needed. Increasingly, however, personnel in these classifications are frequently required, particularly for larger projects within urban areas that are socially and environmentally sensitive. Also, large underground projects frequently require a senior safety supervisor as well as a safety engineer on each shift. On many projects, first aid training given to supervisory personnel and craft foreman may meet the project first aid requirements, but some require one or more individuals assigned exclusively to the provision of first aid services. In ethnically sensitive urban areas, an individual exclusively assigned to administration of the contractor's equal employment opportunity responsibilities is occasionally required by the contract documents.

Increasingly, additional regulatory requirements, ranging from industrial hygiene issues to administration of required environmental mitigation efforts such as waste water discharge control and noise abatement, may require additional contractor job staff personnel. Where required, the costs for such persons would be provided in this section of the salaried payroll indirect cost estimate.

Total Salaried Payroll Costs

As shown in page 2 of Figure 11-1, the salaried payroll costs estimated for the example project total $1,433,192. This figure appears in the labor and total columns of the estimate.

TIME-RELATED OVERHEAD EXPENSE

Recall from Chapter 3 that this section of the indirect cost is utilized for estimating those expenses that recur on a periodic basis, usually monthly (in addition to salaried payroll expense, which obviously recurs on a monthly basis). Referring to that portion of the indirect cost software template illustrated by Figure 11-2, time-related overhead expenses for the example project are broken down into the following subdivisions

DESCRIPTION	QUANTITY	UNIT	EOE U.C.	EOE AMOUNT	REPAIR LABOR U.C.	REPAIR LABOR AMOUNT	RENTAL U.C.	RENTAL AMOUNT	LABOR U.C.	LABOR AMOUNT	EM U.C.	EM AMOUNT	SUBCONTRACT U.C.	SUBCONTRACT AMOUNT	TOTAL AMOUNT
SURVEY:															
Party Chief		MAN-MO													
Inst. Man		MAN-MO													
Rod/Chain Man		MAN-MO													
GENERAL SERVICE EQUIPMENT:															
Boom Truck	19	UNIT-MO	1501.00	28519	1347.00	25593	4688.00	89072	9106.00	173014					316198
Flat Rack		UNIT-MO													
AUTOMOTIVE EQUIPMENT:															
Sedans	48	UNIT-MO	391.00	18768	809.00	38832	407.00	19536							77136
Pickups	58	UNIT-MO	694.00	40252	809.00	46922	493.00	28594							115768
Ambulance		UNIT-MO													
WAREHOUSE OPERATION:															
Teamster		MAN-MO													

FIGURE 11-2 (PAGE 1 OF 3)   Time-related overhead expense.

DESCRIPTION	QUANTITY	UNIT	EOE U.C.	EOE AMOUNT	REPAIR LABOR U.C.	REPAIR LABOR AMOUNT	RENTAL U.C.	RENTAL AMOUNT	LABOR U.C.	LABOR AMOUNT	EM U.C.	EM AMOUNT	SUBCONTRACT U.C.	SUBCONTRACT AMOUNT	TOTAL AMOUNT
SUPPLIES:															
Office	24	MONTH									750.00	18000			18000
Engineering	24	MONTH									250.00	6000			6000
Safety & First Aid	24	MONTH									250.00	6000			6000
Drinking Water Supplies	24	MONTH									250.00	6000			6000
OFFICE UTILITY BILLS:															
Telephone	24	MONTH									625.00	15000			15000
Electricity		MONTH													
Gas/Oil	24	MONTH									938.00	22512			22512
Water		MONTH													
MISC. SERVICES:															
Photo	24	MONTH									163.00	3912			3912
CPM Update		MONTH													
Data Processing	24	MONTH									563.00	13512			13512
Security		MONTH													
Sanitary	24	MONTH									688.00	16512			16512
Trash	24	MONTH									63.00	1512			1512
RENTALS:															
Office		MONTH													
Yard/Warehouse	24	MONTH									375.00	9000			9000

FIGURE 11-2 (PAGE 2 OF 3)   Time-related overhead expense.

DESCRIPTION	QUANTITY	UNIT	EOE U.C.	EOE AMOUNT	REPAIR LABOR U.C.	REPAIR LABOR AMOUNT	RENTAL U.C.	RENTAL AMOUNT	LABOR U.C.	LABOR AMOUNT	EM U.C.	EM AMOUNT	SUBCONTRACT U.C.	SUBCONTRACT AMOUNT	TOTAL AMOUNT
MAINTENANCE EXPENSE:															
Buildings & Facilities	24	MONTH							250.00	6000	375.00	9000			15000
Radios	24	MONTH									250.00	6000			6000
Elect. Distribution System	13	MONTH											7500.00	97500	97500
DISTRIBUTE DRINKING WATER	19	MONTH							4563.00	86697					86697
INTERIM JOB CLEANUP		MONTH													
JOB ELECT. ENERGY CONSUMPTIO	24	MONTH									1500.00	36000			36000
JOB WATER CONSUPTION	24	MONTH									125.00	3000			3000
PROTECTIVE CLOTHING	5853200	LABOR $									0.02	117064			117064
ENTERTAINMENT		MONTH													
SAL. PERSONNEL LIVING ALLOW.		MONTH													
CONCRETE TESTING		MONTH													
MISCELLANEOUS DRAYAGE	24	MONTH									250.00	6000			6000
MISCELLANEOUS PERSONNEL EXP	24	MONTH									250.00	6000			6000
B SUBTOTAL				87539		111347		137202		265711		301024		97500	1000323

FIGURE 11-2 (PAGE 3 OF 3)   Time-related overhead expense.

(many of which contain a series of entries for the typical kinds of costs that are usually incurred):

- Survey
- General service equipment
- Automotive equipment
- Warehouse operations
- Supplies
- Office utility bills
- Miscellaneous services
- Rentals
- Maintenance expense
- Distribute drinking water
- Interim job cleanup
- Job electric energy consumption
- Job water consumption
- Protective clothing
- Entertainment
- Salaried personnel living allowance
- Concrete testing
- Miscellaneous drayage
- Miscellaneous personnel expense

Survey

Normally, contract specifications will require that the contractor provide line and grade control for the project work from baseline monuments established by the owner's engineer. In the case of the example bridge substructure project, the state highway department provided all survey control, so no entries appear in Figure 11-2 for this category of cost. When these services are contractor provided, at least one survey crew would be required for the entire 24 month project duration. During the peak of the work more than one crew probably would be required.

Contractors typically use either a three-person crew consisting of a party chief, an instrument person, and a rod and chain person, or a two-person crew consisting of a party chief (who also does the instrument work) and a rod/chain person. The number of crews and the duration that each crew is likely to be required on the project can be estimated from the construction schedule developed at the time the direct cost estimate is prepared, which determines the number of person months for each survey classification required. This information, coupled with the monthly labor cost for each of the several classifications, enables the total survey costs to be computed in this section of Figure 11-2.

General Service Equipment

During preparation of the direct cost estimate, the required construction equipment is usually included with the work crews when the individual work items are priced. However, there will usually be a need for some additional equipment that provides assistance and service to all parts of the job on an as-needed basis. At a minimum, such equipment would consist of a flat-rack truck driven by a teamster for interjob material transport or delivery or perhaps a larger flat-rack truck equipped with a hydraulically operated boom for loading and unloading material. This latter unit would be operated by a teamster, who would drive the truck and operate the hydraulic boom, and a laborer to assist by hooking and unhooking materials handled by the boom. Larger projects might include a medium-sized motor crane and one or more tractor-trailer combinations, including interchangeable high-bed trailers and a low-bed trailer for transporting heavy track-mounted equipment. Such general service equipment, if required at all, is usually needed for the majority of the project duration.

Another instance of general service equipment need was discussed in Chapter 3, where for water-borne construction the estimating team might elect to carry the cost of required tugboats as an indirect cost rather than including the tugboat cost with the other equipment included with the work crews. That was not the case for the example project but, had it been the case, the cost of the required tugboats would be included in the general service equipment subsection of the time-related overhead expense.

If the full equipment costs, and lubricating and operating personnel costs for service units such as fuel trucks, mechanics repair trucks, etc., are included in the EOE and the RL rates when the equipment rate library is set up for the project (as is commonly the case), these services would not be included in the indirect cost estimate. However, if not provided for in the EOE and RL rates, these vehicle and operating personnel costs would be included in this section of the indirect cost estimate.

In all cases, the number of required general service units and the duration that each unit will be required on the project can be estimated from the construction schedule developed with the direct cost estimate. When the costs for the general service equipment are included in the estimate in this manner, it is convenient to include the labor cost of the teamsters or operators required to run the equipment, as well as the labor cost of oilers and/or laborers that would be assigned to the specific units of general service equipment.

For our example project, the only piece of general service equipment required was a boom truck operated by a teamster assisted by a laborer that would be required for 19 months of the 24 month project duration. The unit costs for equipment operating expense (EOE), repair labor (RL), and rental (R) that appear in Figure 11-2 were determined by multiplying the respective library rate per equipment hour by 173 hr (reflecting a full-time usage of the equipment one shift per day, 5 days per week, 4.33 weeks per month).[1] The unit cost for labor shown in Figure 11-2 was obtained by adding the hourly labor rate for the teamster and the hourly rate for the laborer together and multiplying that sum by 173 hours per month.[2] Unit costs for EOE, RL, R, and the assigned operating crew labor for units of general service equipment that might be required for other types of projects would be determined similarly.

Automotive Equipment

The automotive equipment category of cost is for the EOE, RL, and R costs for vehicles furnished to job management and supervisory personnel. It should be noted that in this case no labor cost category is included because the various persons to whom the vehicles are assigned drive the equipment themselves, and their labor costs were accounted for in the salaried payroll section of the indirect cost estimate.

---

[1] These rates would be doubled if the equipment was expected to operate two shifts per day and tripled if the equipment was expected to operate three shifts per day.

[2] This monthly rate would also be doubled for a two-shift operation and tripled for a three-shift operation.

For the example bridge substructure project, the project manager and project engineer were each furnished with a sedan and each superintendent was furnished with a pickup truck. The number of sedan unit months to be costed is therefore the sum of the person months required for the project manager and project engineer, whereas the number of pickup unit months is the sum of the person months required for the various superintendents. The unit costs for EOE, RL, and R are determined in the same manner as explained for general service equipment.

It should be noted that for projects in remote locations that would require the contractor to provide an on-site ambulance, the number of required unit months would usually be the duration of the project. The unit costs for EOE, RL, and R would be computed on the basis of the hourly rates for each of these costs categories multiplied by a lesser number of hours per month than 173 (perhaps 10% of these hours) because the ambulance would see comparatively little use.

Warehouse Operation

For very large projects the contractor will frequently set up and operate a jobsite warehouse for small tools and supplies, equipment spare parts, and other categories of job required items. Usually, one or more warehouse persons are required to receive, inventory, and issue these types of commodities. The labor costs for these personnel would be reflected in this section of Figure 11-2. The number of persons required per month and the number of months the warehouse would be required are judgment calls driven by the circumstances of the particular project and the schedule of project work developed during direct cost estimate preparation. The monthly labor rate for the personnel required would be the hourly labor rate times 173 hr per month.

Supplies, Office Utilities Bills, Miscellaneous Services, Rentals, and Maintenance Expense

The various specific kinds of recurring monthly expenses typically experienced in each of the above categories are costed in the time-related overhead expense section of the indirect cost on the basis of a monthly use rate determined from the contractor's previous experience on similar projects. Not every category of expense listed in Figure 11-2 will necessarily be required for every project, but most of them usually will be.

For the example substructure project, all of the costs listed under supplies were included. The three items (electricity, gas/oil, and water) listed under office utility bills were also included and costed on the basis of a single use rate ($938 per month) intended to include all three categories of cost.

Costs for a consultant to update a critical path method (CPM) schedule and costs for security services listed under miscellaneous services were not judged to be required. The other four categories (photo costs, data processing charges, sanitary services charges, and trash collection costs) were judged to be required and were costed as shown in Figure 11-2.

Job office space was intended to be provided by means of job trailers that would be purchased, erected on the jobsite, dismantled, and sold at the end of the project (for which the costs are included in the construction plant in-and-out section of the indirect cost estimate), so no expense was required to be costed for office rental included in the rental category. However, a lay-down area outside of the right-of-way furnished by the state was determined to be required, and a monthly cost for renting this area from third parties was included in the rental section.

Under maintenance expense (as distinct from the expense of furnishing the facilities), the costs for maintaining the contractor's jobsite trailers and facilities and for maintaining the job radio communication system were costed on the basis that these services would be required for the full 24 month duration of the project at use rates determined from previous similar projects. It was determined from the job schedule that the electrical distribution system would be in service for 13 months of the 24 month duration. The unit cost of maintenance per month was determined from a prebid quotation from the electrical contractor who had quoted the cost to furnish the electrical distribution system and dismantle it at the end of the project (the furnish and dismantle costs are included separately in the construction plant in-and-out section).

Distribute Drinking Water

Contractors employing large numbers of workers during hot summer working seasons must supply cold drinking water throughout the project at every job location where work is being performed. The cost of obtaining the potable water for this purpose is usually negligible, but the cost of providing ice to cool the water and for providing the insulated containers in which to contain it for distribution and use about the project plus the labor to prepare the drinking water containers at the start of the working day, distribute them to the work locations, collect them from the working locations at the end of the day, and clean them for the next day's use is surprisingly large.

The cost of the containers themselves and the paper drinking cups that are usually utilized was covered for the example project under the supplies category. The labor costs were estimated by developing a unit cost per month equal to the labor cost for two laborers working 4 hr per day 5 days per week, 4.33 weeks per month. This unit cost was extended against the total number of months of warm weather estimated to require drinking water distribution during the 24 month project duration (in this case 19 months). It should be noted that the equipment required for the drinking water distribution and collection operation has already been provided.[3]

Interim Job Cleanup

If the project location was such that a special crew would be required to perform daily cleanup services to meet extra stringent environmental concerns, the cost of providing this crew would be included under this item. It should be noted that general cleanup costs for picking up and disposing of debris from typical construction work operations should be provided by the personnel in the individual work crews as part of normal job housekeeping.

For the example bridge substructure project constructed on the banks (and across) a major river, interim job cleanup services would not be required.

Job Electrical Energy Consumption

This category is the cost for miscellaneous electrical energy required during the performance of the project work. It does not include electrical energy consumed by the job offices or by the electric motors on construction equipment. These latter energy costs

---

[3]Distributing and collecting drinking water cans is one of the many items of work performed by the boom truck provided under general service equipment.

are included in the equipment EOE expense. The cost is usually estimated on the basis of a monthly use rate determined from past experience on similar projects (in this case $1,500 per month) for the full 24 month duration of the project. The monthly use rate for larger projects could be much higher, particularly for projects requiring large lighting loads such as underground projects and/or surface projects operating three shifts per day.

Job Water Consumption

Potable job water consumption is a relatively minor cost determined for the full duration of the project by application of a use rate determined from prior experience.

Protective Clothing

When project workers are exposed to onerous working conditions, the provision of the necessary protective clothing (consisting mainly of waterproof overalls and jackets) is a surprisingly large expense. It is often estimated as an experience factor percentage of the total labor cost in the direct cost, including repair labor.[4] As indicated in Figure 11-2, the total labor figure ($5,853,200) is entered in the quantity column, and the experience factor percentage allowance of 2% (0.02) is entered in the unit cost column.

This cost will vary, depending on the amount of direct labor and on the nature of the project. For a tunneling project of equal time duration where the work is performed under wet conditions, this cost would be much higher.

Miscellaneous Drayage

The cost for miscellaneous drayage is typically incurred for miscellaneous random freight forwarding services and airport freight pickup services provided by a local carrier in lieu of project equipment leaving the project site for this purpose. It is determined on the basis of an experience use rate per month applied to the full 24 month duration of the project.

Miscellaneous Personnel Expense

Miscellaneous personnel expense is intended to reimburse salaried personnel who may incur minor personal out-of-pocket expenses for the benefit of the project. It is estimated by application of an experience use rate for the full duration of the project.

Remaining Time-Related Overhead Expense Items

The three remaining items, **entertainment, salaried personnel living allowances,** and **concrete testing,** are typical of the many kinds of time-related minor expenses for which provision would be made in the estimate by application of a use rate from past experience against the full duration of the project.

For the example bridge substructure project, no significant costs in these categories were expected to be incurred. This would not be the case for larger more complex projects.

---

[4]For the example project, this labor cost is $5,363,400 plus $489,800, for a total of $5,853,200.

Total Time-Related Overhead Expense for Example Project

The total costs estimated for the time-related overhead expense section of the indirect cost appearing on page 3 of Figure 11-2 are

EOE	$	87,539
RL		111,347
R		137,202
L		265,711
EM		301,024
Total		$1,000,323

NON-TIME-RELATED OVERHEAD EXPENSE

Unlike time-related overhead expenses, costs in the non-time-related overhead expense section of the indirect cost occur infrequently during project performance, often only once. Typical cost descriptions that are usually set up in the contractor's estimating software include those listed for the example project in Figure 11-3. With the exception of the salaried personnel expense category, these items are all estimated as single lump sum occurrence expenses (even though a few of them may consist of widely separated individual expenditures made more than once during the life of the project).

For the example bridge substructure project, the non-time-related overhead expense totals on Figure 11-3 are

L	$ 12,500
EM	191,925
Total	$204,425

The only cost items requiring discussion in the non-time-related overhead expense group are the outside engineering and the salaried personnel expense categories. The outside engineering category is for the inclusion of engineering expenses expected to be performed by other than the jobsite engineering organization, usually meaning that they would be performed by a third-party engineering consultant (although in the case of a joint venture, these services could be performed under separately reimbursed arrangements by one of the partners to the joint venture). For the bridge substructure example project, the particular engineering services expected to be required were the design work for the cofferdams and trestle structures, for which an estimated cost of $156,300 was provided. Costs for similar services for off-site engineering work on larger more complex projects could be considerably more.

The salaried personnel expense category is for the cost of relocating salaried personnel to the project site (including their families) from their previous employment location. These expenses can be considerable when a number of employees must be moved, particularly when the move distances are lengthy. Not only are the actual moving expenses involved but usually the employee will receive a per-diem allowance for a number of days after arriving at the project location to cover restaurant and motel expenses until more permanent living arrangements have been made. The average cost of a relocation move is derived from past company experience, as is the necessary per-diem allowance figure.

DESCRIPTION	QUANTITY	UNIT	EOE U.C.	EOE AMOUNT	REPAIR LABOR U.C.	REPAIR LABOR AMOUNT	RENTAL U.C.	RENTAL AMOUNT	LABOR U.C.	LABOR AMOUNT	EM U.C.	EM AMOUNT	SUBCONTRACT U.C.	SUBCONTRACT AMOUNT	TOTAL AMOUNT
OFFICE FURNITURE	1	LS									5000.00	5000			5000
OFFICE EQUIPMENT	1	LS									5000.00	5000			5000
ENGR. EQUIPMENT	1	LS									7500.00	7500			7500
SAFETY & FIRST AID EQUIPMENT	1	LS									11250.00	11250			11250
SET-UP CPM		LS													
OUTSIDE ENGINEERING	1	LS									156300.00	156300			156300
OUTSIDE SURVEYING		LS													
AGC DUES	1	LS									1875.00	1875			1875
MEDICAL FEES		LS													
LEGAL FEES		LS													
PURCHASE OF ICE MACHINE		LS													
PURCHASE W.H. BINS		LS													
PURCHASE OF JOB RADIOS		LS							12500.00	12500	5000.00	5000			17500
PUNCH LIST & FINAL CLEANUP	1	LS													
SALARIED PERSONNEL EXPENSE:	1														
Relocation		MOVES													
Per Diem		MAN-DAYS													
Travel Expense		TRIPS													
C SUBTOTAL										12500		191925			204425

FIGURE 11-3 Non-time-related overhead expense.

If the nature of the employee's engagement is such that incidental trips during employment will be required, costs for the expected number of such longer trips will be included in this section of the estimate also.

In the case of projects located outside the continental limits of the United States, salaried personnel relocation expenses can be particularly sizable.

INSURANCE AND TAXES OTHER THAN PAYROLL

This section of the indirect cost estimate is reserved for recording the costs the contractor will incur for insurance premiums and taxes that are not accounted for by elements included in the labor rates that were set up for the estimate. The major kinds of insurance policies requiring premium payments that are typically costed in this section of the estimate include the following:

- **Automotive liability and casualty insurance** for the automotive equipment operated by the contractor in connection with the project work.
- **Equipment floater insurance** for loss or damage to the contractor's normal land-based construction equipment **when operating on land.**
- **Equipment floater insurance** for loss or damage of the contractor's land-based equipment **when this equipment is operating on the decks** of barges over water.
- **Marine insurance,** which includes protection for loss or damage to floating equipment such as work barges, material storage barges, dredges, cranes, pile drivers, and concrete batch/mix plants when permanently mounted on floating barges (**hull insurance**), and public liability and property damage insurance for marine operations (**protective and indemnity insurance**).
- **Builder's risk insurance** for loss or damage to the permanent construction work of the project prior to its acceptance by the owner.
- **Public liability and property damage insurance** protecting the contractor from monetary damages liability for injury or death and/or property damage to the general public related to the contractor's land-based construction operations, when the labor rate structure set up for the estimate does not include an element for the premium costs for this insurance.[5]
- **Railroad protective insurance** for any type of loss caused to railroad passengers, freight, equipment, or right of way when the project work area includes an active railroad.
- **Deductibles** to cover that portion of casualty losses on public liability and property damage insurance policies that must be paid by the insured contractor before any insurance coverage will apply.
- **Worker's compensation insurance** premiums would also be included in this section of the indirect cost estimate when the labor rate structure set up for the estimate does not include an element for worker's compensation insurance premiums built into the rate.[6]

---

[5]Refer to the discussion in Chapter 5 explaining the optional ways that contractors provide for these insurance premiums.

[6]Refer to the discussion in Chapter 5.

The major tax considerations include

- **Gross receipts taxes** that some government entities (usually municipalities) levy on all gross business revenue.
- **Property taxes** levied by government entities (usually state but sometimes municipalities as well) annually on construction equipment and salvageable construction materials such as steel sheet piling, steel forms, and like items owned by the contractor and in use on the project.
- **Permits and fees** known to be required that, contractually, are for the contractor's account.
- The employer's portion of **FICA, SUI,** and **FUI taxes** are also included in this section when the labor rate structure set up for the estimate does not include elements covering these contractor labor-related costs.[7]

Entries for these typical costs would appear in the contractor's estimating software. Details for determination of these costs for the example bridge substructure project, as shown in Figure 11-4, are discussed as follows.

Automotive Insurance

The premium cost for automotive insurance is usually quoted on the basis of a fixed dollar rate per vehicle per year of project duration. For the example project, the flat annual rate per vehicle was $940. Because six vehicles were involved, the annual cost was $(6)(\$940) = \$5,640$. Extending this annual cost by the 2-year duration of the project results in the $11,280 total premium cost shown in Figure 11-4.

Equipment Floater Insurance: Land-Based Equipment

Premium quotations for this insurance are usually quoted in the form of a fixed dollar rate per $100 of equipment replacement value per year of project duration. For the example project, the quoted figure was $0.70/$100 per year, and the estimated replacement value for the contractor's equipment that would be operating on land was $1,500,000. The premium cost per year was therefore $(0.70)(1,500,000)/100 = \$10,500$ per year. Extending this annual cost by the 2-year project duration yields the $21,000 total cost shown in Figure 11-4.

Equipment Floater Insurance: Water-Based Equipment

The premium costs for water-based equipment insurance are quoted in the same manner as for land-based equipment except that the dollar rate per $100 of equipment replacement value per year is higher. For the example project, this dollar rate was $1.60/$100 per year and the estimated replacement value for equipment that would be operating on barges, or operating on trestles over the water, was $6,000,000. On this basis, the annual cost was $(1.60)(6,000,000)/100 = \$96,000$. The $96,000 annual premium times the 2-year duration of the project is the $192,000 total premium, as shown on Figure 11-4.

---

[7]See discussion in Chapter 5.

DESCRIPTION	QUANTITY	UNIT	EOE U.C.	EOE AMOUNT	REPAIR LABOR U.C.	REPAIR LABOR AMOUNT	RENTAL U.C.	RENTAL AMOUNT	LABOR U.C.	LABOR AMOUNT	EM U.C.	EM AMOUNT	SUBCONTRACT U.C.	SUBCONTRACT AMOUNT	TOTAL AMOUNT
INSURANCE:															
Auto	2	YEAR									5640.00	11280			11280
Eq. Floater-Land	2	YEAR									10500.00	21000			21000
Eq. Floater-Water	2	YEAR									96000.00	192000			192000
Marine	2	YEAR									81250.00	162500			162500
Builder's Risk	1	LS									442150.00	442150			442150
Railroad Protective	1	LS									1875.00	1875			1875
PL & PD	1	$									227050.00	227050			227050
Deductibles	1	LS									12000.00	12000			12000
		$													
		$													
		$													
		$													
TAXES:															
Gross Receipts		YEAR													
Personal Prop. (Equipment)	2	YEAR									35625.00	71250			71250
Personal Prop. (Const. Mat.)	2	YEAR									9375.00	18750			18750
		$													
		$													
		$													
PERMITS & FEES	1	ILS									1875.00	1875			1875
D SUBTOTAL												1161730			1161730

FIGURE 11-4 Insurance and taxes other than payroll.

Marine Insurance

This insurance was quoted as a combined policy covering loss or damage to the contractor's floating equipment (hull insurance) and public liability and property damage to third parties related to the contractor's marine operations (protective and indemnity insurance). For the example bridge substructure project, it was quoted at a rate of $81,250 per year of project duration for all of the tugs and barges as a group. On this basis, the total premium was ($81,250)(2 years) = $162,500, as shown on Figure 11-4.

On other projects, the basis for the premium quotation might vary and could be similar to that explained for equipment floater insurance.

Builder's Risk Insurance

The premium for builder's risk insurance is often quoted on the basis of a fixed dollar rate per $100 of total contract value rather than on the basis of a separate premium charge for each year duration of the project. For the example bridge substructure project, the quoted fixed dollar rate was $1.85. To compute the total builder's risk premium, it is necessary to make an intelligent guess as to the total bid price to establish the total contract value. This is usually fairly easy to do for estimators who have studied the details of the project and who are familiar with their company's past bidding philosophy. For the example project, the direct cost was $15,910,000, and the estimators believed that the total indirect cost might total $5,000,000 and the markup and contingency considerations another $3,000,000, for a total contract value of approximately $23,900,000. On this basis, the builder's risk insurance premium became (1.85)(23,900,000)/100 = $442,150, as shown in Figure 11-4.

Public Liability and Property Damage Insurance

As discussed in Chapters 2 and 5, contractors typically include the premium cost for this insurance as either (1) an element of the labor rates set up for the estimate or (2) on the basis of the total contract value, in which case the premium would be costed in the insurance and taxes section of the indirect cost estimate. Some of the example calculations presented in Chapters 2 and 5 illustrated the method of including the premium cost as an element of the labor rate. In the case of the bridge substructure project, however, the premium was quoted at a rate of $0.95 per $100 of contract value, so the public liability and property damage (PL & PD) insurance premium would be (0.95)(23,900,000)/100 = $227,050, as shown in Figure 11-4.

Railroad Protective Insurance

For the bridge substructure project, there was a railroad passing through the work area on one of the abutments. The premium costs for railroad protective insurance is usually a minor cost and in this case was quoted as $1,875 for the life of the project.

Deductibles

For the example project, the PL & PD policy contained a provision stating that the first $1,000 of loss on any third-party claim must be paid by the insured before the insurance company incurred any payment liability. On this basis, the estimators provided for a total of 12 third-party liability claims during the project duration for a total of $12,000, as indicated on Figure 11-4.

Gross Receipts Taxes

When applicable, gross receipts taxes are usually levied at a stated percentage of gross contract revenue. A common figure is in the vicinity of three-quarters of 1% to 1%. For the example project, gross receipts taxes were not required and no money was provided in Figure 11-4.

Personal Property Taxes

Personal property taxes will almost always be levied if the project is prominent and the duration is long enough to attract the attention of the taxing authorities in whose jurisdiction the project is located. The tax rates are commonly different for construction equipment and for salvageable construction material. Each is commonly levied at a stated dollar rate per $100 on a stated percentage of either the average equipment replacement value over the project duration, per year, or for construction materials, a stated dollar rate per $100 on a stated percentage of the estimated average value of the construction materials during the project, per year.

As stated previously, the total replacement value for construction equipment was $1,500,000 (land-based equipment) plus $6,000,000 (equipment operating on barges), or a total of $7,500,000. Additionally, the average of the purchase price and salvage value at the end of the job of construction materials, consisting of sheet piling and steel forms for the example project, was $750,000. The personal property tax rate for construction equipment was $1.90 per $100 on 25% of the replacement value, per year, whereas on construction materials the rate was $5.00 per $100 on 25% of the average value per year. On this basis, the monies for personal property taxes to be included in the estimate were:

Equipment tax per year	(1.90)(0.25)(7,500,000)/100
	$35,625
Total equipment tax	($35,625 per year)(2 years)
	$71,250
Construction material tax per year	(5.00)(0.25)(750,000)/100
	$9,375
Total tax	($9,375 per year)(2 years)
	$18,750

These amounts are reflected on Figure 11-4.

Permits and Fees

Permits and fees are usually a minor cost for which a nominal allowance is established in the estimate. For the example project, the estimators established a lump sum amount of $1,875 for the life of the project. This figure is reflected in Figure 11-4.

Total Insurance and Taxes Other Than Payroll for Example Project

As indicated in Figure 11-4, the total amount for the example project for this section of the indirect cost is $1,161,730. In this case, all of this money is recorded in the expendable materials column. As this example illustrates, the amount of money required for this section of the indirect cost is sizable.

If worker's compensation premiums and FICA, SUI, and FUI taxes had not been included as elements of the labor rate when the labor rate library was set up for the estimate, the amount of money in this section of the indirect cost would be even higher, as the discussion below will indicate.

Alternate Method of Costing Insurance and Payroll Taxes

There is another way to compute insurance and tax costs when all insurance premiums, FICA, SUI, and FUI taxes are not included in the labor rate library labor rates and public liability insurance is paid on the basis of the contractor's labor exposure rather than on total project revenue. As previously explained in Chapters 2 and 5, many contractors handle the costing of labor-related insurance premiums and labor taxes on this alternate basis. When this is done, the labor rates set up in the labor rate library consist of only the base wage and union fringes (or the equivalent of union fringes for contractors not operating under labor agreements).

To illustrate how the computations for labor-related insurance premiums and taxes would be made under these circumstances, assume that for the example bridge substructure project the following facts apply.

Worker's Compensation Insurance

Six worker's compensation rate classifications were applied to the project at the following premium rates:

Worker	Rate per $100 payroll
Clerical	$ 2.80
Concrete work on land	$18.14
Concrete work on water	$25.64
Pile driving/cofferdam work on land	$38.52
Pile driving/cofferdam work on water	$46.02
Repair and service labor	$18.14

The total base wages included in the direct and indirect cost estimates, separated into the above six worker's compensation categories, were as follows:[8]

Clerical	$ 105,000
Concrete work on land	$ 707,000
Concrete work on water	$ 850,000

---

[8]Such summary information is usually produced by the software employed for computer-aided estimates. For manual estimates, these labor figures must be summarized for the different categories of work by hand.

Pile driving/cofferdam work on land	$ 721,000
Pile driving/cofferdam work on water	$1,140,000
Repair and service labor	$ 406,000

Payroll Taxes

The payroll taxes applying to the project were as follows:

Tax	Tax rate per $100 of base pay
FICA	$6.95
FUI	$0.85
SUI	$5.75

Public Liability and Property Damage Insurance

The quoted rate for PL & PD insurance was $6.45 per $100 of base rate payroll. On the basis of the above information, the worker's compensation insurance premium charge for the project would be

Clerical	$2.80 × $105,000/100	= $ 2,940
Concrete work on land	$18.14 × $707,000/100	= $128,250
Concrete work on water	$25.64 × $850,000/100	= $217,940
Pile driving/cofferdam work on land	$38.52 × $721,000/100	= $277,729
Pile driving/cofferdam work on water	$46.02 × $1,140,000/100	= $524,628
Repair and service labor	$18.14 × $406,000/100	= $ 73,648

The public liability and property damage insurance premium would be calculated as follows:

Clerical payroll	$ 105,000
Concrete work on land payroll	$ 707,000
Concrete work on water payroll	$ 850,000
Pile driving/cofferdam work on land payroll	$ 721,000
Pile driving/cofferdam work on water payroll	$1,140,000
Total base rate payroll (other than RL)	$3,523,000
Base payroll (RL)	$406,000

PL & PD premium (other than RL)	($6.45)($3,523,000)/100	= $227,234
PL & PD premium (on RL)	($6.45)($406,000)/100	= $ 26,187

Finally, payroll taxes would be computed in the following manner:

FICA (other than RL)	($6.95)($3,523,000)/100	= $244,849
FUI (other than RL)	($0.85)($3,523,000)/100	= $ 29,946
SUI (other than RL)	($5.75)($3,523,000)/100	= $202,573
FICA (RL)	($6.95)($406,000)/100	= $ 28,217
FUI (RL)	($0.85)($406,000)/100	= $ 3,451
SUI (RL)	($5.75)($406,000)/100	= $ 23,345

The insurance and taxes section of the indirect cost estimate would appear as shown in Figure 11-5 if the premiums for worker's compensation insurance and PL&PD insurance and payroll taxes were costed into the contractor's estimate by the above methods rather than being included as components of the labor rates or (in the case of PL&PD insurance) being costed on a premium rate per $100 of total contract value. The reader should note that because in this case PL&PD insurance premiums are now labor-related costs, and labor taxes are labor-related costs, they are recorded in the L and RL columns, respectively.

The reader will also note from examination of Figure 11-5 that the total amount of money now included in this section of the indirect cost estimate is much greater than that previously indicated in Figure 11-4. However, if this method of costing labor-related insurance premiums and taxes had been used, the amounts appearing in the repair labor and labor columns of the direct cost and indirect cost estimate would be correspondingly less, and the total estimated cost for the project (direct cost plus indirect cost) would not be appreciably affected.

CONSTRUCTION PLANT IN-AND-OUT

This section of the indirect cost estimate is utilized for costing the expenses involved in providing the temporary on-site facilities that the contractor will require to perform the permanent work, removing those facilities at the completion of the work, and the expense involved in mobilizing the required construction equipment at the site. The computer software template used at California State University, Chico, contains the following major cost sections, each of which has a number of preprinted subitems and blank spaces in which to enter other items:

- Office and yard area site work
- Furnish buildings
- Erect buildings, shops, plants, etc.
- Install utilities
- Install work access and protection
- Remove plant facilities
- Construction equipment in-and-out

For the example bridge substructure project, Figure 11-6 illustrates the various subentries required under the above-listed major sections. A discussion of these entries follows.

Office and Yard Area Site Work

For the example project, the contractor had received a prebid quotation from a local excavation contractor to clear, grade, and surface the areas required for the job office, shop, and change house facilities. Because the surfacing rock was included in the subcontractor's quotation, the only other expense in this section reflects a separate subcontract quotation from a fencing contractor for 600 lf of security fence at a unit price of $15/ft. For larger projects, more facilities would be required, and these costs would be considerably higher.

DESCRIPTION	QUANTITY	UNIT	EOE U.C.	EOE AMOUNT	REPAIR LABOR U.C.	REPAIR LABOR AMOUNT	RENTAL U.C.	RENTAL AMOUNT	LABOR U.C.	LABOR AMOUNT	EM U.C.	EM AMOUNT	SUBCONTRACT U.C.	SUBCONTRACT AMOUNT	TOTAL AMOUNT
INSURANCE:															
Auto	2	YEAR									5640.00	11280			11280
Eq. Floater-Land	2	YEAR									10500.00	21000			21000
Eq. Floater-Water	2	YEAR									96000.00	192000			192000
Marine	2	YEAR									81250.00	162500			162500
Builder's Risk	1	LS									442150.00	442150			442150
Railroad Protective	1	LS									1875.00	1875			1875
PL & PD (Other than RL)	3523000	$							0.0645	227234					227234
PL & PD	406000	$			0.0645	26187									26187
Deductibles	1	LS									12000.00	12000			12000
Worker's Compensation:															
Clerical	105000	$							0.0280	2940					2940
Concrete Land	707000	$							0.1814	128250					128250
Concrete Water	850000	$							0.2564	217940					217940
Pile Dr./Coffer Land	721000	$							0.3852	277729					277729
Pile Dr./Coffer Water	1140000	$							0.4602	524628					524628
TAXES:															
Gross Receipts		YEAR													
Personal Prop. (Equipment)	2	YEAR									35625.00	71250			71250
Personal Prop. (Const. Mat.)	2	YEAR									9375.00	18750			18750
FICA (Other than RL)	3523000	$							0.0695	244849					244849
FUI (Other than RL)	3523000	$							0.0085	29946					29946
SUI (Other than RL)	3523000	$							0.0575	202573					202573
FICA (RL)	406000	$			0.0695	28217									28217
FUI (RL)	406000	$			0.0085	3451									3451
SUI (RL)	406000	$			0.0575	23345									23345
PERMITS & FEES	1	ILS									1875.00	1875			1875
D SUBTOTAL						81200				1856087		934680			2871967

FIGURE 11-5 Insurance and taxes.

DESCRIPTION	QUANTITY	UNIT	EOE U.C.	EOE AMOUNT	REPAIR LABOR U.C.	REPAIR LABOR AMOUNT	RENTAL U.C.	RENTAL AMOUNT	LABOR U.C.	LABOR AMOUNT	EM U.C.	EM AMOUNT	SUBCONTRACT U.C.	SUBCONTRACT AMOUNT	TOTAL AMOUNT
OFFICE/YARD AREA SITEWORK:															
Clear/Grub/Grade	1	LS											18750.00	18750	18750
Surfacing Rock		TON													
Fencing	600	LF											15.00	9000	9000
FURNISH BUILDINGS:															
Main Office		SF													
Supts Office	3	SF									15000.00	45000			45000
First Aid		SF													
Tool Vans	6	SF									3750.00	22500			22500
Mechanic Vans	1	SF									6250.00	6250			6250
Change House	6	SF									6250.00	37500			37500
Change House (miners)		SF													
Compressor House		SF													
Warehouse		SF													
Salvage on Bldgs	1	LS									-33375.00	-33375			-33375

FIGURE 11-6 (PAGE 1 OF 3)   Construction plant in-and-out.

327

DESCRIPTION	QUANTITY	UNIT	EOE U.C.	AMOUNT	REPAIR LABOR U.C.	AMOUNT	RENTAL U.C.	AMOUNT	LABOR U.C.	AMOUNT	EM U.C.	AMOUNT	SUBCONTRACT U.C.	AMOUNT	TOTAL AMOUNT
ERECT BLDGS, SHOPS, PLANTS, ET															
Office Trailers	3	EA							625.00	1875	625.00	1875	1250.00	3750	7500
Office Other		SF													
Mech. Shop	13	EA							250.00	3250	250.00	3250	313.00	4069	10569
Change House		EA													
Change House (Miners)		EA													
Compressor House		SF													
Warehouse		EA													
Saw Yard/Deck		EA													
Fuel Storage		LS													
Oxy. Aceletene Storage		LS													
Magazine		EA													
Batch Plant		EA													
Signs & Bulletin Boards		LS													
INSTALL UTILITIES:															
Telephone Services	1	EA											1250.00	1250	1250
Water Service	1	EA											625.00	625	625
Water Distribution		LS													
Power Drops	2	EA											12500.00	25000	25000
Power Distribution	1	LS											150000.00	150000	150000
Air Distribution		LS													
Pump Discharge		LS													
Sewer Connections		EA													
Septic Tank	1	EA							2500.00	2500	625.00	625			3125

FIGURE 11-6 (PAGE 2 OF 3) Construction plant in-and-out.

DESCRIPTION	QUANTITY	UNIT	EOE U.C.	EOE AMOUNT	REPAIR LABOR U.C.	REPAIR LABOR AMOUNT	RENTAL U.C.	RENTAL AMOUNT	LABOR U.C.	LABOR AMOUNT	EM U.C.	EM AMOUNT	SUBCONTRACT U.C.	SUBCONTRACT AMOUNT	TOTAL AMOUNT
INSTALL WORK ACCESS & PROTEC															
Road Clear/Grub/Grade	1	LS											31250.00	31250	31250
Road Surfacing Rock	9600	TON									8.50	81600			81600
Road Drainage	1	LS							6250.00	6250					6250
Causeway		LF													
Trestle	750	LF	187.50	140625	62.50	46875	62.50	46875	312.50	234375	312.50	234375			703125
Ladders, Platfrms, Stair Rail	1	LF							75000.00	75000	50000.00	50000			125000
Work Area Fencing		LF													
Trestle Concrete	500	LF	31.25	15625	15.00	7500	15.00	7500	125.00	62500	125.00	62500			155625
REMOVE PLANT FACILITIES:															
Trailers, Buildings, Shops	1	LS							1281.00	1281			1956.00	1956	3237
Utilities	1	LS							250.00	250			17688.00	17688	17938
Work Access & Protection	1	LS	62500.00	62500	21750.00	21750	21750.00	21750	148750.00	148750					254750
Restore Site	1	LS													
CONSTRUCTION EQUIPMENT IN &															
Frieght In	1	LS									78125.00	78125			78125
Unload, Assemble	1	LS							40000.00	40000	37500.00	37500			77500
Interim Job Moves		LS													
Dismantle Load Out	1	LS							40000.00	40000	37500.00	37500			77500
Frieght Out	1	LS									24375.00	24375			24375
E SUBTOTAL				218750		76125		76125		616031		689600		263338	1939969

FIGURE 11-6 (PAGE 3 OF 3) Construction plant in-and-out.

Furnish Buildings

The main job office, superintendent's office, and first aid office requirements could be taken care of for the example project by the purchase of three 12 by 60 ft trailers at a cost of $15,000 each. Six tool vans were required at $3,750 each, one mechanics van at $6,250, and six change trailers at $6,250 each. A guaranteed buy-back at the end of the job was available from the trailer supplier at 30% of the purchase price for all trailers. These purchase cost entries are all reflected in the proper places in Figure 11-6, and the guaranteed buy-back at the end of the job is reflected as a lump sum (a credit).

These costs for a larger project, requiring larger offices and extensive repair shop facilities, would be much higher.

Erect Buildings, Shops, Plants, and So On

The buildings furnished as explained above must be erected and connected to electric power and sewage (septic tank) facilities to make them ready for use. These types of costs are often provided on the basis of experience factor allowances for labor, expendable materials, and subcontract costs per trailer, extended against the number of trailers involved. In this case, the contractor applied the following experience factors for the three large trailers:

Labor	$625 each
Expendable materials	$625 each
Subcontracts	$1,250 each

Similarly, the following experience factors per building were applied to the smaller trailers:

Labor	$250 each
Expendable materials	$250 each
Subcontracts	$313 each

The resulting erection costs are reflected in Figure 11-6.

Concrete dams and similar large projects commonly require extensive plant facilities including aggregate producing plants, concrete batch-and-mix plants, aggregate overland conveyor delivery systems, and so on. Tunnel projects require compressor plants, an underground linear plant system consisting of rail, material hoists, personnel lifts, and similar items. For these types of projects, the procurement of the necessary materials (excluding machinery considered to be equipment) and the installation of these plant items would be costed in this section of the indirect cost estimate.

Install Utilities

Out of the nine separate utility services listed on the preprinted computer template, the following would be required for the example project:

- Telephone services
- Domestic water service
- Electric power drops (electrical substations)

- Electric power distribution
- Septic tanks

In most cases, installation of utilities will be performed by specialty subcontractors and are priced on the basis of prebid subcontract quotations. In the case of the example project, the pricing used was

Telephone services (one service at office trailer)	$1,250 each
Water service (one service at office trailer)	$625 each
Power drops (one on each side of river)	$12,500 each
Power distribution system (on both banks and on trestles)	$150,000 (LS)
Septic tanks (one at office trailer)	$2,500 labor
	$625 EM

The extensions for this pricing are shown in Figure 11-6.

The utility systems required for the previously mentioned large dam projects can be very extensive, consisting primarily of electrical substations, electric power and compressed air and water distribution systems, waste-water treatment plants, etc. Tunnel projects require large electrical substations, extensive underground electric power, lighting, compressed air and water distribution systems, fresh air ventilation systems, ground water discharge systems, and water discharge treatment plants. The prebid planning and costing of all of these examples of temporary plant facilities require special expertise, and contractors often engage experienced consultants for this purpose.

In all cases, the costs for these kinds of construction plant facilities would be entered in this section of the indirect cost estimate. It should be noted that only the costs for furnishing the materials and subcontract cost components of plant facilities and the labor and equipment costs to erect them on-site are being discussed here. The operation of plant facilities during the construction of the permanent work is separately costed in the direct cost portion of the estimate.

Install Work Access and Protection

Out of the eight preprinted entries appearing in this section of the computer template, six were used for the example bridge substructure project. They were priced using a combination of experience factors from prior work and (for the grading for the job roads) on a basis of a prebid quotation obtained from a local excavation contractor. Pricing details follow.

Road: Clear, Grub, and Grade

Access roads were required on both sides of the river. In this case, the contractor had received a prebid quotation from a local excavation contractor in the amount of $31,250 to prepare the roadbeds and lay down the surfacing rock for these roads.

Road Surfacing Rock

To provide an adequate surface on the access roads for heavy construction traffic, 9,600 tons of crushed rock (to be purchased by the contractor) were estimated to be required at a unit cost of $8.50 per ton. This cost was carried in the EM column.

Road Drainage

The contractor separately provided an allowance of $6,250 labor to install drainage structures for the access roads.

Work Trestle

A 750-ft-long work trestle was determined to be required for heavy construction equipment access to the pier locations that, for most of the project duration working time, would be located in the water. The contractor had prepared a preliminary design for this trestle for which experience factors for the various categories of installation costs from prior experience were available. The following experience factors were used:

EOE	$187.50/ft
RL	$62.50/ft
R	$62.50/ft
L	$312.50/ft
EM	$312.50/ft

Ladders, Platforms, Stairs, and Railings

The cost for ladders, platforms, stairs, and railings is almost impossible to rationally estimate on the basis of making prebid layouts of these items, as was done for the work trestle. Rather, contractors usually include an allowance for the necessary labor and materials on the basis of experience gained from prior similar projects. In this case, the contractor provided a labor allowance of $75,000 and an expendable material allowance of $50,000. For a large concrete dam, or for a hydroelectric power plant or pumping station, these costs would be much larger.

Concrete Placing Trestle

For the example project, the water was too deep to extend the work trestle to the two most distant piers in the river. Therefore, 500 ft of a much lighter trestle, intended to carry only a personnel walkway and a series of concrete delivery conveyors, was determined to be required. This trestle was priced in a similar manner to the work trestle at the following component cost allowances from past experience:

EOE	$31.25/ft
RL	$15.00/ft
R	$15.00/ft
L	$125.00/ft
EM	$125.00/ft

The above pricing information is reflected in the extensions in the various cost categories appearing in Figure 11-6.

Once again, it should be noted that the provision of work access facilities for larger projects can be far more extensive and expensive than for the bridge substructure project discussed above. The expenditure of several millions of dollars may be required to provide access roads to remote sites in difficult terrain, and concrete dams, powerhouses, and pumping stations often require extensive trestles or other concrete delivery systems. In all cases, however, the costs would be carried in this section of the indirect cost estimate.

PHOTO 11-1 These photos illustrate the kind of construction plant expense associated with large underground projects. Upper left: Erection of 25-ft-diameter tunnel boring machine. Upper right: Erected tunnel muck-handling facilities. Lower left: Tunnel muck dump hopper, trackwork, ventilation, and electric power facilities. Lower right: Personnel elevator in tunnel main access shaft.

PHOTO 11-2 These four photos illustrate the extensive construction plant installations required for a large concrete dam. Upper left: Erection of large electric-powered cranes and high concrete-placing trestles. Upper right: 300 cy/hr concrete batch-and-mix plant. Lower left: 3-1/2-mile-long aggregate delivery conveyor, construction bridge, and access roads. Lower right: 600 ton/hr aggregate screening plant.

Remove Plant Facilities

At the completion of the work, the contractor's temporary facilities must be removed and the site restored to essentially the preconstruction condition. There are four preprinted cost categories in the computer template, the following three of which were utilized for the example bridge substructure project:

- Trailers, buildings, shops
- Utilities
- Work access and protection

The basis of removal costing for the example project was as follows.

Trailers, Buildings, and Shops

Removal of trailers, buildings, and shops is a simple matter, in this case estimated to cost 25% of each cost category required for installation except for the expendable material category for which no additional cost would be required. The procedure used was simply to total the estimated installation cost in each separate cost category (L, EM, and S) for all 13 trailers and then to take 25% of each total as an allowance for removal of all 13.

Utilities

A similar procedure to that employed for trailers, buildings, and shops was utilized for utilities except that the percentage of installation costs taken was 10% rather than 25%.

Work Access and Protection

The same procedure was utilized for the work access and protection category of removal cost, except that the percentage of installation costs used was 40% rather than 25%.

Construction Equipment In-and-Out

The preprinted computer template contains five categories of cost in the construction equipment in-and-out subsection consisting of

- *Freight in:* Considerable cost can be incurred for freight charges (railroad or truck) to transport the heavy construction equipment from its existing location, or locations, to the jobsite. For the example project, most of this equipment consists of large crawler cranes, concrete pumps, conveyors, other placing equipment, and pile-driving equipment. The actual computation of these costs will be explained in Chapter 12 because these computations can be most conveniently performed as part of the overall **equipment adjustment considera- tion** made at the conclusion of the direct cost estimate.
- *Unload, assemble:* Heavy construction equipment usually must be partially disassembled to transport it on railroad cars or trucks. Once the transport vehicles reach the jobsite, the disassembled components must be unloaded and re- assembled and the unit made ready for work. Like the freight costs, these costs are more conveniently estimated at the time of the equipment adjustment per- formed at the completion of the direct cost estimate. They are then simply posted to this section of the indirect cost estimate.
- *Interim job moves:* On some projects extending over a large site, it is sometimes necessary to disassemble equipment previously erected on the site, transport

the equipment to a new location on the site, and re-erect it. If these kinds of costs are anticipated, they probably would be estimated at the time of the equipment adjustment and then posted to this section of the indirect cost estimate.

- *Dismantle, load out:* At the completion of the work, equipment must be broken down into its component parts and loaded for transport off the jobsite onto railroad cars or trucks. These costs are also estimated at the time of the equipment adjustment performed at the direct cost estimate and posted to this section of the indirect cost estimate.

- *Freight out:* The freight charges for equipment owned by the contractor (for the example project either of the joint-venture partners) from the project site to its next location of use usually will be costed into the estimate for the subsequent project on which the equipment will be used rather than the project just completed. However, return freight charges for equipment rented from third parties must be paid by the contractor and are costed into the project estimate.

For the example bridge substructure project, only four of the five categories were used and the following amounts of money were transferred directly to the indirect cost template from the previously completed equipment adjustment sheets prepared at the completion of the direct cost estimate:

Freight in	$78,125 (EM)
Unload, assemble	$40,000 (L)
	$37,500 (EM)
Dismantle, load out	$40,000 (L)
	$37,500 (EM)
Freight out	$24,375 (EM)

These figures are shown in Figure 11-6.

Total Construction Plant In-and-Out Costs for Example Project

The total costs estimated for this section of the indirect costs appearing on page 2 of Figure 11-6 are

EOE	$218,750
Repair labor	76,125
Rental	76,125
Labor	616,031
Expendable materials	689,600
Subcontracts	263,338
Total	$1,939,969

TOTAL INDIRECT COST ESTIMATE

The summary figures for the total indirect cost estimate are shown in Figures 11-7 and 11-8. The summary figures shown in Figure 11-7 reflect the insurance and taxes appearing in Figure 11-4 (reckoned on the basis of worker's compensation insurance premiums

DESCRIPTION	QUANTITY	UNIT	EOE U.C.	EOE AMOUNT	REPAIR LABOR U.C.	REPAIR LABOR AMOUNT	RENTAL U.C.	RENTAL AMOUNT	LABOR U.C.	LABOR AMOUNT	EM U.C.	EM AMOUNT	SUBCONTRACT U.C.	SUBCONTRACT AMOUNT	TOTAL AMOUNT
TOTALS:															
A—SALARIED PAYROLL	PG. 2									1433192					1433192
B—TIME RELATED OH EXPENSE	PG. 5			87539		111347		137202		265711		301024		97500	1000323
C—NON-TIME RELATED OH EXPENS	PG. 6									12500		191925			204425
D—INS & TAXES—OTHER THAN LAB FEES	PG. 7											1161730			1161730
E—GENERAL PLANT IN & OUT	PG. 10			218750		76125		76125		616031		689600		263338	1939969
GRAND TOTAL				306289		187472		213327		2327434		2344279		360838	5739639

FIGURE 11-7 Indirect cost estimate summary.

DESCRIPTION	QUANTITY	UNIT	EOE		REPAIR LABOR		RENTAL		LABOR		EM		SUBCONTRACT		TOTAL
			U.C.	AMOUNT	U.C.	AMOUNT	U.C.	AMOUNT	U.C.	AMOUNT	U.C.	AMOUNT	U.C.	AMOUNT	AMOUNT
TOTALS:															
A—SALARIED PAYROLL	PG. 2									1433192					1433192
B—TIME RELATED OH EXPENSE	PG. 5			87539		111347		137202		265711		301024		97500	1000323
C—NON-TIME RELATED OH EXPENS	PG. 6									12500		191925			204425
D—INS & TAXES—OTHER THAN LAB FEES	PG. 7			218750		81200				1856087		934680			2871967
E—GENERAL PLANT IN & OUT	PG. 10					76125		76125		616031		689600		263338	1939969
GRAND TOTAL				306289		268672		213327		4183521		2117229		360838	7449876

FIGURE 11-8 Indirect costs estimate summary.

and payroll taxes being included as components of the labor rates set up for the estimate and PL & PD insurance premiums paid on the basis of the total contract value), and Figure 11-8 reflects the insurance and taxes figures shown in Figure 11-5 (where worker's compensation insurance premiums and payroll taxes are included in the indirect cost rather than in the labor rates, and PL & PD insurance premiums are also excluded from the labor rates but are paid on the basis of the contractor's total labor exposure). In comparing the total indirect cost for Figures 11-7 and 11-8 ($5,739,639 and $7,523,524, respectively), it should be remembered that for the indirect cost shown in Figure 11-8 ($7,523,524), the corresponding direct cost estimate would be correspondingly lower because it would not include in the labor category totals (RL and L) the premium costs for worker's compensation insurance or for payroll taxes.[9]

CONCLUSION

This chapter has presented in detail, by way of an illustrative example project, the procedures and format for assembling an indirect cost estimate. In doing this, alternate methods were explained for handling worker's compensation insurance premiums, PL&PD insurance premiums, and payroll taxes. As each element of indirect cost was explained, it was pointed out how these costs might vary for projects differing from the example project.

It was also pointed out that in estimating construction equipment-related costs in the construction plant in-and-out section, the figures used were extracted from a separate equipment adjustment analysis that would be performed at the conclusion of the direct cost estimate. Chapter 12 is devoted to an explanation of this equipment adjustment analysis.

QUESTIONS AND PROBLEMS

1. What are the five major sections of the indirect cost vertical format discussed in this chapter?

2. What five major divisions of the salaried payroll section are used in the indirect cost vertical format presented in this chapter? Explain the general kind of duties performed by the personnel costed under each division. What two pieces of information are necessary to cost each personnel classification? How is the number of person-months a particular personnel classification is likely to be required determined?

3. What 19 major divisions constitute the time-related overhead section of the indirect cost estimate? Explain the general nature of each division of expense. How are the monthly rates for EOE, RL, and R for general service and automotive equipment determined? How are the required number of unit months for general service equipment determined? Answer the same questions for automotive equipment.

4. How are contractor-provided survey costs typically determined? What are the usual crew compositions? How are the number of crews and the duration of time each crew is likely to be required determined?

5. To what other element of the contractor's costs does the cost of protective clothing relate? How is past experience tied in or used to determine protective clothing costs on a new project?

---

[9]The reader should note that the PL&PD insurance costs for the example project (paid on the basis of total contract value) were not included in the $15,910,100 direct cost figure and thus *were* included in the $5,739,639 total of Figure 11-7. They are also included in the $7,523,524 total of Figure 11-8 (where they are recorded in the labor and repair labor columns, respectively).

6. For the balance of the individual cost line items in the 19 major divisions of the time-related section of the indirect cost estimate, what two input parameters are required in each case to determine the cost? How does the construction schedule relate to the first parameter? What is the source of the second parameter?

7. What are the 15 line items of the non-time-related section of the indirect cost appearing in Figure 11-3? What two line items are discussed in some detail? Would you expect that it might be possible to obtain quotations from specialist consultants for the first of the line items that was discussed? What two parameters define the costs that would be incurred for the second? How would each of these parameters be determined?

8. What are the nine different kinds of insurance costs that must be considered when pricing the insurance and taxes section of the indirect cost estimate? Explain how the premium quotations are usually structured for each. To which three kinds of these insurance costs is the duration of the construction schedule related? Which one of the nine kinds of insurance costs is always related to the bare labor component of the contractors costs? Is this component also related to the kind of work that comprises the bare labor costs? Which one of the nine kinds of insurance costs may or may not be related to the bare labor component of the contractors costs depending on the way the premium is quoted? With respect to this latter kind of insurance cost, what are the two alternate ways the premium may be quoted? Identify the column in the horizontal format of cost expression in which the premium cost would be recorded, depending on the structure of the premium quotation.

9. What two general methods are explained in this chapter for incorporating worker's compensation insurance premiums and PL & PD insurance premiums (the latter in cases where the premium quotation is based on the contractor's bare labor exposure) into the contractor's bid estimate? In what part of the estimate (i.e. direct cost or indirect cost) would the premium costs for the worker's compensation insurance be carried for each of the two methods? In what column in each case? Answer the same question in the case of PL & PD insurance (in cases where the premium quotation is based on the contractor's bare labor exposure).

10. In what part of the contractor's estimate is the replacement value of construction equipment for pricing equipment floater insurance determined?

11. What six kinds of taxes and fees are discussed in this chapter and how is the cost to be incurred by the contractor determined in each case? Which three are related to the contractor's bare labor costs?

12. What two general methods are explained in this chapter for incorporating the three labor-related taxes into the contractor's bid estimate? In what part of the contractor's estimate, and in which column of the horizontal cost format, are the costs entered in each case?

13. What are the seven major divisions of the construction plant in-and-out section of the indirect cost estimate?

14. What are some of the individual line items where subcontract quotations would likely be sought by the prime contractor (based on the example project discussed in the text)? What item lends itself to a supply/buy-back arrangement? For which of the individual line items requiring special expertise do contractors often engage specialty consultants?

15. For what category of costs carried in the construction plant in-and-out section of the contractor's estimate are the actual costs developed in a separate analysis?

16. The following data are furnished for your use:

Project layout drawing (foldout bridge project layout drawing at the back of the book). This drawing was prepared by Alfred Benesch & Company, Consulting Engineers, Chicago, Illinois, as part of the contract drawings for the Jefferson Barracks Bridge Project across the Mississippi River.

Project time line (at end of this chapter)

Craft labor wage rates

Salaried payroll wage rates

Equipment rental rates

Equipment operating expense rates

Prepare the indirect cost estimate for a project consisting of construction of a series of bridge piers across the Mississippi River between Illinois and Missouri as shown on the project layout drawing. One large pier (pier 12) in the middle of the river controls the schedule of the project. In other words, the time requirements for constructing this pier are such that all the other piers are expected to be completed within the time span required to construct pier 12. The direct cost estimate has been built around the project time line. The direct cost line item totals for the entire project are

EOE	$ 718,800
RL	$ 391,800
R	$ 1,328,100
LABOR	$ 4,290,700
EM	$ 2,171,800
PM	$ 3,797,000
SUBS	$ 28,800
TOTAL	$12,728,000

Use either the indirect cost template furnished in the Appendix or manual methods to prepare the indirect cost estimate, based on the following information obtained from previous experience. Determine the durations of time-related expenses from the project time line.

A. Salaried Payroll

- One project manager is required for duration of project.
- One general superintendent is required for duration of project.
- One pile-driving superintendent is required from the start of pier 12 cofferdam to end of removal of pier 12 cofferdam.
- One carpenter/concrete superintendent (one person) required from the start of construction of all piers other than pier 12 to end of pier 12 structural concrete. (One person supervises both carpenter and concrete work.)
- One project engineer is required for the duration of the project.
- One field engineer is required for the duration of the project. A second field engineer is required for a 12 month period in the middle portion of the job.
- One office manager is required for the duration of the project.
- One secretary clerk is required for the duration of the project.

These personnel (as listed above) will be kept on the payroll during periods of high water and other delays to the work. No other salaried personnel will be required.

B. Time-Related Overhead

Survey: All field surveying work will be performed by the state highway department survey crews at no cost to the contractor.

General service equipment: One boom truck will be required for the duration of the project including setup time for construction plant and move out time. This unit will not be required during high-water periods when there is no active work. Truck will be operated by one teamster and one laborer.

Automotive equipment:

Project manager, sedan assigned

Project engineer, sedan assigned

All superintendents, one pickup assigned to each superintendent

These vehicles are assumed to operate during periods of high water and/or any other delays.

Warehouse operation: Not required.

Supplies: Records of previous similar projects indicate that the following monthly allowances are likely to be expended for all periods personnel are on the project, including high-water periods.

Office supplies	$600/month
Engineering supplies	$200/month
Safety and first aid supplies	$200/month
Drinking water supplies	$250/month

Office utility bills:

Telephone	$500/month
Electricity/gas/oil/water	$750/month (for all four)

Miscellaneous services:

Photo	$130/month
CPM update	By staff (no cost)
Data processing	$450/month
Security	Not required
Sanitary	$550/month
Trash	$ 50/month

Rentals:

Office	Not required
Yard	$300/month

Maintenance expenses:

Buildings and facilities	$200/month (labor)
	$400/month (materials)
Radios	$200/month

Electrical distribution system: A subcontract quotation has been obtained to maintain this system and to provide all hookups and disconnects required for the work crews for $6,000/month. (Not required during move-in, high-water, or move-out periods.)

Other time-related expenses: Distributing drinking water is required during all active work, excluding high-water periods. Requires the time of two laborers, 4 hr per working day each. Working days considered to be 5 days per week, Monday through Friday.

Interim job cleanup	Not required
Job electric energy consumption	$1,200/month (excluding high-water periods)
Job water consumption	$100/month

Protective clothing	$2% of direct labor (include RL)
Miscellaneous overhead personal expense	$250/month
Salaried personnel living allowance	Not required
Concrete testing	By state at no cost
Miscellaneous drayage	$200/month

C. Non-Time-Related Overhead Expense

Office furniture	$4,000 Lump sum (LS)
Office equipment	$4,000 (LS)
Engineering equipment	$6,000 (LS)
Safety and first aid equipment	$9,000 (LS)
Set up CPM	By staff (no cost)
Outside engineering	Allow $125,000 for cofferdam and trestle designs
Outside surveying	Not required
AGC dues	$1,500 (LS)
Medical fees	Not required
Purchase warehouse bins	Not required
Purchase job radios	$20,000 (LS)
Punch list and final cleanup	Allow $10,000 labor and $5,000 materials
Salaried personnel expense	Not required

D. Insurance and Taxes

Insurance quotations have been obtained for the various required insurance policies as follows:

Auto	$750/vehicle/year
Equipment floater	$0.55/$100 of equipment replacement value/year for land-based equipment
	$1.25/$100/year for equipment on barges

The equipment replacement values are $1,250,000 for equipment on land and $5,250,000 for equipment on barges.

Marine hull and marine P&I	$65,000/year for all of the tugs and barges as a group
Builder's risk	$1.45/$100 of contract value[a]
Railroad protective insurance	$1,500 lump sum for job
PL & PD	$0.75/$100 of contract value[a]
Deductibles	Allow $10,000 (LS)

[a]You will have to make a trial guess at the "contract value" (amount of the bid) to compute these premiums. For this purpose, take the total indirect cost at $4,750,000 and mark up and contingency at $2,850,000.

Taxes are as follows:

Gross receipts	Not required
Personal property (equipment)	$1.50/$100 on 30% of equipment replacement value/year
Personal property (construction materials)	$4.00/$100 on 30% of estimated value/year

	Figure this tax on an estimated value of sheet pile on the job of $600,000[a]
Permits, fees	Allow $1500

[a]Average of purchase price and salvage values.

E. General Plant In-and-Out

Office and yard area site work: A subcontract quotation has been obtained for $15,000 to grade and surface with gravel the office and yard area. Six hundred feet of fencing is required and can be bought for an erected price of $12.00/ft.

Furnish buildings: The main office, superintendent's office, and first aid office requirements can be taken care of by three 12 × 60 ft trailers that cost $12,000 each. Six tool vans are required at $3,000 each. One mechanic's van at $4,500, and six change trailers at $5,000 each are also required. A guaranteed buy back at the end of the job is available at 25% of purchase price for all trailers.

Erect buildings, shops, plants: Trailers can be erected for the following:

	Labor, $	Materials, $	Subs, $
Office trailers, each	500	500	1,000
Vans, change trailers, each	200	200	250

Install utilities: Quotations and/or estimates are available for the following required facilities:

	Labor, $	Materials, $	Subs, $
Phone service (1 each)			1,000/each
Water service (1 each)			500/each
Power drops (2 each)			10,000/each
Power distribution (LS)			120,000
Septic tank (1 each)	2000	500	

Install work access and protection: Quotations and/or estimates are available for the following:

All job roads required for project including grading, rock surfacing, and drainage as lump sum allowances:

      Labor $ 5,000
      EM $65,000
      Subs $25,000

Construction trestle across slough on Illinois bank and out to pier 10. Deck at elevation 400, 750 × 25 ft wide (see layout drawing).

      Labor $250/lf
      EOE $150/lf
      RL $ 50/lf
      R $ 50/lf
      EM $250/lf

Concrete conveyor support trestle from pier 10 to pier 12. Elevation 400 to 500 lf × 5 ft wide (see layout drawing). (Write in this line item on the spreadsheet just below "work area fencing.")

Labor	$100/lf
EDE	$ 25/lf
RL	$ 12/lf
R	$ 12/lf
EM	$100/lf

Temporary ladders, platforms, and stairs for construction at all the piers are estimated to cost for life of job (to erect/maintain):

Labor	$60,000
EM	$40,000

Remove plant facilities: Plant facilities can be removed at the end of the job for the following estimated costs:

Office and yard area site work	No removal necessary
Buildings, shops, and plants	20% of installation cost
Utilities	15% of installation cost
Work access and protection	35% of installation cost
	(for every cost category except EM column)

Construction equipment in-and-out: These costs, determined for the total job for the various required pieces of equipment are estimated to be:

	Labor, $	EM, $
Freight in		62,500
Unload, assemble	32,000	30,000
Dismantle, load out	32,000	30,000
Freight out		19,500

Craft Wage Rates:

Classification	Fully loaded labor rate per hour
Laborer	$21.10
Teamster	$21.01

Salaried Labor Rates:

Classification	Fully loaded labor rate per month
Project manager	$8,769/month
General Superintendent	$8,253/month
Carpenter/concrete superintendent	$6,563/month
Pile-driving superintendent	$7,221/month
Project engineer	$6,706/month
Office engineer	$4,598/month
Field engineer	$5,158/month
Office manager	$4,180/month
Secretary/clerk	$2,926/month

Equipment Rates per Hour:

Description	EOE	RL	R
Three-axle boom truck	$6.94	$6.23	$21.68
Sedan	$1.81	$3.74	$ 1.88
Pickup	$3.21	$3.74	$ 2.28

17. Assume that the direct cost labor ($4,290,700) and repair labor ($391,800) components of the $12,728,000 direct cost total in problem 16 had been estimated using labor rates containing base pay and union fringes only. Assume further that the base pay portion of the direct labor and repair labor components break down as follows:

Description	Total, $	Base pay portion, $
Direct labor clerical	127,900	84,000
Direct labor concrete work on land	861,100	565,600
Direct labor concrete work on water	1,035,200	680,000
Direct labor pile-driving/cofferdam on land	878,100	576,800
Direct labor pile-driving/cofferdam on water	1,388,400	912,000
Direct labor total	4,290,700	2,818,400
Repair labor	391,800	257,400

Also, assume the following rates for labor-related insurance premiums and for taxes applied and that all other data remained unchanged:

Worker's compensation insurance	
Clerical	$ 2.24/$100 of base payroll
Concrete work on land	$14.51/$100 of base payroll
Concrete work on water	$20.51/$100 of base payroll
Pile-driving/cofferdam on land	$30.82/$100 of base payroll
Pile-driving/cofferdam on water	$30.82/$100 of base payroll
Repair labor	$14.51/$100 of base payroll

PL & PD insurance, $5.16/$100 of base payroll

Taxes:

FICA	$5.56/$100 of base payroll
FUI	$0.68/$100 of base payroll
SUI	$4.60/$100 of base payroll

Repeat the insurance and taxes portion of the indirect cost estimate in the format of Figure 11-5 on the basis of the above revised information.

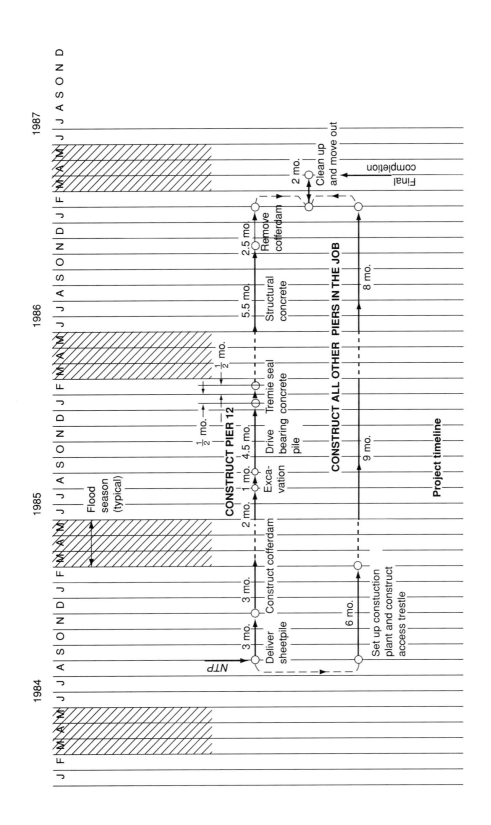

12

Equipment Adjustment and Determination of Equipment Mobilization and Demobilization Costs

Key Words and Concepts

Source of the equipment
Equipment adjustment procedure
Determination of number of units in each equipment class
Least-cost method of obtaining construction equipment
Required equipment adjustment
Selective equipment adjustment approach
Evaluation of alternate equipment adjustment options
Importance of equipment adjustment spreadsheet foot-totals

Total dollars going to third parties
Total dollars going to in-house equipment division
Cash outlay at beginning of project
Total dollars to be realized as equipment salvage value
Freight costs for transport to project site
Unload and make-ready costs
Dismantle and load-out costs
Value to insure
Transfer of cost figures and value to insure to indirect cost estimate

The reader will recall from Chapter 5 that the first step the chief estimator or estimate sponsor must take to determine the library rates for equipment rental (R) is to make a best-guess assumption for the **source of the equipment** that will be used for the project. That is, an assumption must be made that the equipment will be obtained from:

- The contractor's presently owned fleet
- Third-party rental
- New (or used) purchase (or lease-purchase) at the inception of the project and then sold for a salvage price when the project is completed (or absorbed at that time into the contractor's fleet)
- Owner-operators on an operated-and-maintained basis (O&M rental)
- A combination of the above

In the case of rental equipment, the monthly rental rate is likely to be known at the time of setting up the estimate, but a further assumption must be made for the average number of hours of equipment use for each month the equipment is to be rented to derive an hourly rate to set up in the library. Rental charges for equipment rented on an O&M basis are paid only for the hours of actual use, so to set up the proper rate in the library for that type of equipment rental, it is only necessary to canvass the market to obtain the then-current O&M rental rate.

For equipment that is to be obtained on a buy/sell basis, the purchase price is ordinarily reasonably ascertainable, but an estimate must be made for the salvage value at the completion of the work. In addition, two further assumptions must be made:

- The number of months the particular unit of equipment will be required on the project
- The average number of hours the equipment will be used for each month the equipment is on the project

At the time the estimate is set up (requiring all of the above described assumptions to be made), comparatively little is known about the project. At that point, no one has analyzed the details of any part of the actual project work. However, a great deal more information is available after the completion of the direct cost estimate:

- Final choices of particular units of equipment have been made by the estimators for the various work items of the project.
- A construction schedule will have been prepared, defining the time spans over which specific work items (which in turn require specific types of equipment) are planned to be carried out.
- The total number of operating hours for each type of construction equipment that has been priced into the estimate can be summarized and tabulated.[1]
- The total dollars in the rental column for the project direct cost is available from the direct cost summary sheet.

Based on accurate information for the time spans over which the various classes or types of equipment will be used, and the total number of required operating hours in the estimate for each class, it is now possible to determine the number of individual units of equipment that will be required for each class or type of equipment.

---

[1]For computer-aided estimates, the estimating software usually develops these equipment hour summaries by type and size of equipment. For manual estimates, these summaries must be developed by hand.

Once this information has been determined, final decisions can be made about the most economical source for the equipment, and reliable figures can be developed, on a global basis, for the probable costs of providing the equipment. By comparing this revised total dollar figure with the total dollars already in the rental column of the direct cost estimate, an appropriate adjustment can be made, up or down, to the rental total and, thus, to the direct cost total.

The procedure for making this equipment adjustment is the primary subject of this chapter. In addition, the procedures for determining (1) the costs for equipment mobilization on the project, (2) costs for demobilization at the completion of the work, and (3) the replacement value of equipment that should be insured are also described.

EQUIPMENT ADJUSTMENT PROCEDURE

The methodology for making the above described equipment adjustment is best illustrated by considering a greatly simplified hypothetical example project. Assume that for this project only six classes of equipment were required for performance of the project work[2] and that a computer-generated summary sheet listed the following summary figures from the direct cost estimate for each class of equipment.[3]

Description	Hours of use in estimate	Total rental $ in estimate
100 ton crawler crane	1,320	37,000
15 ton center-mount crane	10,200	122,400
1200 cfm compressor	1,936	26,600
185 cfm compressor	8,400	25,200
300 A welding machine	10,200	17,900
Concrete pump	2,900	72,500

Further, by study of the project schedule developed during preparation of the direct cost estimate, the time spans during which each class of equipment would be either continually or intermittently used were determined to be as follows:

100 ton crawler crane	Used for work on a single shift between 6/89 and 6/90 (i.e., 12 months single-shift use)
15 ton center-mount crane	Used for work on two shifts between 1/89 and 6/90 (i.e., 18 months two-shift use)
1200 cfm compressor	Used for work on a single shift between 6/89 and 6/90 (i.e., 12 months single-shift use)
185 cfm compressor	Used for work on two-shifts between 6/89 and 6/90 (i.e., 12 months two-shift use)
300 A welding machine	Used for work on two-shifts between 6/89 and 6/90 (i.e., 12 months two-shift use)
Concrete pump	Used for work on a single shift between 6/89 and 6/90 (i.e., 12 months single-shift use)

---

[2] This is a highly unlikely scenario chosen strictly for purposes of illustration. Typically, 15 or 20 separate classes of equipment (or more) would be required.

[3] Or, in the case of a manual estimate, the information was summarized by hand.

Based on the above information derived from the direct cost estimate, the following key questions can now be answered:

- How many units of each class of equipment will be required?
- How, and from where, can the required equipment be obtained at the lowest possible cost and what is that cost?
- What is the dollar adjustment, if any, that should be made to the monies in the rental column of the direct cost estimate to reflect the benefit of the additional knowledge derived from completion of the direct cost estimate?

Determination of Number of Units within Each Equipment Class

If work requiring the continual or intermittent use of a given class of construction equipment is to be conducted over a known fixed time duration in months, during which a known number of days of work per week and shifts to be worked per day are known, and the total number of required equipment operating hours for the total operation duration is also known, the **theoretical number of units** of equipment can be easily calculated. See, for instance, the determination of the number of excavation load-and-haul units and embankment compactor units that was explained in Chapter 9. In that case, the continuity of operations was sufficiently certain and the number of work locations was sufficiently limited so that there was little possibility of a "stacking up" or "peaking" of simultaneous demands for the same units of equipment in different work locations. The theoretical calculated numbers of required units need not be increased. The same would be true for tunnel excavation and similar work scenarios consisting of just one or two locations of concentrated activity. However, many other types of work, such as structural concrete operations conducted over a large worksite, tend to result in situations where more than the theoretical number of units are required due to a peaking of demand (i.e., the available equipment must satisfy needs in a number of locations at the same time). There is nothing more destructive to profitability than when production efficiency has been severely handicapped because the job was underequipped. This possibility can be averted, at least in part, by use of a **peak factor** to increase the amount of equipment available on the job to accommodate peak demands. A factor ranging between 1.00 and 1.50 is not unreasonable, and for the calculations shown in Figure 12-1 for the hypothetical example project, peak factors have been varied according to the estimator's judgment for the several classes of equipment involved. Generally speaking, the larger units of equipment do not require as large a peak factor as the smaller units (which are more likely to be simultaneously needed in a number of separate locations).

Based on the calculations shown in Figure 12-1, the number of required units for each class of equipment for the hypothetical project are

100 ton crawler crane	One unit required for the full 12 month duration
15 ton center-mount crane	Two units required for the entire 18 month duration; one additional unit required for a period of 10 months
1200 cfm compressor	One unit required for the full 12 month duration; one additional unit required for 2 months
185 cfm compressor	Three units required for the full 12 month duration; one additional unit required for 2 months

Equipment description	Planned work conditions	Months on job	Required equipment hours	Available shift hours	Theoretical no. of units[4]	Peak factor	No. units to procure[5]	Procurement notes
100 ton crawler crane	8.0/1/5	12	1,320	2,078[1]	0.64	1.25	0.80	1 for 12 months
15 ton center-mount crane	7.75/2/5	18	10,200	6,040[2]	1.69	1.50	2.53	2 for 18 months 0.53 × 18 months = 1 additional for 10 months
1200 cfm compressor	7.75/1/5	12	1,936	2,078[1]	0.93	1.25	1.16	1 for 12 months 0.16 × 12 months = 1 additional for 2 months
185 cfm compressor	7.75/2/5	12	8,400	4,027[3]	2.09	1.50	3.14	3 for 12 months 0.14 × 12 months = 1 additional for 2 months
300 amp welder	7.75/2/5	12	10,200	4,027[3]	2.53	1.50	3.79	3 for 12 months 0.79 × 12 months = 1 additional for 10 months
Concrete pump	8.0/1/5	12	2,900	2,078[1]	1.40	1.25	1.75	1 for 12 months 0.75 × 12 months = 1 additional for 9 months

[1]12 months × 4.33 weeks per month × 5 days per week × 1 shift per day × 8.0 hr per shift.

[2]18 months × 4.33 weeks per month × 5 days per week × 2 shifts per day × 7.75 hr per shift.

[3]12 months × 4.33 weeks per month × 5 days per week × 2 shifts per day × 7.75 hr per shift.

[4]Required equipment hours ÷ available shift hours.

[5]Theoretical no. units × peak factor.

FIGURE 12-1 Example calculations for required units of equipment.

| 300 A welding machine | Three units required for the full 12 month duration; one additional unit required for 10 months. |
| Concrete pump | One unit required for the full 12 month duration; an additional unit required for 9 months. |

How, and from Where, Can the Required Equipment Be Obtained at the Least Possible Cost and What Is That Cost?

Once that the project equipment needs have been quantified in specific terms, rational decisions can be made to determine the **least-cost method of obtaining the equipment.** These decisions will vary widely, depending on market conditions, the size and duration of the project, the available cash and credit resources available to the contractor, and the contractor's equipment management philosophy. Assume that for the hypothetical example the contractor had made the following equipment procurement decisions:

100 ton crawler crane	Rent one crane from the company's in-house equipment division at $9,000 per month for the entire 12 month use period.
15 ton center-mount crane	Rent two cranes from the in-house equipment division at $2,900 per month for the full 18 month use period. Rent one crane from in-house equipment division at $2,900 per month for a 10 month period.
1200 cfm compressor	Rent one compressor from the in-house equipment division at $2,400 per month for the entire 12 month use period. Rent one compressor from third-party sources at $3,600 per month for a 2 month period.
185 cfm compressor	Buy three compressors new at a purchase price of $11,000 each. Assume a salvage value at the end of the use period of $6,750 each. Rent one compressor from third-party sources at $800 per month for 2 months.
300 A welding machine	Purchase four welding machines new at a purchase price of $10,000 each. Assume a salvage value at the end of the use period of $3,750 each.
Concrete pump	Purchase two concrete pumps new at a purchase price of $205,000 each. Assume a salvage value at the end of the use period of $143,500 each.

These procurement decisions can be converted into a dollars and cents project equipment rental charge by use of either a manual or computer-aided spreadsheet approach, as shown in Figure 12-2. It should be noted that the calculation method illustrated by Figure 12-1 for determining the number of individual construction equipment

Equipment description	No.	Date required	Release date	Total months	Operator hours	Source & location of equipment	Outside rental Rate	Outside rental Amount	In-house rental Rate	In-house rental Amount	Purchase Rate	Purchase Amount	Salvage amount	Write-off amount	In estimate amount	(+) Amount	(−) Amount
100-ton crawler crane (8.0/1/5)	1	6/89	6/90 '	12	1,320	Rent in-house			9,000	108,000					37,000	71,000	
15-ton center-mount crane (7.75/2/5)	2 / 1	1/89	6/90	18 18 10	10,200	Rent in-house Rent in-house			2,900 2,900	104,400 29,000					122,400	11,000	
1200 cfm compressor (8.0/1/5)	1 / 1	6/89	6/90	12 12 2	1,936	Rent in-house Rent 3rd party	3,600	7,200	2,400	28,800					26,600	9,400	
185 cfm compressor (7.25/2/5)	3 / 1	6/89	6/90	12 12 2	8,400	Buy new Rent 3rd party	800	1,600			11,000	33,000	20,250	12,750	25,200		10,850
300-amp welding machine (7.75/2/5)	4	6/89	6/90	12 12	10,200	Buy new					10,000	40,000	15,000	25,000	17,900	7,100	
Concrete pump	2	6/89	6/90	12 12	2,900	Buy new					205,000	410,000	287,000	123,000	72,500	50,500	
Totals								8,800		270,200		483,000	322,250	160,750	301,600	149,000	10,850
Equipment adjustment															+ 138,150		
Adjusted equipment cost															439,750		

FIGURE 12-2 Equipment adjustment analysis for example project.

units has been reflected in Figure 12-2. With a little practice, a person can perform the necessary arithmetic manipulations of the known information that has been entered onto Figure 12-2 very quickly (using a hand-held calculator) to determine the required number of equipment units of each equipment class and how long each unit will be required at the project. It is not necessary that the spreadsheet illustrated by Figure 12-1 be prepared as an interim step. It was done in this text only to demonstrate the necessary thought processes.

For the hypothetical example project, the cost for meeting the total equipment needs of the project are shown by the foot-totals of Figure 12-2 to be

Outside rental	$ 8,800
In-house rental	$ 270,200
Purchase/salvage write-off	$ 160,750
Total	$ 439,750

Required Equipment Adjustment

Based on the equipment use rates for equipment rental established at the time the estimate was set up, the total amount in the rental column for the project direct cost appears in the in-estimate column of Figure 12-2 as $301,600, the "plus" equipment adjustments as $149,000, the "minus" adjustments as $10,850, for a net required equipment adjustment of $149,000 − $10,850 = $138,150. Adding this to the $301,600 already in the estimate raises that figure to the required $439,750. This latter figure, which is based on the latest and best information available at the time of bid (as contrasted to the information available when the estimate was set up), would be used from that point onward in formulating the bid.

Selective Equipment Adjustment Approach

The above described procedure for the hypothetical example project can be applied for every class of equipment appearing anywhere in the direct cost estimate. However, in the case of larger projects, the list of equipment classes utilized in the direct cost estimate is usually lengthy and will contain a number of equipment classes for which the usage costed into the estimate is minor and highly intermittent. This makes the process of determining specific use periods from the schedule very laborious. It can be done, but the adjustments usually turn out to be so minor that it is not worth the effort.

For the above explained reason, many estimators simply accumulate all of the minor usage equipment rental dollars and enter the total on the equipment adjustment spreadsheet under a group heading entitled "Minor unadjusted equipment." Under this general heading, the only entry made on the equipment adjustment spreadsheet would be the accumulated dollar figure in the in-estimate column. All of the other columns in the spreadsheet would be left blank. In effect, this procedure makes the statement to anyone reading or reviewing the spreadsheet that, although this minor usage equipment class must be procured and utilized for the work, the estimator did not feel it worthwhile to attempt to analyze the details and was willing to ride with the dollars already in the estimate for the cost of providing that class of equipment.

This procedure ensures that the spreadsheet foot-totals in the in-estimate column correctly represent the dollars actually in the estimate prior to the equipment adjustment consideration, even though the spreadsheet does not reflect any of the specific periods of usage or procurement details for this minor usage equipment. The other spreadsheet foot-totals correctly represent the usage, procurement, and cost details for all of the major usage equipment classes that were included in the equipment adjustment.

Evaluation of Alternate Equipment Procurement Options

One of the important fallout advantages of the above described equipment adjustment procedure is the opportunity it affords to study and evaluate different options for equipping the project that would be difficult, if not impossible, to evaluate at the time the use rates in the equipment rental library were set up. As mentioned previously, these procurement decisions are driven by the particular business circumstances the contractor is in at the time of bid, which vary widely. The point is that when considering how to procure the equipment for each major class, the least-cost option should be chosen, dictated by the particular contractor's business circumstances.

For example, suppose that for the hypothetical project concrete pumps could be rented at a monthly rental rate of 5% of the purchase cost, or (0.05) $(\$205,000) =$ $\$10,250$. The contractor would then have three alternate choices for providing the use of this equipment:

Option 1: Third-party rental for both pumps

$$\text{Cost} = (\$10,250)(12 + 9) \text{ pump-months} = \$215,250$$

Option 2: Purchase one pump and third-party rental for the other

$$\text{Cost} = (\text{one pump})(\$205,000 - \$143,500) + (\$10,250)(9 \text{ pump-months})$$

$$= \$61,500 + \$92,250 = \$153,750$$

Option 3: Purchase both pumps

$$\text{Cost} = (\text{two pumps})(\$205,000 - \$143,500) = \$123,000$$

The third option clearly results in the least cost. It would be chosen if the contractor had the cash to purchase the pumps (or had the credit line to borrow the necessary funds) and was confident that the assumed $143,500 salvage figure would stand up. This latter consideration is not solely a question of whether the pumps would be worth the assumed salvage value at the end of the usage period but also involves the question of whether there would be a ready market demand for used concrete pumps. If not, the contractor would either have to regard the continued ownership of the pumps as a good investment for the company's business or at least would have to be willing to wait until market conditions improve so that the estimated salvage value would be recovered at that time.

The consideration of continued ownership of the pumps involves the contractor's equipment management philosophy and the general type of construction projects the contractor normally pursues, both influencing how willing the contractor would be to continue to leave part of the company's net worth tied up in ownership of that particular piece of equipment. For instance, a contractor whose primary line of contracting was structural concrete might favor maintaining ownership of the concrete pumps, whereas another contractor with less interest in concrete work would not.

Decisions of this kind also would depend on the general availability and cost of owner-operators with concrete pumping equipment. Also, decisions involving purchase/salvage options must recognize several other kinds of ownership costs in addition to equipment write-off (depreciation), such as interest on investment, personal property taxes and insurance, storage costs, and so on. This is particularly true for joint venture bids where these types of costs are reflected in the indirect cost portion of the estimate and in the interest consideration part of the vertical format of the estimate leading to the total bid figure (see the discussion in Chapter 3 on this subject). For single-contractor bidding situations, a contractor who prices the bid estimate using the procurement method illustrated by Figure 12-2 (that is, new purchase at the inception of this project and salvage upon completion) would, if the bid was successful, probably immediately incorporate all purchased equipment into the company's equipment division and establish in-house rental rates for charging the project that would recognize the interest on investment and other equipment ownership costs.

IMPORTANCE OF THE EQUIPMENT ADJUSTMENT SPREADSHEET FOOT-TOTALS

The equipment adjustment spreadsheet foot-totals play an important part in the business management of a heavy construction company in addition to their use for developing a monetary adjustment to the rental column of the direct cost estimate. These highly significant foot-totals include:

- Total dollars going to third parties
- Total dollars going to the company's in-house equipment division
- Total cash outlay required at the beginning of the project
- Total dollars that must be realized at the completion of the project as equipment salvage value

Total Dollars Going to Third Parties

This foot-total represents "cash-out" expenditures incurred progressively during the job. From a business management standpoint, a large number in this foot-total is undesirable for two reasons. First, it represents a large progressive depletion of working capital, and second, it represents capital leaving the company and going to some other business entity.

A wise businessperson will minimize this kind of expenditure to the maximum extent possible. Contractors whose business circumstances result in large dollar amounts in this category should probably search for other projects to bid that reflect a better fit to their current business situation. Additionally, third-party rental usually is the most expensive way to procure construction equipment. If equipment usage constitutes a large percentage of the total project costs, as usually is the case in heavy construction, a high figure in this category of cost probably means that the contractor's bid will not be competitive.

Total Dollars Going to the Company's In-House Equipment Division

Most heavy construction contractors maintain an in-house equipment division to manage and maintain the company-owned equipment assets. Monies in a bid estimate for rental charges from such an in-house entity do not represent cash expenditures and do not draw down the company's supply of working capital. To the contrary, they represent revenue to defray equipment ownership costs the company will continue to incur whether the prospective project is obtained or not. From a business standpoint, the healthiest thing a company can do to ensure its continued profitability is to put equipment that is already owned to work in new revenue-producing projects. Therefore, the larger the percentage of the total project equipment cost represented by the in-house rental column foot-total, the better.

Cash Outlay Required at the Beginning of the Project

The magnitude of this foot-total is an important management consideration because of its obvious impact on the company's cash resources and/or credit line. This figure represents a cash investment that will only be made if the bid is successful. Its magnitude is not important if the company possesses the cash or credit line resources to handle it and if the potential equipment usage on the project is large enough so that the purchase costs will be significantly written down by the time of project completion. The most favorable situation is for the purchase cost to be written off on the bid project down to the salvage value of the equipment or near to it. This guarantees that the initial investment will be returned from the combination of the revenues earned on the project at hand plus the salvage value of the equipment.

Total Dollars to Be Realized at Project Completion as Equipment Salvage Value

Depending on its magnitude, this foot-total may represent a significant risk that will be assumed if the bid is successful. To protect the profit expected to be earned on the project, all estimated project costs, in aggregate, must not be exceeded. In heavy construction, the estimated cost in the equipment rental column is usually one of the significant project costs, and every dollar of cost overrun means a dollar lost from the job profit figure that would otherwise be earned. A large number in the salvage value foot-total represents a large risk to the job profit, in the event that, for some reason, the expected salvage value is not realized. The most favorable situation is for this number to be as low as possible.

In situations where a sizable figure for salvage value shows up in the spreadsheet foot-totals, common business prudence demands that the write-off percentages be reviewed to be certain they are realistic so that the remaining salvage value total is a figure that can be relied on at the end of the project.

DETERMINATION OF EQUIPMENT MOBILIZATION AND DEMOBILIZATION COSTS

Immediately following completion of the equipment adjustment consideration by the type of analysis represented by Figure 12-2, it is convenient to compute the costs involved for equipment mobilization and demobilization. This logically follows because, as part of the equipment adjustment process, the details of how and from where the equipment will be procured are determined. These details are the drivers for the computation of mobilization and demobilization costs.

The separate mobilization and demobilization cost elements to be determined include:

- Freight costs for transport of the equipment to the project site
- Costs to unload the equipment components from railroad cars or trucks, to assemble the equipment, and to make the equipment ready to perform work
- Costs to break down the equipment at the completion of the project and load the equipment components onto railroad cars or trucks for shipping the equipment off the jobsite
- Freight cost for transport of the equipment off the jobsite

Another figure that is developed along with the computation of the mobilization/demobilization costs is the dollar figure for the total replacement value of those particular items of equipment that the contractor plans to include in the coverage of an equipment floater insurance policy.

The computations to determine all of the above can be conveniently made in spreadsheet format. Costs for the example project are calculated in this manner on Figure 12-3 based on the following hypothetical information.

Freight Costs for Transport of the Equipment to the Project Site

Assume that the equipment to be furnished in-house from the company's equipment department was located at another job nearing completion at Sacramento, California, and that the freight rate for shipment to the new jobsite was $4.50/cwt. Further, assume that the new equipment purchases were made on the basis of f.o.b. factory prices at Chicago, Illinois, and Pittsburgh, Pennsylvania, for which the freight rates for shipment to the jobsite were $7.50 and $10.50/cwt, respectively. The jobsite is in an urban area and equipment can be delivered to the job from local equipment rental entities in the area for a shipping charge of $1.00/cwt.

The normal bid estimate practice is to price into the estimate all equipment freight costs to the jobsite but to include only those freight costs necessary to return equipment rented from third parties. The reasoning is that the next project where project-owned equipment is to be employed will absorb the freight charges from the present project site to the new project location.

Computations reflecting the above in Figure 12-3 result in total freight-in and freight-out costs of $23,660 and $200, respectively.

Unload and Make-Ready Costs

Heavy construction equipment usually must be disassembled into smaller components before loading onto railroad cars or trucks for shipment to the jobsite. Upon arrival at the jobsite, the equipment must be unloaded, reassembled, and made ready to operate. This usually involves a crew of mechanics assisted by some sort of hoisting equipment which usually can be rented locally. Costs are usually priced on the basis of past experience.

For the example hypothetical project, it would not be unreasonable to price these costs on the basis of the following assumptions:

- Equipment costs will run about $800/shift
- Mechanic labor costs will run $1,200/shift
- The number of shifts required to unload, assemble, and make ready the various units will run as follows:

Equipment mobilization/demobilization: Detail

Description	No.	Source and location of equipment	Gross weight, cwt Unit	Gross weight, cwt Total	Freight in $ Unit	Freight in $ Total	Freight out $ Unit	Freight out $ Total	Unload and make ready — Equipment allowance $ Unit	Total	Labor $ Unit	Total	Dismantle and load out — Equipment allowance $ Unit	Total	Labor $ Unit	Total	Value to insure $ Unit	Total
100-ton crawler crane	1	Sacramento, CA	2,000	2,000	4.50	9,000			3,200	3,200	4,800	4,800	2,400	2,400	3,600	3,600	800,000	800,000
18-ton center-mount crane	3	Sacramento, CA	360	1,080	4.50	4,860			400	1,200	600	1,800	300	900	450	1,350	180,000	540,000
1200 cfm compressor	1	Sacramento, CA	150	150	4.50	675			200	200	300	300	150	150	225	225	30,000	30,000
	1	Locally	150	150	1.00	150	1.00	150	200	200	300	300	150	150	225	225	30,000	30,000
185 cfm compressor	3	Chicago, IL	50	150	7.50	1,125			100	300	150	450	75	225	115	345	11,000	33,000
	1	Locally	50	50	1.00	50	1.00	50	100	100	150	150	75	75	115	115	11,000	11,000
300-amp welding machine	4	Chicago, IL	50	200	7.50	1,500			100	400	150	600	75	300	115	460	10,000	40,000
Concrete pump	2	Pittsburgh, PA	300	600	10.50	6,300		—	400	800	600	1,200	300	600	450	900	205,000	410,000
Totals						23,660		200		6,400		9,600		4,800		7,220		1,894,000

FIGURE 12-3 Mobilization and demobilization costs and value to insure for example project after application of equipment adjustment.

100 ton crawler crane	Four shifts
18 ton center-mount crane	One-half shift
1200 cfm compressor	One-quarter shift
185 cfm compressor	One-eighth shift
400 A welding machine	One-eighth shift
Concrete pump	One-half shift

On this basis, the following foot-totals were developed in Figure 12-3:

| Total unload and make ready equipment costs | $6,400 |
| Total unload and make ready labor costs | $9,600 |

Dismantle and Load-Out Costs

Dismantle and load-out costs can be reasonably estimated to run about 75% of the unload and make-ready costs. The foot-totals in Figure 12-3 therefore are

| Total dismantle and load-out equipment costs | $4,800 |
| Total dismantle and load-out labor costs | $7,220 |

Value to Insure

The values to insure were developed in Figure 12-3 on the basis that all equipment required for the project would be insured at the full replacement value. Such insurance must generally be provided by the renter on the equipment obtained from third parties, and it is optional for project-owned equipment. Prudent business policy would require that most project-owned equipment be insured. Figure 12-3 reflects a total replacement value to insure of $1,894,000.

Transfer of Equipment Mobilization and Demobilization Costs and Value to Insure to the Indirect Cost Estimate

The reader will recall from Chapter 11 that the construction plant in-and-out section of the indirect cost estimate contains a subsection entitled construction equipment in-and-out. It is under this subsection that the foot-totals in Figure 12-3 for freight in, unload and assemble, dismantle and load out, and freight out are actually entered into the estimate.

Also, the replacement value of construction equipment to be insured that is needed to compute the equipment floater insurance policy premiums under the insurance and taxes section of the indirect cost estimate is obtained from the value to insure column of Figure 12-3.

CONCLUSION

This chapter has explained how, through the equipment adjustment process, the costs required to provide the required construction equipment for a project are finalized following the completion of the direct cost estimate. This process is based on all of the fully developed items of information about the project known at that point in time (information that would not be known at the time the estimate was set up).

Additionally, the chapter explained that this process also produces summary figures for the planned equipment usage on the job that are valuable in determining the relative attractiveness of the job, compared to other bid prospects, and which are otherwise necessary for the sound business management of a construction company.

Finally, the chapter explained the details of a concurrent analysis, made at the time of the equipment adjustment, to develop the costs for mobilizing the equipment on the project and demobilization at the end of the job and the replacement value of construction equipment that should be covered under the contractor's equipment floater insurance policy.

The details of the procedure to accomplish the above tasks were explained in terms of a simplified hypothetical project for which the necessary factual material normally developed in a direct cost estimate was assumed. The magnitude of the numbers developed was relatively small. On actual projects in heavy construction, which normally will be much larger, the magnitude of the equipment adjustment and the other developed foot-total numbers will be much larger, sometimes reaching the level of several millions of dollars.

Chapter 13, the final chapter in this text, is devoted to the procedure (and strategy) involved in pricing a schedule-of-bid-items bid form following completion of the entire estimate and determination of the total intended bid figure.

QUESTIONS AND PROBLEMS

1. What five modes of obtaining construction equipment for a project are discussed in this chapter?

2. When pricing equipment costs on a buy/sell basis, what two further assumptions must be made in addition determining the purchase and salvage values?

3. Four items of specific information about construction equipment requirements available at the end of a direct cost estimate are never known at the start of the estimate. What are they as mentioned in this chapter?

4. In addition to explaining the equipment adjustment, what are three additional determinations for which procedures are explained in this chapter?

5. Explain the "selective approach" to making the equipment adjustment discussed in this chapter. Why do many estimators employ this approach?

6. How does the equipment adjustment approach explained in this chapter afford an opportunity for evaluating alternate equipment procurement options?

7. What four highly significant foot-totals generated by the equipment adjustment process are discussed in this chapter? Discuss the significance of each from a business management standpoint.

8. What four equipment mobilization and demobilization cost elements are determined along with the equipment adjustment? In what part of the contractor's indirect cost estimate are these costs carried? What other information required for computation of costs to be entered in the indirect cost estimate is determined along with the equipment adjustment?

9. Complete the equipment adjustment sheet and equipment mobilization and demobilization sheet for the following lock and dam project in the format of

Figures 12-2 and 12-3. The summary figures (determined by others) for the total amount of project equipment (excluding the hoisting equipment) are as follows:

Column	Amount, $
Outside rental	275,000
In-house rental	0
Purchase	7,095,000
Salvage	3,891,000
Write-off	3,204,000
In estimate	3,434,300
Adjustment	+ 44,700
Freight in	75,100
Freight out	10,200
Unload and make ready equipment allowance	116,200
Unload and make ready labor allowance	185,600
Dismantle and load out equipment allowance	108,600
Dismantle and load out labor allowance	169,100
Value to insure	7,095,000

Enter the above values as line totals under the description "Balance of Job" on the first line of each of the two required spreadsheets. Then complete the analysis for the project by adding on succeeding lines your own analysis for the hoisting equipment consisting of one or more units of:

EQ 1: 165 ton crawler crane
EQ 2: 125 ton crawler crane
EQ 3: 25 ton center mount
EQ 4: 15 ton center mount
EQ 5: 225 ton crawler crane

Thus, by addition of these items, you will develop a final line total called "Total Project" for the summary figures for the project as a whole. The following other data also apply:

(a) The computer-generated equipment summary for the direct cost estimate shows the following information for each of the units of hoisting equipment to build the job.

Equipment	Operated hours	Rental in estimate, $
EQ 1	30,555	759,700
EQ 2	23,763	409,100
EQ 3	14,420	88,400
EQ 4	22,461	129,000
EQ 5	10,250	250,600

The bidding entity is a four-company joint venture for which your company is the sponsoring contractor.

(b) Assume that the joint venture principals have firmed up the equipment acquisition plans for the project as follows:

> EQ 1: Purchase new for $905,000 per unit at start of job.
>
> EQ 2: Purchase new for $795,000 per unit at start of job.
>
> EQ 3: Purchase new for $325,000 per unit at start of job.
>
> EQ 4: Purchase new for $265,000 per unit at start of job. (In addition, short-term third-party rental is available for 2.5% of the purchase cost per month.)
>
> EQ 5: Third-party rental at $32,500 per unit per month. (freight both ways for account of contractor renter)

The joint venture principals have also decided that all units are to be written off at the rate of 1.25% of the purchase cost for each month the unit is on the job irrespective of the degree of utilization of the unit.

(c) The freight rate for delivery of the purchased hoisting equipment from the f.o.b. point (factory) to the jobsite is $10.75/cwt. You may assume the weight of each unit as equal to the lifting capacity. The freight rate for rental equipment is $7.50/cwt.

(d) The jobsite crew of mechanics who will unload, assemble, and make ready the equipment will cost

Equipment allowance	$1,000.00 per shift
Labor	$1,600.00 per shift

Each unit will require the following number of mechanic crew shifts to unload, assemble, and make the unit ready to work:

> EQ 1: Six shifts
>
> EQ 2: Four shifts
>
> EQ 3: One-half shift
>
> EQ 4: One-half shift
>
> EQ 5: Seven shifts

The required mechanic crew shifts for dismantle and load-out can be taken as three-quarters of the unload, assemble, and make ready mechanic crew shifts.

(e) Refer to the project timeline. The anticipated utilization of these cranes to perform the project work is as follows:

> EQ 1: Used for excavation, concrete, and stone work
>
> EQ 2: Used for concrete work only
>
> EQ 3: Used for general service throughout the job
>
> EQ 4: Used for general service throughout the job
>
> EQ 5: Used for gate and machinery erection only

(f) The value to insure for each crane is the purchase cost. The purchase cost of the 225 ton crane is $1,273,000.

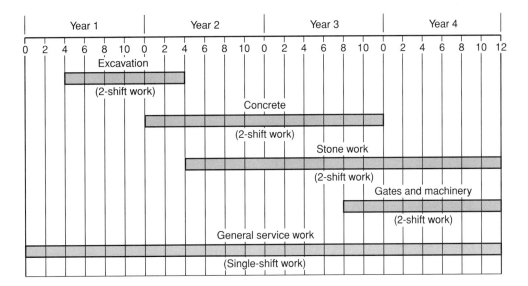

13

Pricing the Bid Form
for Schedule-of-Bid-Item Bids

Key Words and Concepts

Fixed project costs and markup

Distribution based on total estimated direct cost

Distribution based on estimated direct labor

Balanced bid

The effect of inaccurate bid quantities on a balanced bid

Overstating bid quantities to create a hidden contingency

Protection against loss due to overstated bid quantities

Unbalanced bid

Rejection of unbalanced bid as nonresponsive

Front-end loading

Redistribution of fixed costs and markup

When heavy construction project cost estimates have been totally completed and each line item in the vertical estimate format making up the intended bid figure has been finalized, as explained in Chapter 3, the bid form that is to be submitted to the owner must be priced. That is, the intended bid that will determine precisely how the contractor will be paid for performing the project work must be fully and completely set down in writing on the owner's preprinted form, strictly according to the instructions published in the instructions to bidders section of the bid documents.

As Figure 3-5 in Chapter 3 illustrates, there normally is very little the bidder has to do to price the bid form for lump-sum bids (as that term is defined in Chapter 3) other than to enter the intended bid figure in the appropriate blank on the bid form.[1] However, pricing the bid form for schedule-of-bid-item bids (see Figure 3-6, Chapter 3) is a complex matter involving strategic, legal, and ethical issues in addition to the extensive numerical manipulations required. This pricing process is the subject of this chapter.

DISTRIBUTION OF FIXED PROJECT COSTS AND BID MARKUP TO INDIVIDUAL BID ITEMS

Figure 3-6 in Chapter 3 indicates, that **fixed project costs and markup** must be distributed to the individual bid items and then added to the previously estimated direct cost totals to determine the total amount that will constitute the bidder's bid for each bid item. In other words, the total amount to be distributed, D_t in Figure 3-6 consisting of

- The indirect costs
- Labor escalation
- Material escalation
- Equipment escalation
- Interest
- Contingency
- Bond
- Markup

must be distributed in the amounts D_1, D_2, D_3, D_4, and so on. How does one make this distribution? Out of several possible methods, the following two are commonly used:

- Distribution based on total estimated direct cost
- Distribution based on estimated direct labor

Distribution Based on Total Estimated Direct Cost

Consider the general estimate summary sheet depicted in Figure 13-1 for a hypothetical bridge substructure project representing an intended bid figure of $14,395,100. The direct costs totaled $8,348,700, whereas the sum of the fixed[2] line totals to be distributed was

---

[1] Some lump-sum bids may require that additional pricing details be entered on the bid form as well, such as any alternate bid totals, and unit prices for a few specifically designated quantity additions or deletions, but the amount of such supplementary pricing is usually not extensive. Occasional lump-sum bids may involve additional complexity but not to the extent required for schedule-of-bid-item bids.

[2] These line totals are considered to be "fixed" because they are independent from and do not directly relate to the individual bid item quantities.

Bid item no.	Description	Unit	Quantity	EOE	RL	R	L	EM	PM	S	Total direct cost	Distribution	Total bid
1	Earth excavation	cy	13,650							55,100	55,100	39,905	95,005
2	Rock excavation	cy	5,175							84,000	84,000	60,836	144,836
3	Cofferdam excavation	cy	14,150	12,100	4,200	13,200	25,500	1,200			56,200	40,702	96,902
4	Cofferdam #1	LS[(1)]	1	3,400	2,900	21,800	41,500	16,200			85,800	62,139	147,939
5	Cofferdam #2	LS	1	5,000	3,400	30,300	58,700	23,100			120,500	87,270	207,770
6	Cofferdam #3	LS	1	6,300	4,700	32,200	63,900	29,200			136,300	98,713	235,013
7	Cofferdam #4	LS	1	8,600	6,100	72,100	106,200	38,200			231,200	167,443	398,643
8	Cofferdam #5	LS	1	71,700	39,400	88,900	256,300	702,500			1,158,800	839,241	1,998,041
9	Cofferdam #6	LS	1	12,300	7,300	76,400	162,700	187,500			446,200	323,153	769,353
10	Furnish piling	lf	36,475						346,800		346,800	251,164	597,964
11	Drive piling	lf	34,775	141,900	63,300	107,100	290,900	27,200			630,400	456,556	1,086,956
12	Tremie concrete	cy	6,350	36,500	18,700	37,100	188,000	91,200	496,800		818,300	592,639	1,410,939
13	Footing concrete	cy	6,175	68,200	39,900	76,400	493,900	186,200	291,500		1,156,100	837,285	1,993,385
14	Column and web wall concrete	cy	5,325	69,200	40,500	77,600	501,300	189,000	258,600		1,136,200	822,873	1,959,073
15	Column concrete	cy	2,160	32,300	18,900	36,200	234,100	88,300	98,700		508,500	368,272	876,772
16	Cap concrete	cy	1,895	34,800	20,400	38,900	251,900	94,900	84,600		525,500	380,584	906,084
17	Reinforcing steel	lb	2,150,200	24,100	18,200	43,400	248,700	43,600	447,800		825,800	598,071	1,423,871
18	Stone rip rap	ton	1,500							27,000	27,000	19,554	46,554
	Total direct cost			526,400	287,900	751,600	2,873,600	1,718,300	2,024,800	166,100	8,348,700	6,046,400	14,395,100
	Indirect cost			149,100	51,400	55,800	1,017,000	1,183,600	2,024,800	206,900	2,663,800		
	Subtotal			675,500	339,300	807,400	3,890,600	2,901,900	2,024,800	373,000	11,012,500		
	Labor escalation				25,500		291,800				317,300		
	Equipment escalation			40,500							40,500		
	Interest							325,000			325,000		
	High-water contingency			22,800	6,900	6,300	99,000	166,200			301,200		
	Total job costs (w/o bond)			738,800	371,700	813,700	4,281,400	3,393,100	2,024,800	373,000	11,996,500		
	Markup (50% labor)										2,327,000		
	Total bid (w/o bond)										14,323,500		
	Bond (1/2 of 1% of bid)										71,600		
	Total bid										14,395,100		

[(1)]LS = lump sum.

FIGURE 13-1 Example of general estimate summary with distribution of indirects, profit, contingency, etc., to bid items on basis of estimated total direct cost.

Indirect costs	$2,663,800
Labor escalation	$317,300
Equipment escalation	$40,500
Interest	$325,000
High water contingency	$301,200
Bond	$71,600
Markup	$2,327,000
Total	$6,046,400

This amount was distributed to the bid items by the ratio of the total estimated cost of each bid item to the total $8,348,700 direct cost figure. In other words,

$$\text{Distribution to bid item 1} = \left(\frac{55,100}{8,348,700}\right) \times \$6,046,400 = \$39,905$$

$$\text{Distribution to bid item 2} = \left(\frac{84,000}{8,348,700}\right) \times \$6,046,400 = \$60,836$$

and so on. By adding the distributed amounts to the previously estimated total direct cost, a total bid figure is obtained for each bid item. Note that on Figure 13-1 the foot-total of the distributed amounts equals the original $6,046,400 to be distributed, and the foot-total for the total bid column equals the intended bid figure of $14,395,100.

Although widely used, and viewed by some as a "balanced" distribution, this method has the serious disadvantage of assigning the same dollar-for-dollar share of the distributed monies to a subcontracted bid item, or to a bid item consisting largely of the purchase of permanent materials, as it does to bid items requiring a great deal of contractor-provided labor and heavy equipment operation for the performance of significant construction work. This is not realistic because very little of the costs or markup making up the distributed amount is associated with managing subcontracted bid items or administering purchase orders for permanent materials. This method of distribution greatly exaggerates the true value of such bid items compared to bid items requiring sufficient contractor work effort and involving significant monetary risk of performance.

Distribution Based on Estimated Direct Labor

This method is the same as distribution based on total direct costs except that distributed amounts for each bid item are based on the ratio of the estimated direct labor in the bid item to the estimated labor total in the total direct cost. Thus,

$$\text{Distribution to bid item 3} = \left(\frac{25,500}{2,873,600}\right) \times \$6,046,400 = \$53,655$$

$$\text{Distribution to bid item 4} = \left(\frac{41,500}{2,873,600}\right) \times \$6,046,400 = \$87,321$$

and so on. The distribution to any bid item not containing a labor cost component is zero. Figure 13-2 shows all of the distributions to bid items based on estimated direct labor. Once again, note that the foot-totals for the total distribution and total bid columns are $6,046,400 and $14,395,100, respectively.

Bid item no.	Description	Unit	Quantity	EOE	RL	R	L	EM	PM	S	Total direct costs, $	Distribution	Total bid
1	Earth excavation	cy	13,650							55,100	55,100		55,100
2	Rock excavation	cy	5,175							84,000	84,000		84,000
3	Cofferdam excavation	cy	14,150	12,100	4,200	13,200	25,500	1,200			56,200	53,654	109,854
4	Cofferdam #1	LS [1]	1	3,400	2,900	21,800	41,500	16,200			85,800	87,321	173,121
5	Cofferdam #2	LS	1	5,000	3,400	30,800	58,700	23,100			120,500	123,512	244,012
6	Cofferdam #3	LS	1	6,300	4,700	32,200	63,900	29,200			136,300	134,453	270,753
7	Cofferdam #4	LS	1	8,600	6,100	72,100	106,200	38,200			231,200	223,458	454,658
8	Cofferdam #5	LS	1	71,700	39,400	88,900	256,300	702,500			1,158,800	539,286	1,698,086
9	Cofferdam #6	LS	1	12,300	7,300	76,400	162,700	187,500			446,200	342,340	788,540
10	Furnish piling	lf	36,475						346,800		346,800		346,800
11	Drive piling	lf	34,775	141,900	63,300	107,100	290,900	27,200			630,400	612,089	1,242,489
12	Tremie concrete	cy	6,350	36,500	18,700	37,100	138,000	91,200	496,800		818,300	290,369	1,108,669
13	Footing concrete	cy	6,175	68,200	39,900	76,400	493,900	186,200	291,500		1,156,100	1,039,225	2,195,325
14	Column and web wall concrete	cy	5,325	69,200	40,500	77,600	501,300	189,000	258,600		1,136,200	1,054,795	2,190,995
15	Column concrete	cy	2,160	32,300	18,900	36,200	234,100	88,300	98,700		508,500	492,575	1,001,075
16	Cap concrete	cy	1,895	34,800	20,400	38,900	251,900	94,900	84,600		525,500	530,028	1,055,528
17	Reinforcing steel	lb	2,150,200	24,100	18,200	43,400	248,700	43,600	447,800		825,800	523,295	1,349,095
18	Stone rip rap	ton	1,500							27,000	27,000		27,000
	Total direct cost			526,400	287,900	751,600	2,873,600	1,718,300	2,024,800	166,100	8,348,700	6,046,400	14,395,100
	Indirect cost			149,100	51,400	55,800	1,017,000	1,183,600		206,900	2,663,800		
	Subtotal			675,500	339,300	807,400	3,890,600	2,901,900	2,024,800	373,000	11,012,500		
	Labor escalation				25,500		291,800				317,300		
	Equipment escalation			40,500							40,500		
	Interest							325,000			325,000		
	High-water contingency			22,800	6,900	6,300	99,000	166,200			301,200		
	Total job costs (w/o bond)			738,800	371,700	813,700	4,281,400	3,393,100	2,024,800	373,000	11,996,500		
	Markup (50% labor)										2,327,000		
	Total bid (w/o bond)										14,323,500		
	Bond (1/2 of 1% of bid)										71,600		
	Total bid										14,395,100		

[1]LS = lump sum.

FIGURE 13-2 Example general estimate summary with distribution of indirects, profit, contingency, etc., to bid items on basis of estimated direct labor.

Distribution by this method properly reflects a rational apportionment of the various distributed costs to those bid items where the required contractor performance closely relates to the need for most of the various indirect cost line items as well as for the escalation, interest, contingency, and bond costs and for the markup. This is so because the estimated labor figure in each bid item is an accurate indicator of where performance of work with the contractor's own forces will occur on the project. In heavy construction, it is the bid items requiring performance of project work with the contractor's own forces that need the support of the indirect costs and the other distributed costs and that relate to the profit and risk coverage elements reflected in the markup, rather than the bid items for subcontracted work and the furnishing of permanent materials.

Determination of Bid Item Unit Prices and Extended Bid Totals

Figure 13-3 reflects the completed general estimate summary sheet with all bid item unit prices and extensions in the final bid column[3] determined by application of the following described procedure.

For each lump-sum bid item (see Chapter 3), the unit price is simply the previously determined amount in the total bid column for that bid item. Because the quantity is 1, the final bid amount is that same figure. This amount represents the amount of money that the contractor will be paid for performing the total work shown on the bid drawings and/or described in the specifications for that particular lump-sum bid item (again, see Chapter 3 for examples for how lump-sum bid items are described on bid forms).

However, each unit price bid item will require the determination of a discrete unit bid price that, when multiplied by the owner's bid quantity, will extend to the intended bid figure for that quantity of work.[4]

The procedure used to determine the unit price and final bid figures on Figure 13-3 can be illustrated by considering a typical unit price bid item, say bid item 3, for example. The first step is to obtain a trial unit price:

$$\text{Total bid (including distributed amounts)} = \$109,854$$

$$\text{Bid quantity} = \$14,150 \text{ cy}$$

$$\text{Trial unit price } \frac{\$109,854}{14,150 \text{ cy}} = \$7.7635/\text{cy}$$

This trial unit price is not an integer (an even ratio of two whole numbers) and when multiplied by the bid quantity will not produce an extension in even dollars:[5]

$$(14,150 \text{ cy})(\$7.7635) = \$109,853.5250 \ldots$$

---

[3]The distributed amounts in Figure 13-3 were determined by distribution based on direct labor. From this point onward in this chapter, no further consideration will be given to distribution to bid items by the total direct cost method.

[4]These quantities will not necessarily be the quantities of work performed when the project is built, but if they have been accurately determined from the bid drawings by the owner's engineer when the bid documents are created, and the work of the project is not changed, the differences in the quantities of work actually performed from the bid quantities will not be large.

[5]For general convenience in determining final bid totals for comparing competitive bids, all final bid extensions should be kept in even dollars.

Bid item no.	Description	Unit	Quantity	EOE	RL	R	L	EM	PM	S	Total direct cost	Distribution	Total bid[1]	Final bid[1] Unit price	Final bid
1	Earth excavation	cy	13,650							55,100	55,100		55,100	4.04	55,146
2	Rock excavation	cy	5,175							84,000	84,000		84,000	16.24	84,042
3	Cofferdam excavation	cy	14,150	12,100		13,200	25,500	1,200			56,200	53,654	109,854	7.76	109,804
4	Cofferdam #1	LS[2]	1	3,400	2,900	21,800	41,500	16,200			85,800	87,321	173,121	173,121.00	173,121
5	Cofferdam #2	LS	1	5,000	3,400	30,300	58,700	23,100			120,500	123,512	244,012	244,012.00	244,012
6	Cofferdam #3	LS	1	6,300	4,700	32,200	63,900	29,200			136,300	134,453	270,753	270,753.00	270,753
7	Cofferdam #4	LS	1	8,600	6,100	72,100	106,200	38,200			231,200	223,458	454,658	454,658.00	454,658
8	Cofferdam #5	LS	1	71,700	39,400	88,900	256,300	702,500			1,158,800	539,286	1,698,086	1,698,086.00	1,698,086
9	Cofferdam #6	LS	1	12,300	7,300	76,400	162,700	187,500			446,200	342,340	788,540	788,540.00	788,540
10	Furnish piling	lf	36,475						346,800		346,800		346,800	9.52	347,242
11	Drive piling	lf	34,775	141,900	63,300	107,100	290,900	27,200			630,400	612,089	1,242,489	35.72	1,242,163
12	Tremie concrete	cy	6,350	36,500	18,700	37,100	138,000	91,200	496,800		818,300	290,369	1,108,669	174.60	1,108,710
13	Footing concrete	cy	6,175	68,200	39,900	76,400	493,900	186,200	291,500		1,156,100	1,039,225	2,195,325	355.52	2,195,336
14	Column and web wall concrete	cy	5,325	69,200	40,500	77,600	501,300	189,000	258,600		1,136,200	1,054,795	2,190,995	411.44	2,190,918
15	Column concrete	cy	2,160	32,300	18,900	36,200	234,100	88,300	98,700		508,500	492,575	1,001,075	463.45	1,001,052
16	Cap concrete	cy	1,895	34,800	20,400	38,900	251,900	94,900	84,600		525,500	530,028	1,055,528	557.00	1,055,515
17	Reinforcing steel	lb	2,150,200	24,100	18,200	43,400	248,700	43,600	447,800		825,800	523,295	1,349,095	0.63	1,354,626
18	Stone rip rap	ton	1,500							27,000	27,000		27,000	18.00	27,000
	Total direct cost			526,400	287,900	751,600	2,873,600	1,718,300	2,024,800	166,100	8,348,700	6,046,400	14,395,100		14,400,724
	Indirect cost			149,100	51,400	55,800	1,017,000	1,183,600		206,900	2,663,800				14,395,100
	Subtotal			675,500	339,300	807,400	3,890,600	2,901,900	2,024,800	373,000	11,012,500				
	Labor escalation				25,500		291,800				317,300				
	Equipment escalation			40,500							40,500				
	Interest							325,000			325,000				
	High-water contingency			22,800	6,900	6,300	99,000	166,200			301,200				
	Total job costs (w/o bond)			738,800	371,700	813,700	4,281,400	3,393,100	2,024,800	373,000	11,996,500				
	Markup (50% labor)										2,327,000				
	Total bid (w/o bond)										14,323,500				
	Bond (1/2 of 1% of bid)										71,600				5,624
	Total bid										14,395,100				

Corrected bid figures

	Total bid
Bid item 8	14,400,724
1,698,086	−5,624
−5,624	14,395,100
1,692,462	

[1] Balanced on labor.
[2] LS = lump sum.

FIGURE 13-3 Example of general estimate summary with balanced bid item unit prices and extended bid totals (distribution based on estimated labor).

Therefore, the estimator finds by trial and error a unit price as close as possible to the trial unit price (as determined above) that is an integer and that will produce an extension in even dollars. For instance, for $7.76

$$(14,150 \text{ cy})(\$7.76) = \$109,804$$

The extension $109,804 is in even dollars, so the unit price of $7.76 and the $109,804 extension figures are entered in the final bid columns.

When all the bid item final bid figures have been determined in this manner and totaled, the resulting figure will usually not equal the intended bid figure exactly. For instance, the discrepancy on Figure 13-3 amounts to $14,400,724 − $14,395,100, or $5,624. If the bidder is not concerned by raising the bid total to the $14,400,724 figure, nothing further need be done. Alternately, a correction of $−5,624 can be made to one of the lump-sum bid items and to the bid total, restoring the latter to the intended bid figure of $14,395,100 as is shown in the lower right-hand corner of Figure 13-3.

When all unit prices and final bid totals have been worked out in this manner, the unit prices and final bid figures and final bid total on Figure 13-3 (after any necessary adjustments necessary to reflect any changes in assumed permanent material and sub-contract prices due to final quotations received just prior to bid) are ready to be hand-written into the owner's bid form, *provided* that the contractor is willing to submit the bid on the above described "balanced" basis.[6]

THE EFFECT OF INACCURATE BID QUANTITIES ON A BALANCED BID

Unfortunately, the bid-item quantities shown on the owner's bid form are not always accurate. That is, quantities for important bid items constituting significant portions of the project work may differ from the pay quantities determined by the contractor's takeoff from the bid drawings that was performed strictly in accordance with the measurement and payment provisions in the specifications. When these differences are large, the impact on the contractor's financial position can be devastating if the contractor has submitted a **balanced bid.**

Why would the bid-item quantities stated on the bid form be inaccurate? One reason is that the design of heavy construction projects normally proceeds through a series of iterations—from preliminary design, to 50% design, to 75% design, to 90%, and then to the final design advertised for bids. Bid quantities associated with each iteration may or may not be revised each time a new iteration emerges in the design process during the preparation of the bidding documents. If quantity provisions are not carefully kept up to date, and quantities corresponding to the final design are not carefully checked, the advertised bid quantities can be seriously in error. Also, some owners and/or their engineers who intend to serve as the construction manager for the owner during construction, artificially create a **hidden contingency,** or **"cushion"** in the total bid price by **deliberately overstating** important quantities. This cushion provides funding for unexpected change order costs during construction for items in the design that have either been inadvertently omitted or inadequately designed.

The reader should remember that when the work of a bid item is actually performed, the contractor will be paid the bid unit price times the quantity of work that is

---

[6]There is no universal understanding that is accepted in the heavy construction industry as to exactly what a "balanced bid" is. However, the writer believes that most practitioners would consider the bid represented by Figure 13-3 to be a balanced bid.

measured for payment in the field. So, if the bid quantity has been overstated, the contractor will not receive full payment of the amount of money that was distributed to the bid item before the unit price was worked out. Because this distributed money is not directly related to the bid quantities and represents costs that are going to be expended by the contractor whether the bid quantities develop or not, the contractor will suffer an out-of-pocket loss. Fortunately, the practice of deliberately overstating bid quantities for the purpose explained above is relatively rare.

Assume, for example, that the contractor's takeoff for the pay quantities in Figure 13-3 compared to the bid quantities was as follows:

Bid item	Description	Unit	Bid quantity	Takeoff quantity
1	Earth excavation	cy	13,650	13,625
2	Rock excavation	cy	5,175	5,205
3	Cofferdam excavation	cy	14,150	14,165
4	Cofferdam 1	LS	1	1
5	Cofferdam 2	LS	1	1
6	Cofferdam 3	LS	1	1
7	Cofferdam 4	LS	1	1
8	Cofferdam 5	LS	1	1
9	Cofferdam 6	LS	1	1
10	Furnish piling	lf	36,475	32,512
11	Drive piling	lf	34,775	30,512
12	Tremie concrete	cy	6,350	6,055
13	Footing concrete	cy	6,175	5,820
14	Column and web wall concrete	cy	5,325	4,710
15	Column concrete	cy	2,160	1,892
16	Cap concrete	cy	1,895	1,515
17	Reinforcing steel	lb	2,150,200	1,705,525
18	Stone rip rap	ton	1,500	1,490

Based on the contractor's assessment of the pay quantities, the company would suffer the out-of-pocket loss shown by Figure 13-4 even if all the estimated costs were achieved in project performance. In other words, the $2,327,000 markup the contractor would otherwise have received would be decreased by $866,799 to $1,460,201. Although bid quantities that are overstated to this extent may not be common, the writer has occasionally encountered situations similar to the above hypothetical example. Of course, bid quantities can also be understated, resulting in the opposite effect on the contractor's bottom line.

How Can the Contractor Protect Against Monetary Loss Due to Overstated Bid Quantities?

If the contractor has performed a careful takeoff of the pay quantities from the drawings for all important bid items, according to the measurement and payment provisions in the specifications, any bid quantities that are seriously overstated will be readily apparent. Because part of any monies (representing fixed costs and the markup) distributed to such bid items will be irretrievably lost when the actual work is performed, the contractor should bid such bid items no higher than the direct cost level. The same rule

Bid item	Description	Unit	Bid quantity	Final pay quantity	Distributed amount, $	Contractor will lose, $		
11	Drive piling	lf	34,775	30,512	612,089	$(612,089)(1 - 30,512/34,775)$	=	75,035
12	Tremie concrete	cy	6,350	6,055	290,369	$(290,369)(1 - 6,055/6,350)$	=	13,490
13	Footing concrete	cy	6,175	5,820	1,039,225	$(1,039,225)(1 - 5,820/6,175)$	=	59,745
14	Column and web wall concrete	cy	5,325	4,710	1,054,795	$(1,054,795)(1 - 4,710/5,325)$	=	121,821
15	Column concrete	cy	2,160	1,892	492,575	$(492,575)(1 - 1,892/2,160)$	=	382,202
16	Cap concrete	cy	1,895	1,515	530,028	$(530,028)(1 - 1,515/1,895)$	=	106,285
17	Reinforcing steel	lb	2,150,200	1,705,525	523,295	$(523,295)(1 - 1,705,505/2,150,200)$	=	108,221
						Total loss	=	866,799

FIGURE 13-4 Impact of overstated bid-item quantities for example project.

should apply to bid items that, for one reason or another, the contractor suspects may be subject to deletion by the owner.

The above in effect means that monies that would be distributed by a balanced bid to overstated bid items (or to bid items that may be deleted) should instead be distributed to other items, either to unit price items where the quantities are not overstated or to lump-sum items. This in turn means that the submitted bid will no longer be balanced and, semantically speaking, will therefore be unbalanced.

Rejection of Unbalanced Bids as Nonresponsive

Bidding documents often contain provisions to the effect that unbalanced bids may be rejected as nonresponsive. Sometimes these provisions contain words like "grossly unbalanced" or "substantially unbalanced." What do these provisions mean in practice? The following questions logically arise:

1. Does the redistribution of the contractor's fixed costs and markup to protect out-of-pocket loss constitute "unbalancing" that might result in an otherwise responsive low bid being rejected as nonresponsive?
2. Does the uneven distribution of fixed costs and markup favoring bid items that are to be performed early in the job constitute "unbalancing" that might result in an otherwise low bid being rejected as nonresponsive?
3. Are there other contractor motivations for uneven distribution that are likely to result in the bid being rejected as nonresponsive?

In considering these questions the reader should first realize that the owner normally will never see the contractor's general estimate summary sheet showing the distributed fixed costs and the development of the bid-item unit prices and extended totals. The owner will see only the bid form on which the bid-item unit prices and the extended totals have been written in. Further, these unit prices and extended bid totals will vary widely, even among bidders whose total bids are nearly the same. Also, they usually vary widely from those in the engineer's estimate. Thus, unless the bidder's distribution of fixed costs and markup has been made in a manner that results in such grossly distorted unit prices and extensions that they "really stand out in a crowd," the owner will have a difficult time in concluding how particular bidders made their distributions.

In answer to question 1 above, the writer has never seen or heard of a bid rejected as nonresponsive because the contractor had redistributed fixed costs and markup to avoid monetary loss that would otherwise have occurred because of overstated bid quantities. Nor is he aware of a case where such a bid was even questioned by an owner. Also, owners are usually reasonably sympathetic to contractor's distributing higher proportions of distributed monies to bid items expected to be completed early in the construction schedule than to those occurring later[7] (question 2). Owners naturally resist, however, obvious attempts by bidders to reap a financial windfall by grossly overloading with distributed monies a bid item that is seriously understated (question 3).

To ensure that the bid will not be declared nonresponsive and thus rejected, the distribution of fixed costs and the markup must be reasonable, as more fully explained below.

Example Redistribution of Fixed Costs and Markup

The first thing a bidder should do after encountering major differences in bid quantities when compared to their takeoff is to notify the owner of the differences and request that the bid quantities be corrected by addendum. If the owner declines to issue an addendum to correct the bid quantities,[8] the bidder must either decline to bid or deal with the situation in a manner that will protect the bidder's ability to recover fixed costs and markup when the actual work is performed. The procedure to accomplish this latter objective, as well as the reasoning supporting this procedure, can best be illustrated by an example.

Assume the major work in the example bridge substructure project consists of the following sequential work activities:

- Constructing six sheet-pile cofferdams in the river.
- Underwater excavation within the cofferdams to the bottom elevation of the underwater concrete tremie seals (to be placed later).
- Driving bearing piling inside the cofferdams to tip-bearing on rock. The tops of the piling are required to be cut off at a specified elevation to support the permanent bridge piers.
- Pouring the underwater concrete tremie seals.
- Dewatering the cofferdams.
- Constructing the permanent concrete pier within the cofferdam. The permanent pier components are
  Footing concrete
  Column and web wall concrete
  Column concrete
  Cap concrete
- Removing the temporary cofferdams.

Also, assume that

- Earth and rock excavation operations will be performed early in the job.
- Cofferdams will be built and removed in pairs. Cofferdams 5 and 6 will be built together and then removed (at a point when the column and web wall concrete

---

[7]This practice, known as **front-end loading,** is commonly employed to improve contractor cash flow and thus reduce interest costs.

[8]The owner may be reluctant to make those kinds of addendum changes due to limited time before bid opening (or for other reasons).

have been placed), and the sheet pile and bracing material will then be used to build cofferdams 3 and 4. When cofferdams 3 and 4 are removed (following completion of column and web wall concrete), the sheet pile and bracing material will be used to build cofferdams 1 and 2.

Finally, within each cofferdam, cofferdam excavation will be performed first, followed by driving piling, followed by tremie concrete, followed by the remaining concrete items.

As previously discussed, the bid quantities have been seriously overstated for bid items 10 through 17. The balance of the bid-item quantities compare very closely with the contractor's takeoff. To avoid the out-of-pocket loss indicated by Figure 13-4, the contractor should bid bid items 11 through 17 at a level no higher than the estimated direct cost for these items. In this manner, none of the fixed costs and markup that would normally have been distributed to these items will be lost when the project is actually built.[9] Therefore, the difference between the sum of the direct cost estimates for bid items 10 through 17 and the previously determined total bid figures based on labor (see Figure 13-3) should be redistributed to the bid items where the quantities are not overstated (or at least not seriously overstated).

In the hypothetical bridge project, the total number of bid items is not large (18 bid items), so the redistribution can easily and quickly be made to all of the remaining bid items. For projects with more bid items, the redistribution would not necessarily have to be made to all of the remaining items. It would be equally satisfactory to select a limited number of the major remaining bid items (i.e., those with large dollar bid extensions), particularly including the larger lump-sum items.

The redistribution calculations for the hypothetical bridge project are shown in Figure 13-5, which indicates a total of $4,547,962 to be redistributed to bid items 1 through 9 and to bid item 18. The redistribution calculations are also shown.

BI	Final bid balanced on labor, $[1]	Direct cost, $[1]	BI	Original total bid, $[1]	Redistributed, $	New total bid, $
10	347,242	346,800	1	55,146	64,223	119,369
11	1,242,163	630,400	2	84,042	97,876	181,918
12	1,108,710	818,300	3	109,804	127,878	237,682
13	2,195,336	1,156,100	4	173,121	201,617	374,738
14	2,190,918	1,136,200	5	244,012	284,177	528,189
15	1,001,052	508,500	6	270,753	315,320	586,073
16	1,055,515	525,500	7	454,658	529,496	984,154
17	1,354,626	825,800	8	1,698,086	1,977,595	3,675,681
Totals	10,495,562	5,947,600	9	788,540	918,336	1,706,876
	−5,947,600		18	27,000	31,444	58,444
	4,547,962		Totals	3,905,162	4,547,962	8,453,124
	(To be redistributed)					

[1]From Figure 13-3.

FIGURE 13-5 Redistribution calculations for example project.

[9]Bid item 10 is already at the direct cost level because no labor was required for the work of that bid item and no distribution of fixed costs or markup was made to that item.

Bid item no.	Description	Unit	Quantity	Total direct costs	Bid balanced on labor				Adjusted bid		
					Distribution	Total bid	Unit price	Final bid	Total after redistribution[1]	Unit price	Final bid
1	Earth excavation	cy	13,650	55,100		55,100	4.04	55,146	119,369	8.74	119,301
2	Rock excavation	cy	5,175	84,000		84,000	16.24	84,042	181,918	35.12	181,746
3	Cofferdam excavation	cy	14,150	56,200	53,654	109,854	7.76	109,804	237,682	16.80	237,720
4	Cofferdam #1	LS [2]	1	85,800	87,321	173,121	173,121.00	173,121	374,738	374,740.00	374,740
5	Cofferdam #2	LS	1	120,500	123,512	244,012	244,012.00	244,012	528,189	528,192.00	528,192
6	Cofferdam #3	LS	1	136,300	134,453	270,753	270,753.00	270,753	586,073	586,076.00	586,076
7	Cofferdam #4	LS	1	231,200	223,458	454,658	454,658.00	454,658	984,154	984,159.00	984,159
8	Cofferdam #5	LS	1	1,158,800	539,286	1,698,086	1,698,086.00	1,698,086	3,675,681	3,675,701.00	3,675,701
9	Cofferdam #6	LS	1	446,200	342,340	788,540	788,540.00	788,540	1,706,876	1,706,884.00	1,706,884
10	Furnish piling	lf	36,475	346,800		346,800	9.52	347,242	346,800	9.52	347,242
11	Drive piling	lf	34,775	630,400	612,089	1,242,489	35.72	1,242,163	630,400	17.60	612,040
12	Tremie concrete	cy	6,350	818,300	290,369	1,108,669	174.60	1,108,710	818,300	128.86	818,261
13	Footing concrete	cy	6,175	1,156,100	1,039,225	2,195,325	355.52	2,195,336	1,156,100	187.20	1,155,960
14	Column and web wall concrete	cy	5,325	1,136,200	1,054,795	2,190,995	411.44	2,190,918	1,136,200	213.36	1,136,142
15	Column concrete	cy	2,160	508,500	492,575	1,001,075	463.45	1,001,052	508,500	235.40	508,464
16	Cap concrete	cy	1,895	525,500	530,028	1,055,528	557.00	1,055,515	525,500	277.40	525,673
17	Reinforcing steel	lb	2,150,200	825,800	523,295	1,349,095	0.63	1,354,626	825,800	0.40	860,080
18	Stone rip rap	ton	1,500	27,000		27,000	18.00	27,000	58,444	39.00	58,500
	Total direct cost			8,348,700	6,046,400	14,395,100		14,400,724			14,416,881
	Indirect cost			2,663,800				14,395,100			−14,395,100
								5,624			21,781
	Subtotal			11,012,500							
	Labor escalation			317,300							
	Equipment escalation			40,500							
	Interest			325,000							
	High-water contingency			301,200							
	Total job costs (w/o bond)			11,996,500							
	Markup (50% labor)			2,327,000							
	Total bid w/o bond			14,323,500							
	Bond (1/2 of 1% of bid)			71,600							
	Total bid			14,395,100							

Corrected bid figures (Bid balanced on labor):

Bid item 8	Total bid
1,698,086	14,400,724
−5,624	−5,624
1,692,462	14,395,100

Corrected bid figures (Adjusted bid):

Bid item 8	Total bid
3,675,701	14,416,881
−21,781	−21,781
3,653,920	14,395,100

[1] From Figure 13-5.

[2] LS = lump sum.

FIGURE 13-6 Revised bid item for example project.

378

The adjusted bid unit prices and bid extensions for all of the bid items determined after the redistribution are shown in Figure 13-6, as are the original unit prices and bid extensions balanced on labor.

In the writer's experience, a bid submitted in accordance with the adjusted bid figures shown on Figure 13-6 would not be questioned by the owner. If the bid were questioned, the contractor would easily be able to justify it by revealing and discussing the information shown on Figures 13-3, 13-4, 13-5, and 13-6 with the owner.

CONCLUSION

This final chapter in this book has presented the principles involved in pricing out the bid form for schedule-of-bid-item bids, emphasizing the importance of reviewing the accuracy of the owner's bid quantities to protect distributed fixed costs and bid markup.

Once this process has been completed, the final bid-item unit prices and extensions are ready to be entered into the bid form that will be submitted to the owner, subject only to any final adjustments necessary to reflect changes in assumed permanent material and subcontract prices after receipt of final quotations just prior to bid.

The bid form submitted to an owner calling for bids is an important legal document. In a sense, this entire book has been devoted to explaining the construction operations analysis and cost estimating procedures involved in its preparation for heavy construction projects.

QUESTIONS AND PROBLEMS

1. What does it mean to "price" a bid form? What does pricing the bid form normally amount to for lump-sum bids? What three kinds of issues are involved in pricing schedule-of-bid-item bid forms?

2. What seven categories of project costs (in addition to the markup) must be distributed to the various bid items as the first step in the pricing process? What two methods of making this distribution were explained in this chapter? What is the disadvantage or objection discussed in this chapter in regard to the first of these methods?

3. How does the determination of the final bid item extension differ between lump-sum and schedule-of-bid-item bid items? What particular requirement applies to the latter?

4. Is there a universal understanding that is accepted in the heavy construction industry for the definition of a "balanced bid," or does the term have a different meaning to different individuals?

5. What are some of the reasons that were discussed in this chapter why the bid quantities on the bid form might be inaccurate?

6. How is the contractor actually paid for the work of a unit-price bid item during contract performance?

7. Explain how, at the time of bid, the contractor can protect against monetary loss due to overstated bid quantities.

8. What do bid documents frequently have to say about "unbalanced" bids? What do such statements actually mean in practice? As discussed in this chapter, why is it difficult for an owner reviewing bids that have been received to rationally conclude how particular bidders distributed their fixed costs and markup?

9. What is "front-end" loading? Why is it done? What is the usual owner attitude regarding front-end loading? What is one bidding practice that, when recognized by an owner receiving bids, is almost certain to result in rejection of a bid?

10. What is the first thing a bidder should do upon encountering major differences between takeoff figures for pay quantities and the bid quantities stated in the bid schedule? If the owner declines to issue an addendum correcting the bid quantities, what two options are left to the bidding contractor?

11. Repeat the work explained in this chapter for the example bridge project on the basis of the different estimate figures shown on the general estimate summary sheet. Specifically,

    (a) Distribute the fixed costs and markup based on labor and determine final bid-item unit prices and extensions. Use the format of Figure 13-3.

    (b) On the basis of the following comparison of takeoff vs. the bid quantities stated in the bid schedule, determine how much money the contractor would lose if awarded the contract on the basis of the bid developed in (a) above. Use the format of Figure 13-4.

Bid item	Description	Unit	Bid quantity	Takeoff quantity
1	Earth excavation	cy	13,650	13,725
2	Rock excavation	cy	5,175	5,170
3	Cofferdam excavation	cy	14,150	14,210
4	Cofferdam 1	LS	1	1
5	Cofferdam 2	LS	1	1
6	Cofferdam 3	LS	1	1
7	Cofferdam 4	LS	1	1
8	Cofferdam 5	LS	1	1
9	Cofferdam 6	LS	1	1
10	Furnish bearing piling	lf	36,475	31,500
11	Drive bearing piling	lf	34,775	29,450
12	Tremie concrete	cy	6,350	6,100
13	Footing concrete	cy	6,175	5,790
14	Column and web wall concrete	cy	5,325	4,950
15	Column concrete	cy	2,160	1,875
16	Cap concrete	cy	1,895	1,610
17	Reinforcing steel	lb	2,150,200	1,775,600
18	Stone rip rap	ton	1,500	1,495

    (c) Redistribute fixed costs and markup and determine new bid prices so that the contractor will not lose any part of distributed monies when the work is performed. Use formats of Figures 13-5 and 13-6.

Bid item no.	Description	Unit	Quantity	EOE	RL	R	L	EM	PM	S	Total direct cost	Distribution	Total bid(1)	Unit price	Final bid
1	Earth excavation	cy	13,650							68,875	68,875				
2	Rock excavation	cy	5,175							105,000	105,000				
3	Cofferdam excavation	cy	14,150	15,125	5,250	16,500	31,875	1,500			70,250				
4	Cofferdam #1	LS(2)	1	4,250	3,625	27,250	51,875	20,250			107,250				
5	Cofferdam #2	LS	1	6,250	4,250	37,875	73,375	28,875			150,625				
6	Cofferdam #3	LS	1	7,875	5,875	40,250	79,875	36,500			170,375				
7	Cofferdam #4	LS	1	10,750	7,625	90,125	132,750	47,750			289,000				
8	Cofferdam #5	LS	1	89,625	49,250	111,125	320,375	878,125			1,448,500				
9	Cofferdam #6	LS	1	15,375	9,125	95,500	203,375	234,375			557,750				
10	Furnish piling	lf	36,475						433,500		433,500				
11	Drive piling	lf	34,775	177,375	79,125	133,875	363,625	34,000			788,000				
12	Tremie concrete	cy	6,350	45,625	23,375	46,375	172,500	114,000	621,000		1,022,875				
13	Footing concrete	cy	6,175	85,250	49,875	95,500	617,375	232,750	364,375		1,445,125				
14	Column & web wall concrete	cy	5,325	86,500	50,625	97,000	626,625	236,250	323,250		1,420,250				
15	Column concrete	cy	2,160	40,375	23,625	45,250	292,625	110,375	123,375		635,625				
16	Cap concrete	cy	1,895	43,500	25,500	48,625	314,875	118,625	846,001		1,397,126				
17	Reinforcing steel	lb	2,150,200	30,125	22,750	54,250	310,875	54,500	559,750		1,032,250				
18	Stone rip rap	ton	1,500							33,750	33,750				
	Total direct cost			658,000	359,875	939,500	3,592,000	2,147,875	3,271,251	207,625	11,176,126				
	Indirect cost			198,303	68,362	74,214	1,352,610	1,574,188		275,177	3,542,854				
	Subtotal			856,303	428,237	1,013,714	4,944,610	3,722,063	3,271,251	482,802	14,718,960				
	Labor escalation				35,700		408,520				444,220				
	Equipment escalation			56,700							56,700				
	Interest							455,000			455,000				
	High-water contingency			31,920	9,660	8,820	138,600	232,680			421,680				
	Total job costs (w/o bond)			944,923	473,597	1,022,534	5,491,730	4,409,743	3,271,251	482,802	16,096,580	Corrected bid figures			
	Markup (50% labor)										2,982,664				
	Total bid (w/o bond)										19,079,244	Bid item 8	Total bid		
	Bond (1/2 of 1% of bid)										206,056				
	Total bid										19,285,300				

(1) Balance on labor.
(2) LS = lump sum.

381

Appendix

UNIT PRICE LIBRARIES USED FOR TEXT EXAMPLE EXERCISES AND AS INCLUDED IN CSUC, CHICO, COMPUTER ESTIMATING PROGRAM

Labor (fully loaded hourly rates)

Code	Description	Rate	Code	Description	Rate
OE1	Crane operator > 150 ft	50.160	TM4	Boom truck driver	35.140
OE2	Crane operator < 150 ft	48.170	TM5	12 cy transit mix driver	35.030
OE3	Oiler	39.890	TM6	Dumpcrete driver	35.030
OE4	Dozer operator	44.090	TM7	Water truck driver	32.910
OE5	Loader operator	43.280	TM8	Warehouseman	32.410
OE6	Backhoe operator	44.780	LAB1	Common laborer	29.740
OE7	Hydraulic shovel operator	44.780	LAB2	Dumpman	28.010
OE8	Scraper operator	43.280	LAB3	Concrete laborer	30.080
OE9	Compactor operator	44.090	LAB4	Air tool operator	30.080
OE10	Motor grader operator	44.090	LAB5	Driller	29.840
OE11	Gradall operator	43.290	LAB6	Chucktender	28.140
OE12	Compressor operator	38.890	LAB7	Powderman	29.840
OE13	Generator operator	38.890	LAB8	Labor foreman	33.070
OE14	Pump operator	38.880	LAB9	Powder foreman	31.070
OE15	Concrete pump operator	42.880	CARP1	Journeyman carpenter	38.270
OE16	Conveyor operator	42.870	CARP2	Carpenter foreman	40.290
OE17	Operator foreman	50.510	PD1	Piledriver	40.450
OE18	Heavy-duty mechanic	41.840	PD2	Piledriver foreman	42.580
OE19	Lube/tire worker	41.840	CM1	Concrete finisher	33.890
OE20	Grade checker	41.840	CM2	Concrete finisher foreman	35.240
TM1	Off-highway truck driver	34.080	IW1	Reinforcing ironworker	37.890
TM2	On-highway truck driver	32.790	IW2	Structural ironworker	37.890
TM3	Flat-rack truck driver	32.410	IW3	Ironworker foreman	39.910

Equipment (hourly rates) [1]

Code	Description	EOE	RL	Rental
EQ1	200 ton crawler crane	45.240	10.370	53.490
EQ2	165 ton crawler crane	40.960	10.370	46.030
EQ3	120 ton crawler crane	33.780	8.940	38.960
EQ4	100 ton crawler crane	23.250	8.940	27.950
EQ5	165 ton truck crane	47.520	4.910	59.620
EQ6	90 ton truck crane	32.460	5.600	45.980
EQ7	75 ton truck crane	24.390	5.600	50.330
EQ8	30 ton center-mount crane	15.190	6.900	20.920
EQ9	18 ton center-mount crane	13.200	7.020	14.330
EQ10	Boom truck	22.600	5.580	17.080
EQ11	15 cy wheel loader	56.450	8.310	93.590
EQ12	12 cy wheel loader	73.880	8.310	93.590
EQ13	7 cy wheel loader	37.230	8.350	45.330
EQ14	4 cy wheel loader	18.690	3.960	19.710
EQ15	4.5 cy crawler loader	41.120	6.570	27.840
EQ16	2.5 cy crawler loader	26.430	6.570	17.440
EQ17	11 cy Hyd. backhoe	154.00	15.400	91.790
EQ18	6 cy Hyd. backhoe	94.630	9.460	53.130
EQ19	3 cy Hyd. backhoe	50.020	5.000	29.720
EQ20	1 cy wheel loader/hoe	10.750	6.350	24.070
EQ21	1 cy track loader/hoe	24.440	7.290	21.210
EQ22	G660 gradall (wheel)	12.000	6.350	21.650
EQ23	Caterpillar 12 motor grader	13.780	2.870	12.520
EQ24	Caterpillar 14 motor grader	21.360	6.920	25.980
EQ25	Caterpillar 16 motor grader	30.580	6.920	38.080
EQ26	D4 dozer (dirt)	6.860	8.170	9.080
EQ27	D6 dozer (dirt)	18.750	8.440	18.010
EQ28	D8 dozer (dirt)	33.310	10.500	32.040
EQ29	D9 dozer (dirt)	42.440	10.010	36.870
EQ30	D8 dozer (rock)	41.640	13.130	40.05
EQ31	D9 dozer (rock)	53.010	12.510	46.090
EQ32	D9 dozer/ripper	63.620	15.200	55.310
EQ33	D9 push cat	36.050	8.510	31.340
EQ34	824 wheel dozer	27.980	7.590	29.890
EQ35	834 wheel dozer	42.390	7.590	46.450
EQ36	Caterpillar 815 compactor	42.880	9.710	26.140
EQ37	Caterpillar 825 compactor	41.070	15.330	39.200
EQ38	84 in. drum vibrator compactor	7.210	3.500	17.820
EQ39	36 in. drum vibrator compactor	1.840	3.500	3.240
EQ40	Caterpillar 621E scraper (14 cy)	46.540	12.390	32.090
EQ41	Caterpillar 627E scraper (14 cy)	53.080	12.390	29.230

[1]These equipment hourly rates were derived from equipment rates published in the 1991 *Contractor's Equipment Cost Guide* published by Dataquest-MID (now Primedia Information, Inc.).

Code	Description	EOE	RL	Rental
EQ42	Caterpillar 631E scraper (21 cy)	71.370	12.390	23.010
EQ43	Caterpillar 637E scraper (21 cy)	95.430	12.390	64.680
EQ44	Caterpillar 651E scraper (32 cy)	92.270	7.990	66.980
EQ45	Caterpillar 657E scraper (32 cy)	122.41	12.390	81.150
EQ46	Caterpillar 769 end dump (35 T)	47.660	5.690	34.870
EQ47	Caterpillar 773B end dump (50 T)	66.100	5.690	47.860
EQ48	Caterpillar 777B end dump (85 T)	93.300	5.690	68.500
EQ49	Dart 5130 bottom dump (130 T)	88.890	15.050	56.190
EQ50	10 cy on-highway end dump	17.160	3.230	9.410
EQ51	20 cy on-highway bottom dump	20.560	3.220	10.490
EQ52	5,000 gal water truck	25.040	7.930	24.060
EQ53	10,000 gal water truck	70.220	7.930	68.570
EQ54	3 cy clamshell bucket	2.730	1.070	5.310
EQ55	5 cy clamshell bucket	4.730	1.070	9.210
EQ56	3 cy dragline bucket	1.460	0.990	3.250
EQ57	5 cy dragline bucket	3.070	0.990	6.830
EQ58	Jackhammer	0.100	0.430	0.260
EQ59	Air trac (pneumatic)	17.610	2.320	10.110
EQ60	Track-mounted hoe ram	50.800	4.610	31.650
EQ61	185 cfm diesel compressor	5.460	1.610	2.430
EQ62	600 cfm diesel compressor	15.300	1.610	7.430
EQ63	1200 cfm diesel compressor	26.300	1.610	13.680
EQ64	1600 cfm diesel compressor	32.660	1.610	16.000
EQ65	400 A diesel welder	3.200	1.030	1.000
EQ66	400 A electric welder	0.330	0.450	0.340
EQ67	3 kW light plant	0.570	5.830	0.620
EQ68	12 kW light plant	1.920	11.660	2.240
EQ69	40,000 ft·lb diesel hammer	22.230	5.170	16.140
EQ70	32,500 ft·lb diesel hammer	19.150	6.760	31.370
EQ71	250 hp vibro/extractor	37.550	3.080	19.540
EQ72	60 ft swinging leads			10.000
EQ73	80 ft leads/spotter	5.000	5.000	30.000
EQ74	Truck-mounted concrete pump	30.240	4.540	33.350
EQ75	1 cy concrete bucket	0.040	0.400	0.370
EQ76	2 cy concrete bucket	0.070	0.400	0.610
EQ77	4 cy concrete bucket	0.130	0.400	1.040
EQ78	Vibrator with generator	1.650	7.210	1.460
EQ79	Sedan	6.620	0.500	1.170
EQ80	3/4 ton pickup truck	5.000	0.500	1.910
EQ81	4 × 2 flatbed truck	7.520	1.390	5.530
EQ82	6 × 4 flatbed truck	13.900	1.390	11.180
EQ83	Highbed tractor/trailer	17.670	3.580	13.260
EQ84	Spare trailer	2.390	2.190	2.080
EQ85	5 hp electric submersible pump	0.100	0.240	0.300
EQ86	15 hp electric submersible pump	0.300	0.240	0.950
EQ87	25 hp diesel contractor pump	5.000	1.750	5.000
EQ88	50 hp diesel contractor pump	10.000	2.000	10.000

EQ89	150 hp diesel jet pump	15.000	2.500	15.000
EQ90	Paving breaker	0.100	0.430	0.260
EQ91	Backfill tamper	0.100	0.430	0.260
EQ100	Concrete batch/mix plant 100 cy/hr	28.020	7.310	48.740
EQ101	8 cy transit mix truck	24.310	4.040	11.930
EQ102	8 cy dumpcrete truck	18.880	3.560	10.350

Expendable and Permanent Materials

Code	Description	Unit	Expendable materials	Permanent materials
EM1	3000 psi waste/OB Conc.	cy	45.00	
EM2	4000 psi waste/OB Conc.	cy	50.00	
EM3	Dimension lumber	MBF	400.00	
EM4	$\frac{3}{4}$ in. form plywood	sf	600.00	
EM5	Buy steel forms (flat)	sf	30.00	
EM6	Buy steel forms (complex)	sf	50.00	
EM7	Salvage steel forms (flat)	sf	−15.00	
EM8	Salvage steel forms (complex)	sf	−25.00	
EM9	Shoring rental (allowance)	cf	0.038	
EM10	Form ties and miscellaneous (allowance)	sfca	0.25	
EM11	Sandblast materials and supplies (allowance)	sf	0.15	
EM12	Finish materials and supplies (allowance)	sf	0.02	
EM13	Cure materials and supplies (allowance)	sf	0.01	
EM14	Cement	ton	65.00	
EM15	Concrete sand	ton	7.50	
EM16	Concrete aggregate	ton	10.00	
EM17	Buy HP steel piling	ton	500.00	
EM18	Buy steel sheet piling	ton	550.00	
EM19	Buy steel deck beams	ton	500.00	
EM20	Buy steel struts/wales	ton	500.00	
EM21	Buy miscellaneous braces/plates	ton	1000.00	
EM22	Buy heavy timber	MBF	400.00	
EM23	Buy lagging timber	MBF	400.00	
EM24	Trench shores (allowance)	cf	0.08	
EM25	Salvage HP steel piling	ton	−200.00	
EM26	Salvage steel sheet piling	ton	−220.00	
EM27	Salvage deck beams/wales/struts	ton	−220.00	
EM28	Salvage heavy timber	MBF	−80.00	
EM29	Granular backfill	cy	4.75	
EM30	Crushed rock	cy	12.00	
EM31	Rip rap stone	ton	9.00	
EM32	3 in. bits	Each	175.00	
EM33	2 in. bits	Each	75.00	
EM34	16 ft drill steels	Each	275.00	
EM35	Couplings	Each	60.00	
EM36	Shanks	Each	150.00	

Code	Description	Unit	Expendable materials	Permanent materials
EM37	Gelatin dynamite	lb	1.35	
EM38	Ammonium Nitrate Fuel Oil Mixture (ANFO)	lb	0.22	
EM39	Primacord	lf	0.30	
EM40	Presplit powder	lf	10.00	
EM41	Electric Blasting Caps (EBC)	Each	1.25	
EM42	Non pay rebar	lb	0.25	
PM1	3000 psi concrete	cy		45.00
PM2	4000 psi concrete	cy		50.00
PM3	Type I cement	ton		65.00
PM4	Concrete sand	ton		7.50
PM5	Coarse aggregate	ton		10.00
PM6	Fabricated rebar	lb		0.25
PM7	18 in. corrugated culvert	lf		8.00
PM8	24 in. corrugated culvert	lf		12.00
PM9	24 in. reinforced concrete pipe	lf		12.00
PM10	36 in. reinforced concrete pipe	lf		16.00
PM11	3 ft concrete precast MH	vf		30.00
PM12	3 ft manhole frame and cover	Each		400.00

Subcontracts

Code	Description	Unit	Subcontracts
SUB1	Install reinforcing	lb	0.20
SUB2	2 in. wellpoint with header	Each	100.00
SUB3	24 in. × 100 ft deepwell	Each	5000.00
SUB4	Operate wellpoint system (100 points)	month	45000.00
SUB5	Operate deepwell system (10 wells)	month	15000.00
SUB6	50 kip tieback anchor	Each	1500.00
SUB7	100 kip tieback anchor	Each	3000.00
SUB8	6 in. compacted base	sy	2.50
SUB9	6 in. asphalt paving	sy	11.00
SUB10	6 in. concrete paving	sy	15.00
SUB11	6 in. curb and gutter	lf	8.00
SUB12	6 in. asphalt sawcut	lf	3.00
SUB13	6 in. concrete sawcut	lf	6.00
SUB14	Light clear and grub	Acre	3000.00
SUB15	Heavy clear and grub	Acre	5000.00

INSTRUCTIONS FOR USE OF CSUC, CHICO, COST-ESTIMATING PROGRAM

The CSUC, Chico, program is written in Lotus 1-2-3 but will run equally well in Microsoft Excel using Windows 95 or 98 software by making a simple change in keyboard commands.

The program is intended to familiarize students with the estimating pricing process for a single line item, which in the context of the instructional approach taken in this text, is usually a single discrete bid item for a schedule-of-bid-items bid. It is not an integrated program intended for pricing and assembling multiple bid items into a complete general estimate summary and/or to compile labor, equipment, and material summary information for a complete project bid. There are several excellent PC-driven comprehensive heavy construction project estimating programs available in the current commercial market for this purpose.

This single bid-item program is a teaching tool intended for classroom instruction. Students trained by its use in practice estimating exercises should have no difficulty after graduation in adapting to one of the commercially available PC-driven programs or mainframe computer estimating programs in use by their employers.

It is suggested that students acquaint themselves with the estimating program by replicating one or more of the printout solutions in this book such as Figures 9-17 and/or 10-5.

The master program file name is 209mtr00.wk1. This program does not require an extensive knowledge of spreadsheet software but some basics should be understood. **Rule 1: Save your work every 10 min.** Most people learn this rule through the unfortunate experience of losing a great deal of work that was being performed in the **RAM** and only infrequently being saved to a disk.

The entire program is self-contained within the 209mtr00.wk1 file and, when saved to another file name, is also self-contained within that new file. Operating procedures will be discussed below. The most important point to remember is that you should **never** do any work in or save to the 209mtr00.wk1 file. Once an estimate is begun in a new file, it cannot be cleaned, erased, or reset to be used again. The structure of the file and the programming that drives it is nonreversible. Think of each estimate file as being disposable—you use it once and throw it away. (No, don't throw your disk away, just don't reuse an old file on the disk.) For this reason, your first step in any new project is to **save from the 209mtr00.wk1 file to a new file** name of your choice. File naming will be covered in the operating procedures section of these instructions. As an added precaution, you should back up your disk containing the 209mtr00.wk1 master file to another disk before doing any work.

No modifications should be performed on the spreadsheet file using any of the Lotus 1-2-3 menu commands. The 209mtr00.wk1 program is driven by macros written in 1-2-3 command language and can be rendered useless by modifying the file. The program was written for a specific purpose and is a powerful and useful tool when used within its limitations. Doing any of the following may cause part or all of the program to quit functioning:

1. Inserting rows and columns
2. Deleting rows and columns
3. Making entries in other than indicated cells
4. Erasing labels, headings, asterisks, or formulas in protected cells
5. Increasing or decreasing column widths
6. Copying or moving entries within the file or parts of the file to different locations on the spreadsheet
7. Setting up print specifications other than those in the print macro
8. Adding or deleting range names

Inserting, deleting, erasing, copying, moving, and making random entries will break the continuity of the program and cause specific and absolute cell addresses to become meaningless. The asterisks that may seem to appear in random locations perform a very specific function. They act as control gates to direct the flow of data and also act as range indicators to tell the macros the actual size of the estimate. Do not erase them.

If some of these terms and explanations do not make sense at this point, don't worry about it. You don't need to understand them to use the program. Follow the procedures as they are given and the program will work as it should.

If you have inadvertently done one or more of the listed **don't do's** or if something else seems to have gone wrong, don't panic. Because you have been saving your work every 10 min, you're in good shape. **Don't save the mistake.** Just remove your disk from the drive, quit Lotus (or Excel if running in Excel) and then come back in and retrieve your working file. The status of your file will be as it was when you last saved it. To find out where you were, simply view the estimate (covered in the operating procedures section of these instructions) and pick up again where you left off.

Important Things to Know about Macros

Many of the functions of the program are controlled by small, independent programs written in 1-2-3's command language. These small programs are called macros. Once access has been made to the proper working file, executing the correct sequence of macros will enable you to build an estimate. Some macros can be used randomly to perform various tasks or to view different parts of the file. Other macros must be used only in the proper sequence and at the proper time in the estimate building process. When a macro has been activated, **it must be allowed to finish its task, with your only activity being to input information requested by the macro.** A macro is in process when the letters **CMD** appear at the bottom of the screen. Interfering with an operating macro may destroy the format of the estimate. More specific information about individual macros is covered in the following operating procedures sections.

The nine macros that drive this program are listed in the on-screen macro menu, which can be accessed at any time by activating the **ALT R (return to menu)** macro (when running in Excel, this command is **CONTROL R).** The macros are activated by depressing and holding down the ALT (or CONTROL in Excel) key and then typing in the letter of the appropriate macro. Think of the ALT key (or CONTROL key in Excel) as working similarly to the shift key. They must be held down while typing the letter. **Once a macro is in progress, it must be allowed to run its course.** Some macros are completed within a second or two, and others may take up to 20 sec. A second macro cannot be activated while one is in progress. Remember, the letters **CMD** will appear at the bottom of the screen while a macro is in progress.

Operating Procedures

1: Power On

Turn power on to the CPU and monitor. These procedures will assume you are using a self-booting IBM computer with a hard drive.

2: Booting and Preparing to Read Disk Drive

After the computer has booted, follow whatever procedure is necessary to read disks, depending on the way your PC is set up. It is a good idea at this point to hit the **CAPS LOCK** key. Subsequent entries made in the estimate file stand out better in caps than in lowercase.

3: Access Lotus 1-2-3 or Excel

You must first load Lotus 1-2-3 or Excel software into the RAM of your computer by whatever procedure is necessary, depending on how your computer is set up (many computer users maintain either or both of these software programs on the hard drives of their computers).

4: Retrieve Master File

Insert the program disk furnished with this text into the floppy disk drive on your computer. Bring up the program menu up by hitting the **/** key. This is the forward slash key, which is very different in function than the backward slash, ****. Move the cursor to **File** and **ENTER.** The cursor will then be on **Retrieve. ENTER** again. A list of the four files on the disk will appear in the control panel. Move the cursor to the **209mtr00.wk1** file and **ENTER.** Observe the **WAIT** indicator in the upper right corner. No keyboard entries can be made while **WAIT** is flashing. A title screen will appear for a few seconds and then the macro menu will appear.

5: Macro Menu

Read the menu options to get an idea of what is available to you. Notice the first option, **ALT R.** Activating **ALT R** (or **CONTROL R** in Excel) will always return you to the menu that you now see on the screen. Remember, to activate any of the macros, hold the **ALT** key (or the **CONTROL** key in Excel) down and type the letter of the macro that you want to activate.

6: Create Working File

It is **extremely important** that you perform this step before doing anything else. At this point, you have retrieved the master file. **You must not do any work in the master file.** It is to be preserved as a clean, uncontaminated template from which new files are made. **Remove your 209mtr00.wk1 disk from the floppy disk drive** and replace it with a clean disk to which you will save your work under your working file name. Activate **ALT S** (or **CONTROL S** in Excel). The heading **Enter save file name: B: 209mtr00.wk1** will appear in the control panel. Type in a new file name over **209mtr00.wk1.** The file name can be up to eight characters long and must begin with a letter. Do not type **WK1.** The program will append that automatically. After typing your new file name, **ENTER.** Wait for the new file to be created and saved. You are now in the new file where you can begin working. The master file is preserved out of reach on your master floppy disk.

7: Retrieve Working File

Step 6 is performed only when you are beginning a **NEW** estimate. If you wish to continue working on an estimate that was started and **SAVED** in a previous working session, follow through to step 4 and retrieve your working file instead of the master file. If you are starting another new file, follow step 6 again.

Please note that your floppy disk will have a limited capacity. Trying to store too many files on one disk may cause a **Disk Full** error message. As a precaution, you should make at least one backup copy of the disk you received with this text on a separate disk. Disks are fragile and data can be lost. Disks are also cheap compared to the amount of time you may have to spend obtaining new master files. Back up your working files on other disks. For normal saving procedures and to make backups, follow step 8.

8: Saving and Backing Up

You're probably tired of reading this by now, but here it is again: **Save your work every 10 min.** When saving to a new file name from the master, follow step 6. When saving during the process of building your estimate (every 10 min), your working file is in **RAM.** Activate **ALT S** (or **CONTROL S** in Excel). It does not matter where you are in the estimate file; however, another macro cannot be in progress. The name of your working file will appear in the control panel. If **209mtr00.wk1** appears, you did not follow step 6. When you see your file name, **ENTER.** The words **Cancel and Replace** will appear. Type **R** (and **ENTER**) to replace the original work in the working file with the updated work that you have just done. Be aware that whenever you do this, your original file is wiped out and everything that you have done since your last save replaces it. For this reason, if you've made a mistake, **don't save it.** See the discussion in the introduction for this situation.

When saving for the first time to a new file or when backing up a working file to a new disk, the words **Cancel** and **Replace** as noted earlier will appear because there is no file to replace.

9: Beginning a New Estimate

To begin a new estimate for a bid item, activate **ALT B** (or **CONTROL B** in Excel). The estimate information input screen will appear with the cursor in position for the first entry. Type the appropriate entries and **ENTER** after each one.

Important notice: You are interacting with a macro (notice **CMD** at the bottom of the screen). **Do not** use the cursor keys to move the cursor. Hitting the **ENTER** key takes you to the next input cell. If you scroll back up to correct an error, the macro does not recognize that and will not take the remaining needed entries. When you enter the last item, the message, **Are your entries correct? (Y/N):** will appear in the control panel. If you need to correct an error, just type **N** and **ENTER.** Continue hitting **EN-TER** until the macro moves you to the cell with the error and type in the correct entry. **ENTER** until you see the entry message again and type **Y** if everything is correct. You will then be returned to the menu. Step 9 is normally performed only once for an estimate (bid item); however, it may be repeated to change or correct label entries, bid quantity, or takeoff quantity in the estimate heading.

The required entry cells in the begin estimate macro will request the following information:

Cursor position 1: What is the bid-item number?
Cursor position 2: What is the bid-item description?
Cursor position 3: What is the unit of measurement for the bid-item quantity?
Cursor position 4: What is the contractor's takeoff figure for the pay quantity?
Cursor position 5: What is the bid quantity established by the owner?

10: Configure a New Operation

This step configures and loads each operation within the bid-item estimate. Most bid-item estimates will have multiple operations. **Activate ALT C** (or **CONTROL C** in Excel). Remember, if you make an error in your entries, **don't backtrack** to fix it. You will be given the opportunity to go back through your entries after you make the last entry. If everything is correct, type **Y** at the prompt and **ENTER.** If your entries are not correct, type **N** and **ENTER.** The cursor will automatically jump back to the first entry so that you can advance through the macro again to correct your error. When you finally enter **Y,** indicating your entries are now correct, the program will display a message and ask you to wait while it configures the operation according to your input information.

The configuration process expands the size of the estimate to accommodate each new operation. Space for eight line items is provided within each operation. After configuration is complete, a **Ready** message will appear with a prompt in the control panel asking which data library you would like to enter. Type **L, E, M, or S** and **ENTER** for library access. The requested library will appear on the screen.

The required entry cells in the configure macro will request the following information:

Cursor position 1: What is the operation number?

Cursor position 2: What is the operation description?

Cursor position 3: What is the operation work quantity?

Cursor position 4: What are the work quantity units of measure?

Cursor position 5: What is the work crew productivity?

Cursor position 6: What is the small tools and supplies percentage?

Cursor position 7: What is the labor premium package?

After each query is answered and **ENTER** is struck, the cursor will automatically advance until the prompt **are your entries correct?** appears.

For operations involving only the entry of materials and/or subcontracts (that is, when there is no work crew performing some quantity of work), only the queries for the first two cursor positions should be answered. Simply hit **ENTER** at each of the following cursor positions to advance through the macro.

11: Selecting and Transferring Data

With the new operation configured, individual line items may be transferred from the data libraries to the estimate. Remember from step 10, up to eight line items may be transferred to an operation. An error message will appear if you try to transfer more. To recover from any error message, hit the **ESC** (escape) key. If you have more than eight line items for an operation, simply configure a new operation with the same input as the current one and add the extra items to the new one.

To select and transfer data, move the cursor down the quantity column in the data library until you locate a needed item. Note that when viewing a data library that you have accessed, the **CMD** symbol does not appear at the bottom of the screen, so you are not in a macro. You can (and must) use the cursor key to move up or down in the quantity column. Type in the desired quantity of that item and **ENTER.** With the cursor still on the entered quantity, activate **ALT T** (or **CONTROL T** in Excel). The line item item, with your entered quantity, will be transferred to the current (last) configured operation. You will know that this has happened when your line item quantity disappears from the screen.

Important note: Line items can only be transferred to the last configured operation. It is not possible to transfer to previous operations. Only one line item may be transferred at a time (in any quantity). The items will appear in the estimate in the order they are transferred. **ALT T** (or **CONTROL T** in Excel)) should be activated only when a data library is on screen. Strange things will result if it is activated from other locations.

12: View or Modify Estimate

During the selection and transfer process, the actual estimate can be viewed at any time to check or modify entries. To view the estimate, activate **ALT V** (or **CONTROL V** in Excel). This can be done from any of the four data libraries or from the macro menu. Note that when viewing the estimate, the **CMD** symbol does not appear at the bottom of the screen, so you are not in a macro. You may freely use your cursor keys as you would when working with any spreadsheet. With the estimate on the screen, the cursor or **PgDn** and **PgUp** keys can be used to see all the operations and the bottom line totals. If changes are needed in quantities, productivity, or unit prices, move the cursor to the appropriate cell and type in the correct entry. All extended values will be instantly updated. New line item descriptions may also be typed in at the end of any existing operation provided there is room. Enter the new item's quantity, unit of measurement, and applicable unit price or unit prices (if more than one) in the appropriate unit price column (or columns).

Important note: Entries of this type may be entered only in unprotected cells. On a color monitor, unprotected cells are green. On a monochrome monitor, they are highlighted. Do not attempt to make entries in white, protected cells.

To continue adding items to the current operation from the data libraries (up to a total of eight entries under the last operation you have configured), return to the data libraries by activating **ALT D** (or **CONTROL D** in Excel). To configure a following operation, return to the macro menu, **ALT R** (or **CONTROL R**), and activate **ALT C** (or **CONTROL C**). Once you configure a new operation you cannot return to a previously configured operation. The only way that you can alter entries in a previously configured operation is through the process of making additions or changes while viewing the estimate spreadsheet (**ALT V or CONTROL V**). **Save your work every 10 min.**

13: Printing the Estimate

After configuring and loading all the operations in an estimate, you can print the estimate (displayed by **ALT V or CONTROL V**) by following the printing procedure required by your particular computer/printer setup.

14: Printing the Data Libraries

The data libraries may be printed by bringing them up and following the printing procedure required by your particular computer/printer setup. The data libraries in the 209mtr00.wk1 program are the same as listed in the first section of this Appendix.

INSTRUCTIONS FOR USE OF CSUC, CHICO, LOAD-AND-HAUL PROGRAMS

Both CSUC, Chico, programs (**ald-scra.wk1** and **ald-truc.wk1**) are macro-driven Lotus 1-2-3 spreadsheet templates and are on the disk furnished with this book. They will also operate on Microsoft Excel in Windows 95 or 98 software. User options are selected from menu screens and required data input is prompted with required units indicated. Both programs require Lotus 1-2-3 version 2.0 or higher or Microsoft Excel. Both programs can be run from either floppy drives or they can be loaded to the hard drive. Retrieval and saving times are much reduced if the hard drive option is used. **Care should be taken to avoid altering the original template files (ald-scra.wk1 and ald-truc.wk1)** by inadvertently saving a working file for a haul being analyzed over them. **Backup copies of the original template files (ald-scra.wk1 and ald-truc.wk1) should be made and retained in a safe place.**

Once the master file (either **ald-scra.wk1** or **ald-truc.wk1** has been loaded into the CPU RAM, remove the master disk from the computer drive to ensure that it will not become contaminated. Then name a working file (see the file naming procedure explained for the 209mtr00.wk1 estimating program). Save all subsequent work to the working file name.

It is suggested that students familiarize themselves with using these programs by replicating the printouts of solutions shown in this book such as Figures 9-8, 9-11, and 9-12.

GENERAL DIRECTIONS FOR USING BOTH PROGRAMS

Each program is organized into a number of subroutines, which are selected from the main menu. To execute a subroutine hold down the **ALT** key (the **CONTROL** key in Excel) and press the indicated letter key at the same time.

All required data input is prompted (the cursor is automatically located and the required data is indicated by the message in the cell to the left of the cursor or by a column heading). When entering a value use the **ENTER** key so that the cursor will automatically advance to the next data input location. Provision to review and edit input data is included. Be sure to input data with the required units (i.e., 80% is 80, not .80). Menu choices and prompts may appear in the 1-2-3 panel at the top of the worksheet, and these may be executed as for any 1-2-3 menu option.

The suggested sequence of execution of the main menu options for both programs is

1. Select **ALT I** (**CONTROL I** in Excel) and execute the main menu option to enter project information. Following input, a prompt in the 1-2-3 panel at the top of the screen will inquire whether data is correct. If data is correct, respond as prompted to continue. If data requires correction, respond as prompted and use the **ENTER** key to advance the cursor to the field requiring correction and enter the correct data. The same error correction routine is included at the end of all main menu options that require data input from the user.

2. Select **ALT L** (**CONTROL L** in Excel) and execute the main menu option to enter loaded haul information. When more than one haul road segment is required, respond exactly as prompted in the 1-2-3 panel to expand the worksheet and enter the second and subsequent segments. Any number of haul road segments can be included.

3. Select **ALT R** (**CONTROL R** in Excel) and execute the main menu option to enter return haul information. When more than one haul road segment is required, respond exactly as prompted in the 1-2-3 panel to expand the worksheet and enter the second and subsequent segments. Any number of haul road segments can be included. The return haul information is independent and must be separately entered. The return haul does not have to be the mirror image of the loaded haul.

   If errors are made in entering loaded or return haul information, it usually is more time effective to quit the program, then call it up again, and start over, saving your work to the working file name.

4. Select **ALT S** (**CONTROL S** in Excel) and execute the main menu option to elect haul equipment. Information regarding loading equipment selection and cycle constants is also entered in this subroutine.

5. Select **ALT C** (**CONTROL C** in Excel) and execute the main menu option to calculate and view the worksheet. Review the worksheet by moving the cursor with the cursor control keys.

6. Select **ALT P** (**CONTROL P** in Excel) and execute the main menu option to print the worksheet. If your printer is not included in the menu options, follow the necessary printing procedures dictated by your computer/printer setup.

7. Repeat steps 4 through 6 (press **ALT M,** or **CONTROL M** in Excel, to return to the main menu) for each combination of load-and-haul equipment to be investigated for the same haul road configuration and conditions that you have previously loaded into the program. You do not have to reenter the haul road information again.

8. If speed restrictions are desired, they can be accomplished by directly entering the required maximum speed in the **Sus Spd column** for affected haul road segments. After entering the restricted speeds, the worksheet must be recalculated by pressing the **F-9 function key.** Because the entry of the speed limit destroys the original formulas, restrictions should not be imposed until a final equipment selection has been made.

DETAILED DOCUMENTATION FOR ALD-TRUC.wk1— LOAD-AND-HAUL ANALYSIS PROGRAM FOR OFF-HIGHWAY END-DUMP TRUCKS

Project Information Subroutine

1. **Project name, estimator, and load haul number** identification information is self-explanatory.
2. **Efficiency (%)** = number of minutes worked per hour/60 min/hr × 100%.

Loaded and Return Haul Subroutines

1. For each indicated haul segment, enter the segment distance in feet(**Seg Dist**), the estimated rolling resistance (**R.R.**) in lb/ton and the grade resistance in % (**G.R.**) in the indicated columns. Total haul road rolling resistance (**THRR**) is then automatically calculated by the program as a percentage, based on 20 lb/ton of rolling resistance = 1% grade resistance.

2. The basic template provides for one haul road segment in addition to travel segments in the cut and in the fill. To input additional haul road segments, answer the menu query displayed in the 1-2-3 panel area at the top of the worksheet after entering all required information for the first haul road segment. Repeat this procedure for each additional haul segment required except for the travel-on-fill segment.

3. The haul unit speed factor (**Spd Factor**), in miles per hour, corrects the speed for acceleration/deceleration and is automatically determined by the program, using a look-up table based on the manufacturer's rimpull/gradeability/speed curves for the haul unit selected. The sustained speed automatically reflects the manufacturer's continuous grade retarding curves when the total haul road rolling resistance is favorable.

4. The average segment speed (**Av Spd**), in miles per hour, is calculated by the program and is equal to the sustained speed times the speed factor (**Spd Factor**).

5. Travel time in minutes (**Travel**) is calculated from the segment distance and average speed, with 1 mph = 88 fpm.

Equipment Selection Subroutine

1. **Haul Vehicle** is selected from menu choices appearing in 1-2-3 panel area at top of worksheet.

2. **Mtrl. Wt. (#/Bcy)** is in situ density of material to be hauled, including in situ water.

3. **Mtrl Swell (%)** is a factor to account for the increasing volume resulting from loosening of material during excavation. It is equal to

    (number of loose cubic yards per bank cubic yard $-$ 1.0) $\times$ 100%

4. **Load Bckt (cy)** is the nominal rated heaped capacity of the bucket selected.

5. **Bucket Factor** is a factor to convert the nominal rated heaped capacity of the bucket in cubic yards to the volume in bank cubic yards (Bcy) of the material being loaded actually delivered per bucket. It is whatever the estimator judges it to be for the material being loaded.

6. **Loader Cycle (min)** is the loader cycle time in minutes. It is whatever the estimator judges it to be for the loader and conditions visualized.

7. **Spot & Mnvr (min)** is the estimated time required for the haul unit to maneuver to its loading position in minutes as determined by the estimator's judgment.

8. **Turn & Dump (min)** is the estimated time required for the haul unit to turn and dump its load at the fill in minutes as determined by the estimator's judgment.

Worksheet Calculations

When **ALT C or CONTROL C** is activated, all worksheet calculations are instantaneously performed according to the following formulas.

1. **Haul unit payload, tons** = manufacturer's rated maximum capacity

2. **Haul unit payload, Bcy** $= \dfrac{\text{(haul unit payload, tons)(2000 lb/ton)}}{\text{(material weight, lb/Bcy)}}$

3. **Loader bucket payload, Bcy** $= \dfrac{\text{(rated bucket capacity, cy)(bucket fill factor)}}{(1 + \text{material swell \%}/100)}$

4. **Number of loader buckets** $= \dfrac{\text{haul unit payload, Bcy}}{\text{loader bucket payload, Bcy}}$

5. **Haul unit load time, min:**

    (a) When spot and maneuver time is greater than loader cycle time,

    $$= \text{(number of loader buckets)(loader cycle, min)} + \text{(spot and maneuver, min} - \text{loader cycle, min)}$$

    (b) When spot and maneuver time is less than loader cycle time, and the number of haul units is greater or equal to the balance number,

    $$= \text{(number of loader buckets)(loader cycle, min)}$$

6. **Haul unit cycle time, min** = haul unit load time + total travel time + turn & dump time.

7. **Haul unit productivity, Bcy/hr**

$$= \frac{(\text{haul unit payload, Bcy})(60 \text{ min/hr})(\text{efficiency } \%/100)}{(\text{haul unit cycle time, min})}$$

8. **Balanced number of haul units** $= \dfrac{(\text{haul unit cycle time})}{(\text{haul unit load time})}$

9. Spread productivity (**Spread prod**), in bank cubic yards per hour, is governed by haul unit productivity (**Haul unit prod**) when the number of haul units is less than the balance number and by loading unit productivity when the number of haul units is greater than or equal to the balance number. The following formulas apply:

   (a) Less than the balanced number of haul units:

   Spread productivity, Bcy/hr = (number of haul units)(haul unit productivity, Bcy/hr)

   (b) Greater than or equal to the balanced number of haul units:

   Spread productivity, Bcy/hr = (balanced number of haul units)(haul unit productivity, Bcy/hr)

DETAILED DOCUMENTATION FOR ALD-SCRA.wk1— LOAD-AND-HAUL ANALYSIS PROGRAM FOR PUSH-LOADED SCRAPERS

Project Information Subroutine

The input material is the same as for **ald-truc.wk1,** except that the material loaded is identified (**Ld. material**) and that the **Efficiency (%)** is input in the equipment selection subroutine rather than in the project information subroutine.

Equipment Selection Subroutine

1. **Haul vehicle** (scraper) is selected from the menu choices appearing in the 1-2-3 panel area at the top of worksheet.
2. **Enter Efficiency (%)** as for **ald-truc.wk1.**
3. Use **Load equip.** field to identify selected load equipment. This entry is for information purposes only—it does not directly affect worksheet calculations.
4. **Load factor (Bcy/cy)** is a dimensionless factor to convert nominal struck capacity to actual bowl capacity in bank cubic yards for the material being loaded. Material swell and loading characteristics must be taken into account by the estimator when making the judgment for the load factor to use in any given case.
5. **Load & boost (min)** is the estimated time required to push load and boost the scraper, as determined by the estimator's judgment.
6. **P.C. return (min)** is the estimated time after completion of scraper load and boost for the push cat(s) to back up to pushing position for the following scraper, as determined by the estimator's judgment.
7. **P.C. maneuver (min)** is the estimated time for the push cat(s) to maneuver and engage the following scraper in pushing position, as determined by the estimator's judgment.

8. **Turn & dump (min)** is the estimated time for the scraper to dump its load at the fill and turn back on the return haul as determined by the estimator's judgment.

Loaded and Return Haul Subroutines

These are the same as for **ald-truc.wk1.**

Worksheet Calculations

1. **Haul unit struck capacity (cy)** from the manufacturer's handbook is programatically entered for the equipment selected.
2. **Haul unit payload (Bcy)** = (haul unit struck capacity, cy)(load factor, bcy/cy).
3. **P.C. cycle (min)** = load and boost + P.C. return + P.C. maneuver time.
4. **Haul unit cycle time (min)** = P.C. maneuver + load and boost + total travel + turn and dump time.
5. **Haul unit productivity (Bcy/hr)** is the same as for **ald-truc.wk1.**
6. **Balanced # haul units** = (Haul unit cycle time, min)/(P.C. cycle time, min).
7. **Spread prod (Bcy/hr)** is the same as for **ald-truc.wk1.**

INSTRUCTION FOR USE OF CSUC, CHICO, 209IC-98.wkI INDIRECT COST COMPUTER SPREADSHEET TEMPLATE

This computer spreadsheet template (contained in the disk furnished with this text) is in the indirect cost estimate format presented and explained in Chapter 11. The master file name is **209IC-98.wk1.** It will run and respond to the normal spreadsheet commands using either Lotus 1-2-3 (in which it was written) or Microsoft Excel.

This template is not macro-driven, so the user must make all required cursor movements using the cursor keys. All cells are protected except those that might receive input entries necessary for a particular indirect cost estimate. Once that **quantities** and **unit rates** are entered in the appropriate cells, the formulas built into the spreadsheet will make proper extensions and enter appropriate line and column totals **for each of the five major sections of the indirect cost. A summary sheet totaling the five section totals into a grand total** is also built into the spreadsheet. User changes can be accomplished by exercise of the appropriate Lotus 1-2-3 spreadsheet commands.

Operating Suggestions

It is recommended that after loading the program into the **RAM,** the following preliminary steps be taken:

1. With the cursor on **cell B5,** freeze both the horizontal and vertical titles.
2. Suppress all spreadsheet zeros.
3. When the estimate has been completed, clear both horizontal and vertical titles prior to printing.
4. Print the spreadsheet following the printing procedures dictated by your particular computer/printer setup.

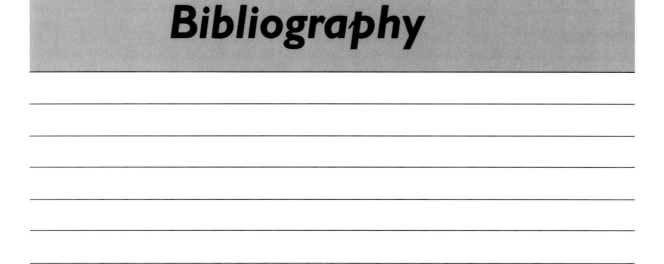

Bibliography

Bickell, John O., Thomas R. Kuesel, and Elwyn H. King. *Tunnel Engineering Handbook*. New York: Chapman & Hall, 1996.

Bonny, J. B., and Joseph P. Frein. *Handbook of Construction Management and Organization*. New York: Van Nostrand Reinhold Company, 1973.

Church, Horace K. *Excavation Handbook*. New York: McGraw-Hill, 1981.

Hurd, M. K. *Formwork for Concrete*. Detroit: American Concrete Institute, 1995.

Parker, Albert D. *Planning and Estimating Heavy Construction*. New York: McGraw-Hill, 1984.

Peurifoy, R. L., W. B. Ledbetter, and C. J. Schexnayder. *Construction Planning, Equipment, and Methods*. New York: McGraw-Hill, 1996.

Ratay, Robert T. *Handbook of Temporary Structures in Construction*. New York: McGraw-Hill, 1996.

Ringwald, Richard C., and Francis Hopcroft. *Means Heavy Construction Handbook*. Kingston, MA: R. S. Means Company, Inc., 1993.

Singh, Jagman. *Heavy Construction—Planning, Equipment, and Methods*. Rotterdam: A. A. Balkema, 1993.

Smith, Ronald C. *Principles and Practices of Heavy Construction*. Englewood Cliffs, NJ: Prentice-Hall, 1976.

Stubbs, Frank W., and John A. Havens. *Handbook of Heavy Construction*. New York: McGraw-Hill, 1971.

Index

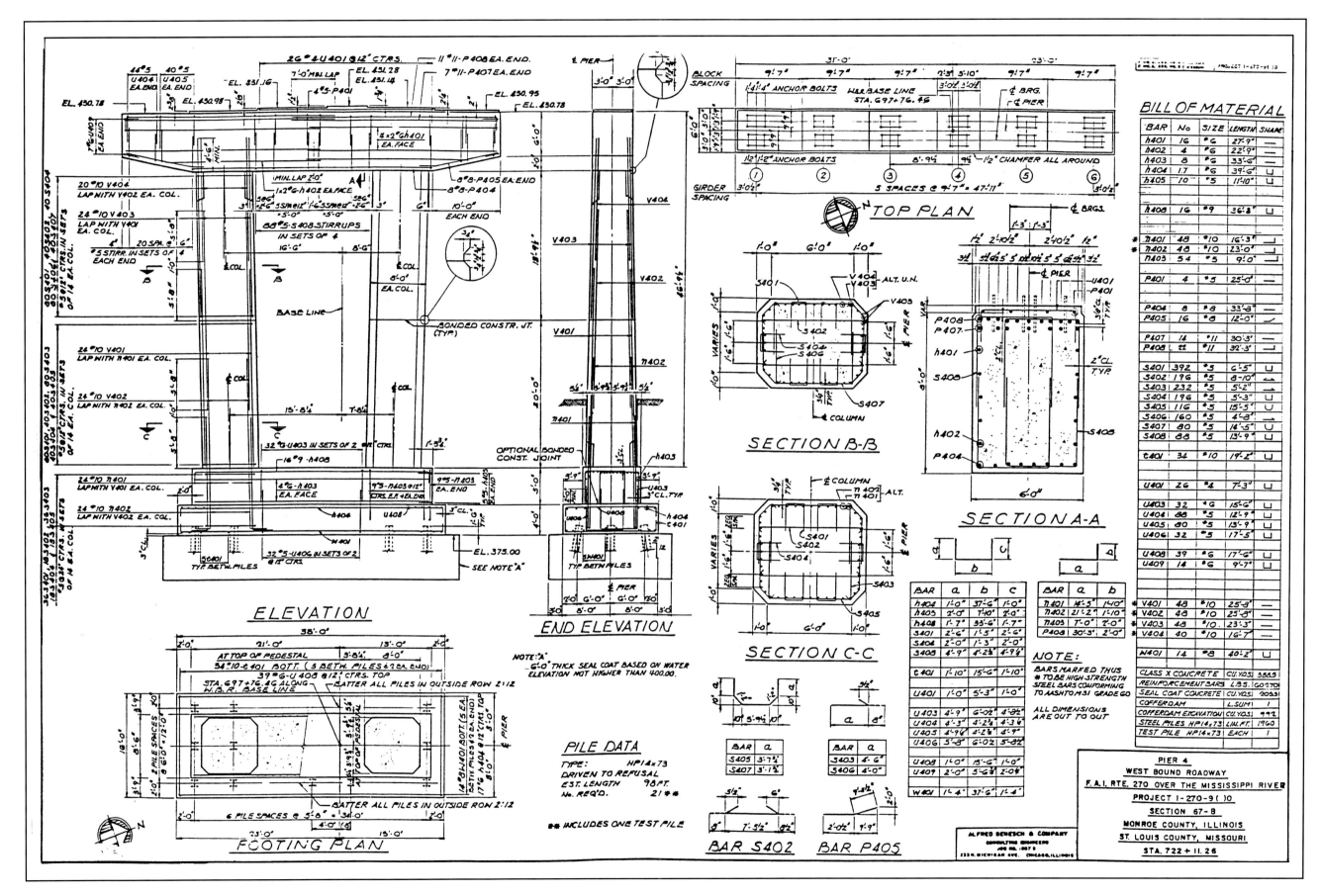

FOLDOUT 1